MARPOL
Consolidated Edition 2017

International Convention for the Prevention of Pollution from Ships

Supplement
February 2019

Since the publication of MARPOL Consolidated Edition 2017, the Marine Environment Protection Committee (MEPC) has adopted resolutions amending MARPOL Annex VI. This supplement presents, in chronological order of their adoption, those amendments that either have entered into force or will have entered into force before the next consolidated edition has been published.*

Resolution	Amends	Date of entry into force	Page
MEPC.286(71)	**Annex VI** Chapter 3, Regulation 13 Appendix V	1 January 2019	2
MEPC.301(72)	**Annex VI** Chapter 3, Regulation 13 Chapter 4, Regulation 21	1 September 2019†	4
MEPC.305(73)	**Annex VI** Chapter 3, Regulation 14 Appendix I	1 March 2020†	5

* Refer to the consolidated text of MARPOL Annex VI including resolutions MEPC.271(69) and MEPC.278(70) which are now in force (see Prospective amendments to MARPOL Annexes under *Additional Information* of *MARPOL Consolidated Edition 2017*, pages 399-465).

† Subject to acceptance.

Resolution MEPC.286(71)
adopted on 7 July 2017

MARPOL Annex VI

Regulations for the prevention of air pollution from ships

Chapter 3 – Requirements for control of emissions from ships

Regulation 13
Nitrogen oxides (NO_x)

Tier III

1 *In paragraph 5.1, after the words* "an emission control area designated for Tier III NO_x control under paragraph 6 of this regulation", *insert the words* "(NO_x Tier III emission control area)".

2 *The existing text of paragraph 5.1.2 is replaced by the following:*

"**.2** that ship is constructed on or after:

.1 1 January 2016 and is operating in the North American Emission Control Area or the United States Caribbean Sea Emission Control Area;

.2 1 January 2021 and is operating in the Baltic Sea Emission Control Area or the North Sea Emission Control Area;"

3 *Between paragraphs 5.1.2 and 5.1.3, the word* "when" *is deleted.*

4 *In paragraph 5.1.3, the words* "an emission control area designated for Tier III NO_x control under paragraph 6 of this regulation" *are replaced by* "a NO_x Tier III emission control area".

5 *In paragraph 5.2.3, the word* "convention" *is replaced by* "Convention".

6 *Insert new paragraphs 5.4 and 5.5, as follows*:

"**5.4** Emissions of nitrogen oxides from a marine diesel engine subject to paragraph 5.1 of this regulation that occur immediately following building and sea trials of a newly constructed ship, or before and following converting, repairing, and/or maintaining the ship, or maintenance or repair of a Tier II engine or a dual fuel engine when the ship is required to not have gas fuel or gas cargo on board due to safety requirements, for which activities take place in a shipyard or other repair facility located in a NO_x Tier III emission control area are temporarily exempted provided the following conditions are met:

.1 the engine meets the Tier II NO_x limits; and

.2 the ship sails directly to or from the shipyard or other repair facility, does not load or unload cargo during the duration of the exemption, and follows any additional specific routing requirements indicated by the port State in which the shipyard or other repair facility is located, if applicable.

5.5 The exemption described in paragraph 5.4 of this regulation applies only for the following period:

.1 for a newly constructed ship, the period beginning at the time the ship is delivered from the shipyard, including sea trials, and ending at the time the ship directly exits the NO_x Tier III emission control area(s) or, with regard to a ship fitted with a dual fuel engine, the ship directly exits the NO_x Tier III emission control area(s) or proceeds directly to the nearest gas fuel bunkering facility appropriate to the ship located in the NO_x Tier III emission control area(s);

.2 for a ship with a Tier II engine undergoing conversion, maintenance or repair, the period beginning at the time the ship enters the NO_x Tier III emission control area(s) and proceeds directly to the shipyard or other repair facility, and ending at the time the ship is released from the shipyard or other repair facility and directly exits the NO_x Tier III emission control area (s) after performing sea trials, if applicable; or

.3 for a ship with a dual fuel engine undergoing conversion, maintenance or repair, when the ship is required to not have gas fuel or gas cargo on board due to safety requirements, the period beginning at the time the ship enters the NO_x Tier III emission control area(s) or when it is degassed in the NO_x Tier III emission control area(s) and proceeds directly to the shipyard or other repair facility, and ending at the time when the ship is released from the shipyard or other repair facility and directly exits the NO_x Tier III emission control area(s) or proceeds directly to the nearest gas fuel bunkering facility appropriate to the ship located in the NO_x Tier III emission control area(s)."

Emission control area

7 *The existing text of paragraph 6 is replaced by the following:*

"**6** For the purposes of this regulation, a NO_x Tier III emission control area shall be any sea area, including any port area, designated by the Organization in accordance with the criteria and procedures set forth in appendix III to this Annex. The NO_x Tier III emission control areas are:

.1 the North American Emission Control Area, which means the area described by the coordinates provided in appendix VII to this Annex;

.2 the United States Caribbean Sea Emission Control Area, which means the area described by the coordinates provided in appendix VII to this Annex;

.3 the Baltic Sea Emission Control Area as defined in regulation 1.11.2 of Annex I of the present Convention; and

.4 the North Sea Emission Control Area as defined in regulation 1.14.6 of Annex V of the present Convention."

Appendix V

Information to be included in the bunker delivery note (regulation 18.5)

8 *The items listed in the Appendix are numbered from 1 to 9.*

9 *In item 7, the comma after "15°C" is deleted and the expression* "kg/m^3" is replaced by "(kg/m^3)".

10 *Item 9 is replaced with the following:*

"A declaration signed and certified by the fuel oil supplier's representative that the fuel oil supplied is in conformity with regulation 18.3 of this Annex and that the sulphur content of the fuel oil supplied does not exceed:

☐ the limit value given by regulation 14.1 of this Annex;

☐ the limit value given by regulation 14.4 of this Annex; or

☐ the purchaser's specified limit value of _____ (% m/m), as completed by the fuel oil supplier's representative and on the basis of the purchaser's notification that the fuel oil is intended to be used:

.1 in combination with an equivalent means of compliance in accordance with regulation 4 of this Annex; or

.2 is subject to a relevant exemption for a ship to conduct trials for sulphur oxides emission reduction and control technology research in accordance with regulation 3.2 of this Annex.

The declaration shall be completed by the fuel oil supplier's representative by marking the applicable box(es) with a cross (x)."

Resolution MEPC.301(72)

adopted on 13 April 2018

MARPOL Annex VI

Regulations for the prevention of air pollution from ships

Chapter 3 – Requirements for control of emissions from ships

Regulation 13

Nitrogen oxides (NO_x)

Tier III

1 *In paragraph 5.3, the words* "an emission control area designated under paragraph 6 of this regulation" *are replaced with the words* "a NO_x Tier III emission control area".

Chapter 4 – Regulations on energy efficiency for ships

Regulation 21

Required EEDI

2 *In "***Table 2** *– Parameters for determination of reference values for the different ship types", referred to in paragraphs 3 and 4, rows 2.34 and 2.35 for ro-ro cargo ships and ro-ro passenger ships are replaced by the following:*

"
<table>
<tr><td rowspan="2">2.34 Ro-ro cargo ship</td><td>1,405.15</td><td>DWT of the ship</td><td rowspan="2">0.498</td></tr>
<tr><td>1,686.17*</td><td>DWT of the ship where DWT $\leq$ 17,000*
17,000 where DWT $>$ 17,000*</td></tr>
<tr><td rowspan="2">2.35 Ro-ro passenger ship</td><td>752.16</td><td>DWT of the ship</td><td rowspan="2">0.381</td></tr>
<tr><td>902.59*</td><td>DWT of the ship where DWT $\leq$ 10,000*
10,000 where DWT $>$ 10,000*</td></tr>
</table>

* To be used from phase 2 and thereafter."

Resolution MEPC.305(73)
adopted on 26 October 2018

MARPOL Annex VI

Regulations for the prevention of air pollution from ships

Chapter 3 – Requirements for control of emissions from ships

Regulation 14
Sulphur oxides (SO_x) and particulate matter

General requirements

1 *Paragraph 1 is replaced by the following:*

"**1** The sulphur content of fuel oil used or carried for use on board a ship shall not exceed 0.50% m/m."

Requirements within emission control areas

2 *Paragraph 3 is replaced by the following:*

"**3** For the purpose of this regulation, an emission control area shall be any sea area, including any port area, designated by the Organization in accordance with the criteria and procedures set forth in appendix III to this Annex. The emission control areas under this regulation are:

.1 the Baltic Sea area as defined in regulation 1.11.2 of Annex I of the present Convention;

.2 the North Sea area as defined in regulation 1.14.6 of Annex V of the present Convention;

.3 the North American Emission Control Area, which means the area described by the coordinates provided in appendix VII to this Annex; and

.4 the United States Caribbean Sea Emission Control Area, which means the area described by the coordinates provided in appendix VII to this Annex."

3 *Paragraph 4 is replaced by the following:*

"**4** While a ship is operating within an emission control area, the sulphur content of fuel oil used on board that ship shall not exceed 0.10% m/m."

4 *The subtitle "***Review provision***" and paragraphs 8, 9 and 10 are deleted.*

Appendix I

Form of International Air Pollution Prevention (IAPP) Certificate (regulation 8)

SUPPLEMENT TO INTERNATIONAL AIR POLLUTION PREVENTION CERTIFICATE (IAPP CERTIFICATE)

RECORD OF CONSTRUCTION AND EQUIPMENT

2 Control of emissions from ships

2.3 *Sulphur oxides (SO_x) and particulate matter* (regulation 14)

5 *Paragraphs 2.3.1 and 2.3.2 are replaced by the following and a new paragraph 2.3.3 is added as follows:*

"2.3.1 When the ship operates outside of an emission control area specified in regulation 14.3, the ship uses:

.1 fuel oil with a sulphur content as documented by bunker delivery notes that does not exceed the limit value of 0.50% m/m, and/or . . . . ☐

.2 an equivalent arrangement approved in accordance with regulation 4.1 as listed in paragraph 2.6 that is at least as effective in terms of SO_x emission reductions as compared to using a fuel oil with a sulphur content limit value of 0.50% m/m. . . . . ☐

2.3.2 When the ship operates inside an emission control area specified in regulation 14.3, the ship uses:

.1 fuel oil with a sulphur content as documented by bunker delivery notes that does not exceed the limit value of 0.10% m/m, and/or. . . . . ☐

.2 an equivalent arrangement approved in accordance with regulation 4.1 as listed in paragraph 2.6 that is at least as effective in terms of SO_x emission reductions as compared to using a fuel oil with a sulphur content limit value of 0.10% m/m . . . . ☐

2.3.3 For a ship without an equivalent arrangement approved in accordance with regulation 4.1 as listed in paragraph 2.6, the sulphur content of fuel oil carried for use on board the ship shall not exceed 0.50% m/m as documented by bunker delivery notes . . . . ☐"

MARPOL
Consolidated Edition 2017

International Convention for the Prevention of Pollution from Ships

Errata

March 2018

Chapter 4 – Regulations on energy efficiency for ships

On pages 293 and 294, and page 433, replace regulations 20 and 21 with the following:

"Regulation 20
Attained Energy Efficiency Design Index (attained EEDI)

1 The attained EEDI shall be calculated for:

.1 each new ship;

.2 each new ship which has undergone a major conversion; and

.3 each new or existing ship which has undergone a major conversion, that is so extensive that the ship is regarded by the Administration as a newly-constructed ship,

which falls into one or more of the categories in regulations 2.25 to 2.35, 2.38 and 2.39 of this Annex. The attained EEDI shall be specific to each ship and shall indicate the estimated performance of the ship in terms of energy efficiency, and be accompanied by the EEDI technical file that contains the information necessary for the calculation of the attained EEDI and that shows the process of calculation. The attained EEDI shall be verified, based on the EEDI technical file, either by the Administration or by any organization duly authorized by it.*

2 The attained EEDI shall be calculated taking into account guidelines† developed by the Organization.

Regulation 21
Required EEDI

1 For each:

.1 new ship,

.2 new ship which has undergone a major conversion, and

.3 new or existing ship which has undergone a major conversion that is so extensive that the ship is regarded by the Administration as a newly-constructed ship,

which falls into one of the categories in regulations 2.25 to 2.31, 2.33 to 2.35, 2.38 and 2.39 and to which this chapter is applicable, the attained EEDI shall be as follows:

$$\text{Attained EEDI} \leq \text{Required EEDI} = \left(1 - \frac{X}{100}\right) \cdot \text{Reference line value}$$

where X is the reduction factor specified in table 1 for the required EEDI compared to the EEDI reference line.

---

* Refer to Guidelines for the authorization of organizations acting on behalf of the Administration (resolution A.739(18), as amended by resolution MSC.208(81)), and Specifications on the survey and certification functions of recognized organizations acting on behalf of the Administration (resolution A.789(19), as may be amended).

† Refer to 2014 Guidelines on the method of calculation of the Energy Efficiency Design Index for new ships (resolution MEPC.245(66), as amended by resolutions MEPC.263(68) and MEPC.281(70)).

2 For each new and existing ship that has undergone a major conversion which is so extensive that the ship is regarded by the Administration as a newly constructed ship, the attained EEDI shall be calculated and meet the requirement of paragraph 21.1 with the reduction factor applicable corresponding to the ship type and size of the converted ship at the date of the contract of the conversion, or in the absence of a contract, the commencement date of the conversion."

Appendix VII
Emission control areas (regulations 13.6 and 14.3)

On pages 313 to 321, and pages 452 to 460, replace appendix VII with the following:

"Appendix VII
Emission control areas (regulations 13.6 and 14.3)

1 The boundaries of emission control areas designated under regulations 13.6 and 14.3, other than the Baltic Sea and the North Sea areas, are set forth in this appendix.

2 The North American area comprises:

.1 the sea area located off the Pacific coasts of the United States and Canada, enclosed by geodesic lines connecting the following coordinates:

Point	Latitude	Longitude
1	32°32′10″ N	117°06′11″ W
2	32°32′04″ N	117°07′29″ W
3	32°31′39″ N	117°14′20″ W
4	32°33′13″ N	117°15′50″ W
5	32°34′21″ N	117°22′01″ W
6	32°35′23″ N	117°27′53″ W
7	32°37′38″ N	117°49′34″ W
8	31°07′59″ N	118°36′21″ W
9	30°33′25″ N	121°47′29″ W
10	31°46′11″ N	123°17′22″ W
11	32°21′58″ N	123°50′44″ W
12	32°56′39″ N	124°11′47″ W
13	33°40′12″ N	124°27′15″ W
14	34°31′28″ N	125°16′52″ W
15	35°14′38″ N	125°43′23″ W
16	35°43′60″ N	126°18′53″ W
17	36°16′25″ N	126°45′30″ W
18	37°01′35″ N	127°07′18″ W
19	37°45′39″ N	127°38′02″ W
20	38°25′08″ N	127°52′60″ W
21	39°25′05″ N	128°31′23″ W
22	40°18′47″ N	128°45′46″ W
23	41°13′39″ N	128°40′22″ W

Point	Latitude	Longitude
24	42°12′49″ N	129°00′38″ W
25	42°47′34″ N	129°05′42″ W
26	43°26′22″ N	129°01′26″ W
27	44°24′43″ N	128°41′23″ W
28	45°30′43″ N	128°40′02″ W
29	46°11′01″ N	128°49′01″ W
30	46°33′55″ N	129°04′29″ W
31	47°39′55″ N	131°15′41″ W
32	48°32′32″ N	132°41′00″ W
33	48°57′47″ N	133°14′47″ W
34	49°22′39″ N	134°15′51″ W
35	50°01′52″ N	135°19′01″ W
36	51°03′18″ N	136°45′45″ W
37	51°54′04″ N	137°41′54″ W
38	52°45′12″ N	138°20′14″ W
39	53°29′20″ N	138°40′36″ W
40	53°40′39″ N	138°48′53″ W
41	54°13′45″ N	139°32′38″ W
42	54°39′25″ N	139°56′19″ W
43	55°20′18″ N	140°55′45″ W
44	56°07′12″ N	141°36′18″ W
45	56°28′32″ N	142°17′19″ W
46	56°37′19″ N	142°48′57″ W
47	58°51′04″ N	153°15′03″ W

.2 the sea areas located off the Atlantic coasts of the United States, Canada and France (Saint-Pierre-et-Miquelon), and the Gulf of Mexico coast of the United States enclosed by geodesic lines connecting the following coordinates:

Point	Latitude	Longitude
1	60°00′00″ N	64°09′36″ W
2	60°00′00″ N	56°43′00″ W
3	58°54′01″ N	55°38′05″ W
4	57°50′52″ N	55°03′47″ W
5	57°35′13″ N	54°00′59″ W
6	57°14′20″ N	53°07′58″ W
7	56°48′09″ N	52°23′29″ W
8	56°18′13″ N	51°49′42″ W
9	54°23′21″ N	50°17′44″ W
10	53°44′54″ N	50°07′17″ W
11	53°04′59″ N	50°10′05″ W
12	52°20′06″ N	49°57′09″ W
13	51°34′20″ N	48°52′45″ W
14	50°40′15″ N	48°16′04″ W
15	50°02′28″ N	48°07′03″ W

Point	Latitude	Longitude
16	49°24′03″ N	48°09′35″ W
17	48°39′22″ N	47°55′17″ W
18	47°24′25″ N	47°46′56″ W
19	46°35′12″ N	48°00′54″ W
20	45°19′45″ N	48°43′28″ W
21	44°43′38″ N	49°16′50″ W
22	44°16′38″ N	49°51′23″ W
23	43°53′15″ N	50°34′01″ W
24	43°36′06″ N	51°20′41″ W
25	43°23′59″ N	52°17′22″ W
26	43°19′50″ N	53°20′13″ W
27	43°21′14″ N	54°09′20″ W
28	43°29′41″ N	55°07′41″ W
29	42°40′12″ N	55°31′44″ W
30	41°58′19″ N	56°09′34″ W
31	41°20′21″ N	57°05′13″ W
32	40°55′34″ N	58°02′55″ W
33	40°41′38″ N	59°05′18″ W
34	40°38′33″ N	60°12′20″ W
35	40°45′46″ N	61°14′03″ W
36	41°04′52″ N	62°17′49″ W
37	40°36′55″ N	63°10′49″ W
38	40°17′32″ N	64°08′37″ W
39	40°07′46″ N	64°59′31″ W
40	40°05′44″ N	65°53′07″ W
41	39°58′05″ N	65°59′51″ W
42	39°28′24″ N	66°21′14″ W
43	39°01′54″ N	66°48′33″ W
44	38°39′16″ N	67°20′59″ W
45	38°19′20″ N	68°02′01″ W
46	38°05′29″ N	68°46′55″ W
47	37°58′14″ N	69°34′07″ W
48	37°57′47″ N	70°24′09″ W
49	37°52′46″ N	70°37′50″ W
50	37°18′37″ N	71°08′33″ W
51	36°32′25″ N	71°33′59″ W
52	35°34′58″ N	71°26′02″ W
53	34°33′10″ N	71°37′04″ W
54	33°54′49″ N	71°52′35″ W
55	33°19′23″ N	72°17′12″ W
56	32°45′31″ N	72°54′05″ W
57	31°55′13″ N	74°12′02″ W
58	31°27′14″ N	75°15′20″ W

Point	Latitude	Longitude
59	31°03′16″ N	75°51′18″ W
60	30°45′42″ N	76°31′38″ W
61	30°12′48″ N	77°18′29″ W
62	29°25′17″ N	76°56′42″ W
63	28°36′59″ N	76°47′60″ W
64	28°17′13″ N	76°40′10″ W
65	28°17′12″ N	79°11′23″ W
66	27°52′56″ N	79°28′35″ W
67	27°26′01″ N	79°31′38″ W
68	27°16′13″ N	79°34′18″ W
69	27°11′54″ N	79°34′56″ W
70	27°05′59″ N	79°35′19″ W
71	27°00′28″ N	79°35′17″ W
72	26°55′16″ N	79°34′39″ W
73	26°53′58″ N	79°34′27″ W
74	26°45′46″ N	79°32′41″ W
75	26°44′30″ N	79°32′23″ W
76	26°43′40″ N	79°32′20″ W
77	26°41′12″ N	79°32′01″ W
78	26°38′13″ N	79°31′32″ W
79	26°36′30″ N	79°31′06″ W
80	26°35′21″ N	79°30′50″ W
81	26°34′51″ N	79°30′46″ W
82	26°34′11″ N	79°30′38″ W
83	26°31′12″ N	79°30′15″ W
84	26°29′05″ N	79°29′53″ W
85	26°25′31″ N	79°29′58″ W
86	26°23′29″ N	79°29′55″ W
87	26°23′21″ N	79°29′54″ W
88	26°18′57″ N	79°31′55″ W
89	26°15′26″ N	79°33′17″ W
90	26°15′13″ N	79°33′23″ W
91	26°08′09″ N	79°35′53″ W
92	26°07′47″ N	79°36′09″ W
93	26°06′59″ N	79°36′35″ W
94	26°02′52″ N	79°38′22″ W
95	25°59′30″ N	79°40′03″ W
96	25°59′16″ N	79°40′08″ W
97	25°57′48″ N	79°40′38″ W
98	25°56′18″ N	79°41′06″ W
99	25°54′04″ N	79°41′38″ W
100	25°53′24″ N	79°41′46″ W
101	25°51′54″ N	79°41′59″ W

Point	Latitude	Longitude
102	25°49′33″ N	79°42′16″ W
103	25°48′24″ N	79°42′23″ W
104	25°48′20″ N	79°42′24″ W
105	25°46′26″ N	79°42′44″ W
106	25°46′16″ N	79°42′45″ W
107	25°43′40″ N	79°42′59″ W
108	25°42′31″ N	79°42′48″ W
109	25°40′37″ N	79°42′27″ W
110	25°37′24″ N	79°42′27″ W
111	25°37′08″ N	79°42′27″ W
112	25°31′03″ N	79°42′12″ W
113	25°27′59″ N	79°42′11″ W
114	25°24′04″ N	79°42′12″ W
115	25°22′21″ N	79°42′20″ W
116	25°21′29″ N	79°42′08″ W
117	25°16′52″ N	79°41′24″ W
118	25°15′57″ N	79°41′31″ W
119	25°10′39″ N	79°41′31″ W
120	25°09′51″ N	79°41′36″ W
121	25°09′03″ N	79°41′45″ W
122	25°03′55″ N	79°42′29″ W
123	25°02′60″ N	79°42′56″ W
124	25°00′30″ N	79°44′05″ W
125	24°59′03″ N	79°44′48″ W
126	24°55′28″ N	79°45′57″ W
127	24°44′18″ N	79°49′24″ W
128	24°43′04″ N	79°49′38″ W
129	24°42′36″ N	79°50′50″ W
130	24°41′47″ N	79°52′57″ W
131	24°38′32″ N	79°59′58″ W
132	24°36′27″ N	80°03′51″ W
133	24°33′18″ N	80°12′43″ W
134	24°33′05″ N	80°13′21″ W
135	24°32′13″ N	80°15′16″ W
136	24°31′27″ N	80°16′55″ W
137	24°30′57″ N	80°17′47″ W
138	24°30′14″ N	80°19′21″ W
139	24°30′06″ N	80°19′44″ W
140	24°29′38″ N	80°21′05″ W
141	24°28′18″ N	80°24′35″ W
142	24°28′06″ N	80°25′10″ W
143	24°27′23″ N	80°27′20″ W
144	24°26′30″ N	80°29′30″ W

Point	Latitude	Longitude
145	24°25′07″ N	80°32′22″ W
146	24°23′30″ N	80°36′09″ W
147	24°22′33″ N	80°38′56″ W
148	24°22′07″ N	80°39′51″ W
149	24°19′31″ N	80°45′21″ W
150	24°19′16″ N	80°45′47″ W
151	24°18′38″ N	80°46′49″ W
152	24°18′35″ N	80°46′54″ W
153	24°09′51″ N	80°59′47″ W
154	24°09′48″ N	80°59′51″ W
155	24°08′58″ N	81°01′07″ W
156	24°08′30″ N	81°01′51″ W
157	24°08′26″ N	81°01′57″ W
158	24°07′28″ N	81°03′06″ W
159	24°02′20″ N	81°09′05″ W
160	23°59′60″ N	81°11′16″ W
161	23°55′32″ N	81°12′55″ W
162	23°53′52″ N	81°19′43″ W
163	23°50′52″ N	81°29′59″ W
164	23°50′02″ N	81°39′59″ W
165	23°49′05″ N	81°49′59″ W
166	23°49′05″ N	82°00′11″ W
167	23°49′42″ N	82°09′59″ W
168	23°51′14″ N	82°24′59″ W
169	23°51′14″ N	82°39′59″ W
170	23°49′42″ N	82°48′53″ W
171	23°49′32″ N	82°51′11″ W
172	23°49′24″ N	82°59′59″ W
173	23°49′52″ N	83°14′59″ W
174	23°51′22″ N	83°25′49″ W
175	23°52′27″ N	83°33′01″ W
176	23°54′04″ N	83°41′35″ W
177	23°55′47″ N	83°48′11″ W
178	23°58′38″ N	83°59′59″ W
179	24°09′37″ N	84°29′27″ W
180	24°13′20″ N	84°38′39″ W
181	24°16′41″ N	84°46′07″ W
182	24°23′30″ N	84°59′59″ W
183	24°26′37″ N	85°06′19″ W
184	24°38′57″ N	85°31′54″ W
185	24°44′17″ N	85°43′11″ W
186	24°53′57″ N	85°59′59″ W
187	25°10′44″ N	86°30′07″ W

Point	Latitude	Longitude
188	25°43′15″ N	86°21′14″ W
189	26°13′13″ N	86°06′45″ W
190	26°27′22″ N	86°13′15″ W
191	26°33′46″ N	86°37′07″ W
192	26°01′24″ N	87°29′35″ W
193	25°42′25″ N	88°33′00″ W
194	25°46′54″ N	90°29′41″ W
195	25°44′39″ N	90°47′05″ W
196	25°51′43″ N	91°52′50″ W
197	26°17′44″ N	93°03′59″ W
198	25°59′55″ N	93°33′52″ W
199	26°00′32″ N	95°39′27″ W
200	26°00′33″ N	96°48′30″ W
201	25°58′32″ N	96°55′28″ W
202	25°58′15″ N	96°58′41″ W
203	25°57′58″ N	97°01′54″ W
204	25°57′41″ N	97°05′08″ W
205	25°57′24″ N	97°08′21″ W
206	25°57′24″ N	97°08′47″ W

.3 the sea area located off the coasts of the Hawaiian Islands of Hawai′i, Maui, Oahu, Moloka′i, Ni′ihau, Kaua′i, Lāna′i and Kaho′olawe, enclosed by geodesic lines connecting the following coordinates:

Point	Latitude	Longitude
1	22°32′54″ N	153°00′33″ W
2	23°06′05″ N	153°28′36″ W
3	23°32′11″ N	154°02′12″ W
4	23°51′47″ N	154°36′48″ W
5	24°21′49″ N	155°51′13″ W
6	24°41′47″ N	156°27′27″ W
7	24°57′33″ N	157°22′17″ W
8	25°13′41″ N	157°54′13″ W
9	25°25′31″ N	158°30′36″ W
10	25°31′19″ N	159°09′47″ W
11	25°30′31″ N	159°54′21″ W
12	25°21′53″ N	160°39′53″ W
13	25°00′06″ N	161°38′33″ W
14	24°40′49″ N	162°13′13″ W
15	24°15′53″ N	162°43′08″ W
16	23°40′50″ N	163°13′00″ W
17	23°03′20″ N	163°32′58″ W
18	22°20′09″ N	163°44′41″ W
19	21°36′45″ N	163°46′03″ W
20	20°55′26″ N	163°37′44″ W

Point	Latitude	Longitude
21	20°13′34″ N	163°19′13″ W
22	19°39′03″ N	162°53′48″ W
23	19°09′43″ N	162°20′35″ W
24	18°39′16″ N	161°19′14″ W
25	18°30′31″ N	160°38′30″ W
26	18°29′31″ N	159°56′17″ W
27	18°10′41″ N	159°14′08″ W
28	17°31′17″ N	158°56′55″ W
29	16°54′06″ N	158°30′29″ W
30	16°25′49″ N	157°59′25″ W
31	15°59′57″ N	157°17′35″ W
32	15°40′37″ N	156°21′06″ W
33	15°37′36″ N	155°22′16″ W
34	15°43′46″ N	154°46′37″ W
35	15°55′32″ N	154°13′05″ W
36	16°46′27″ N	152°49′11″ W
37	17°33′42″ N	152°00′32″ W
38	18°30′16″ N	151°30′24″ W
39	19°02′47″ N	151°22′17″ W
40	19°34′46″ N	151°19′47″ W
41	20°07′42″ N	151°22′58″ W
42	20°38′43″ N	151°31′36″ W
43	21°29′09″ N	151°59′50″ W
44	22°06′58″ N	152°31′25″ W
45	22°32′54″ N	153°00′33″ W

3 The United States Caribbean Sea area includes:

.1 the sea area located off the Atlantic and Caribbean coasts of the Commonwealth of Puerto Rico and the United States Virgin Islands, enclosed by geodesic lines connecting the following coordinates:

Point	Latitude	Longitude
1	17°18′37″ N	67°32′14″ W
2	19°11′14″ N	67°26′45″ W
3	19°30′28″ N	65°16′48″ W
4	19°12′25″ N	65°06′08″ W
5	18°45′13″ N	65°00′22″ W
6	18°41′14″ N	64°59′33″ W
7	18°29′22″ N	64°53′51″ W
8	18°27′35″ N	64°53′22″ W
9	18°25′21″ N	64°52′39″ W
10	18°24′30″ N	64°52′19″ W
11	18°23′51″ N	64°51′50″ W
12	18°23′42″ N	64°51′23″ W

Point	Latitude	Longitude
13	18°23′36″ N	64°50′17″ W
14	18°23′48″ N	64°49′41″ W
15	18°24′11″ N	64°49′00″ W
16	18°24′28″ N	64°47′57″ W
17	18°24′18″ N	64°47′01″ W
18	18°23′13″ N	64°46′37″ W
19	18°22′37″ N	64°45′20″ W
20	18°22′39″ N	64°44′42″ W
21	18°22′42″ N	64°44′36″ W
22	18°22′37″ N	64°44′24″ W
23	18°22′39″ N	64°43′42″ W
24	18°22′30″ N	64°43′36″ W
25	18°22′25″ N	64°42′58″ W
26	18°22′26″ N	64°42′28″ W
27	18°22′15″ N	64°42′03″ W
28	18°22′22″ N	64°40′60″ W
29	18°21′57″ N	64°40′15″ W
30	18°21′51″ N	64°38′23″ W
31	18°21′22″ N	64°38′16″ W
32	18°20′39″ N	64°38′33″ W
33	18°19′15″ N	64°38′14″ W
34	18°19′07″ N	64°38′16″ W
35	18°17′23″ N	64°39′38″ W
36	18°16′43″ N	64°39′41″ W
37	18°11′33″ N	64°38′58″ W
38	18°03′02″ N	64°38′03″ W
39	18°02′56″ N	64°29′35″ W
40	18°02′51″ N	64°27′02″ W
41	18°02′30″ N	64°21′08″ W
42	18°02′31″ N	64°20′08″ W
43	18°02′03″ N	64°15′57″ W
44	18°00′12″ N	64°02′29″ W
45	17°59′58″ N	64°01′04″ W
46	17°58′47″ N	63°57′01″ W
47	17°57′51″ N	63°53′54″ W
48	17°56′38″ N	63°53′21″ W
49	17°39′40″ N	63°54′53″ W
50	17°37′08″ N	63°55′10″ W
51	17°30′21″ N	63°55′56″ W
52	17°11′36″ N	63°57′57″ W
53	17°04′60″ N	63°58′41″ W
54	16°59′49″ N	63°59′18″ W
55	17°18′37″ N	67°32′14″ W

″

Appendix IX

Information to be submitted to the IMO Ship Fuel Oil Consumption Database

On page 406, replace the third footnote with the following:

"‡ DWT means the difference in tonnes between the displacement of a ship in water of relative density of 1,025 kg/m^3 at the summer load draught and the lightweight of the ship. The summer load draught should be taken as the maximum summer draught as certified in the stability booklet approved by the Administration or an organization recognized by it."

Articles, Protocols, Annexes and Unified Interpretations of the International Convention for the Prevention of Pollution from Ships, 1973, as modified by the 1978 and 1997 Protocols

Incorporating all amendments in force on 1 January 2017

London, 2017

First published in 1991 by the
INTERNATIONAL MARITIME ORGANIZATION
4 Albert Embankment, London SE1 7SR
www.imo.org

Sixth edition 2017

Printed and bound by CPI Group (UK) Ltd,
Croydon, CRO 4YY

ISBN 978-92-801-1657-1

IMO PUBLICATION
Sales number: IE520E

Summary of contents

Contents

MARPOL Annex II
Regulations for the control of pollution by noxious liquid substances in bulk

Appendices to Annex II

MARPOL Annex III
Regulations for the prevention of pollution by harmful substances carried by sea in packaged form

Chapter 1 – General

Chapter 2 – Verification of compliance with the provisions of this Annex

Appendix to Annex III

MARPOL Annex IV
Regulations for the prevention of pollution by sewage from ships

Chapter 1 – General

Chapter 2 – Surveys and certification

MARPOL Annex VI

Regulations for the prevention of air pollution from ships

Appendices to Annex VI

Additional information

Introduction

The International Convention for the Prevention of Pollution from Ships, 1973 (MARPOL Convention), was adopted by the International Conference on Marine Pollution convened by the International Maritime Organization (IMO) from 8 October to 2 November 1973. Protocol I (Provisions concerning reports on incidents involving Harmful Substances) and Protocol II (Arbitration) were adopted at the same Conference. This Convention was subsequently modified by the Protocol of 1978 relating thereto, which was adopted by the International Conference on Tanker Safety and Pollution Prevention (TSPP Conference) convened by IMO from 6 to 17 February 1978. The Convention, as modified by the 1978 Protocol, is known as the "International Convention for the Prevention of Pollution from Ships, 1973, as modified by the Protocol of 1978 relating thereto", or, in short form, "MARPOL 73/78". Regulations covering the various sources of ship-generated pollution are contained in the five Annexes of the Convention. The Convention has also been modified by the Protocol of 1997, whereby a sixth Annex was added. It may be noted that the Marine Environment Protection Committee (MEPC), at its fifty-sixth session, decided that, when referring to the Convention and its six Annexes as a whole, the term "MARPOL" should be preferred to "MARPOL 73/78", as the latter would leave Annex VI on Prevention of air pollution from ships, which had been adopted by the 1997 Protocol, outside its scope.

The MEPC, since its inception in 1974, has reviewed various provisions of the MARPOL Convention that have been found to require clarification or have given rise to difficulties in implementation. In order to resolve such ambiguities and difficulties in a uniform manner, the MEPC agreed that it was desirable to develop unified interpretations. In certain cases, the MEPC recognized that there was a need to amend existing regulations or to introduce new regulations with the aim of reducing even further operational and accidental pollution from ships. These activities by the MEPC have resulted in a number of unified interpretations and amendments to the Convention.

The purpose of this publication is to provide an easy reference to the up-to-date provisions and unified interpretations of the articles, protocols and Annexes of the MARPOL Convention, including the incorporation of all amendments in force on 1 January 2017.

The footnotes contained in this publication are not part of the authentic text of the MARPOL Convention. They were agreed by the MEPC when the various amendments to the Convention were adopted, and inserted in the publication by the Secretariat, taking into account the provisions of the uniform wording for referencing IMO instruments adopted by resolution A.911(22). Footnotes inserted or updated refer to codes, guidelines, manuals or decisions of the MEPC and, in some cases, a specific edition of the industry guide or standards relating to a particular text. The reader should make use of the latest versions of the referenced texts, bearing in mind that such texts may have been revised or superseded since the publication of this Consolidated Edition of MARPOL.

Protocol I – Provisions concerning reports on incidents involving harmful substances

This Protocol was adopted on 2 November 1973 and subsequently amended by:

- 1985 amendments (resolution MEPC.21(22)) by which the Protocol was replaced by a revised text: entered into force on 6 April 1987; and
- 1996 amendments (resolution MEPC.68(38)) on amendments to article II(1): entered into force on 1 January 1998.

Protocol II – Arbitration

This Protocol was also adopted on 2 November 1973 and there have been no amendments to it.

Annex I – Regulations for the prevention of pollution by oil

Annex I entered into force on 2 October 1983 and, as between the Parties to MARPOL 73/78, supersedes the International Convention for the Prevention of Pollution of the Sea by Oil, 1954, as amended in 1962 and 1969, which was then in force. A number of amendments to Annex I have been adopted by the MEPC and have entered into force as summarized below:

- 1984 amendments (resolution MEPC.14(20)) on control of discharge of oil; retention of oil on board; pumping, piping and discharge arrangements of oil tankers; subdivision and stability: entered into force on 7 January 1986;
- 1987 amendments (resolution MEPC.29(25)) on designation of the Gulf of Aden as a special area: entered into force on 1 April 1989;
- 1990 amendments (resolution MEPC.39(29)) on the introduction of the harmonized system of survey and certification: entered into force on 3 February 2000;
- 1990 amendments (resolution MEPC.42(30)) on designation of the Antarctic area as a special area: entered into force on 17 March 1992;
- 1991 amendments (resolution MEPC.47(31)) on new regulation 26, Shipboard Oil Pollution Emergency Plan, and other amendments to Annex I: entered into force on 4 April 1993;
- 1992 amendments (resolution MEPC.51(32)) on discharge criteria of Annex I: entered into force on 6 July 1993;
- 1992 amendments (resolution MEPC.52(32)) on new regulations 13F and 13G and related amendments to Annex I: entered into force on 6 July 1993;
- 1994 amendments (resolution 1 adopted on 2 November 1994 by the Conference of Parties to MARPOL 73/78) on port State control on operational requirements: entered into force on 3 March 1996;
- 1997 amendments (resolution MEPC.75(40)) on designation of North West European waters as a special area and new regulation 25A: entered into force on 1 February 1999;
- 1999 amendments (resolution MEPC.78(43)) to regulations 13G and 26 and the IOPP Certificate: entered into force on 1 January 2001;
- 2001 amendments (resolution MEPC.95(46)) to regulation 13G: entered into force on 1 September 2002;
- 2003 amendments (resolution MEPC.111(50)) to regulation 13G, new regulation 13H and related amendments to Annex I: entered into force on 5 April 2005;
- 2004 amendments (resolution MEPC.117(52)) on the revised Annex I: entered into force on 1 January 2007;
- 2006 amendments (resolution MEPC.141(54)) to regulations 1 and 21, addition of regulation 12A and related amendments to Annex I: entered into force on 1 August 2007;
- 2006 amendments (resolution MEPC.154(55)) on designation of the Southern South African waters as a special area: entered into force on 1 March 2008;
- 2007 amendments (resolution MEPC.164(56)) to regulation 38: entered into force on 1 December 2008;
- 2009 amendments (resolution MEPC.186(59)) on addition of a new chapter 8 concerning transfer of oil cargo between oil tankers at sea and amendments to the IOPP Certificate; and resolution MEPC.187(59) on amendments to regulations 1, 12, 13, 17 and 38, the IOPP Certificate and Oil Record Book: both entered into force on 1 January 2011;
- 2010 amendments (resolution MEPC.189(60)) on addition of a new chapter 9 concerning the use or carriage of oils in the Antarctic area: entered into force on 1 August 2011;

- 2012 amendments (resolution MEPC.216(63)) on regional arrangements for port reception facilities: entered into force on 1 August 2013;
- 2013 amendments (resolution MEPC.235(65)) to Form A and Form B of Supplements to the IOPP Certificate: entered into force on 1 October 2014;
- 2013 amendments (resolution MEPC.238(65)) to make the RO Code mandatory: entered into force on 1 January 2015;
- 2014 amendments (resolution MEPC.246(66)) to make the use of the III Code mandatory: entered into force on 1 January 2016;
- 2014 amendments (resolution MEPC.248(66)) on mandatory carriage requirements for a stability instrument: entered into force on 1 January 2016;
- 2014 amendments (resolution MEPC.256(67)) to regulation 43: entered into force on 1 March 2016;
- 2015 amendments (resolution MEPC.265(68)) to make the use of environment-related provisions of the Polar Code mandatory: entered into force on 1 January 2017; and
- 2015 amendments (resolution MEPC.266(68)) to regulation 12: entered into force on 1 January 2017.

Annex II – Regulations for the control of pollution by noxious liquid substances in bulk

To facilitate implementation of the Annex, the original text underwent amendments in 1985, by resolution MEPC.16(22), in respect of pumping, piping and control requirements. At its twenty-second session, the MEPC also decided that, in accordance with article II of the 1978 Protocol, "Parties shall be bound by the provisions of Annex II of MARPOL 73/78 as amended from 6 April 1987" (resolution MEPC.17(22)). Subsequent amendments have been adopted by the MEPC and have entered into force as summarized below:

- 1989 amendments (resolution MEPC.34(27)), which updated appendices II and III to make them compatible with chapters 17/VI and 18/VII of the IBC Code and BCH Code, respectively: entered into force on 13 October 1990;
- 1990 amendments (resolution MEPC.39(29)) on introduction of the harmonized system of survey and certification: entered into force on 3 February 2000;
- 1992 amendments (resolution MEPC.57(33)) on designation of the Antarctic area as a special area and lists of liquid substances in appendices to Annex II: entered into force on 1 July 1994;
- 1994 amendments (resolution 1 adopted on 2 November 1994 by the Conference of Parties to MARPOL 73/78) on port State control on operational requirements: entered into force on 3 March 1996;
- 1999 amendments (resolution MEPC.78(43)) on addition of new regulation 16: entered into force on 1 January 2001;
- 2004 amendments (resolution MEPC.118(52)) on the revised Annex II: entered into force on 1 January 2007;
- 2012 amendments (resolution MEPC.216(63)) on regional arrangements for port reception facilities: entered into force on 1 August 2013;
- 2013 amendments (resolution MEPC.238(65)) to make the RO Code mandatory: entered into force on 1 January 2015;
- 2014 amendments (resolution MEPC.246(66)) to make the use of the III Code mandatory: entered into force on 1 January 2016; and
- 2015 amendments (resolution MEPC.265(68)) to make the use of environment-related provisions of the Polar Code mandatory: entered into force on 1 January 2017.

Annex III – Regulations for the prevention of pollution by harmful substances carried by sea in packaged form

Annex III entered into force on 1 July 1992. However, long before this entry into force date, the MEPC, with the concurrence of the Maritime Safety Committee (MSC), agreed that the Annex should be implemented through the IMDG Code. The IMDG Code had amendments covering marine pollution prepared by the MSC (Amendment 25-89) and these amendments were implemented from 1 January 1991. Subsequent amendments have been adopted by the MEPC and have entered into force as summarized below:

- 1992 amendments (resolution MEPC.58(33)), which totally revised Annex III as a clarification of the requirements in the original version of Annex III rather than a change of substance, and incorporated the reference to the IMDG Code: entered into force on 28 February 1994;
- 1994 amendments (resolution 2 adopted on 2 November 1994 by the Conference of Parties to MARPOL 73/78) on port State control on operational requirements: entered into force on 3 March 1996;
- 2000 amendments (MEPC.84(44)), deleting a clause relating to tainting of seafood: entered into force on 1 January 2002;
- 2006 amendments (resolution MEPC.156(55)) on the revised Annex III: entered into force on 1 January 2010;
- 2010 amendments (resolution MEPC.193(61)) on the revised Annex III: entered into force on 1 January 2014;
- 2014 amendments (resolution MEPC.246(66)) to make the use of the III Code mandatory: entered into force on 1 January 2016; and
- 2014 amendments (resolution MEPC.257(67)) to the appendix on criteria for the identification of harmful substances in packaged form: entered into force on 1 March 2016.

Annex IV – Regulations for the prevention of pollution by sewage from ships

Annex IV entered into force on 27 September 2003. Subsequent amendments have been adopted by the MEPC and have entered into force as summarized below:

- 2004 amendments (resolution MEPC.115(51)) on the revised Annex IV: entered into force on 1 August 2005;
- 2006 amendments (resolution MEPC.143(54)) on new regulation 13 concerning port State control on operational requirements: entered into force on 1 August 2007;
- 2007 amendments (resolution MEPC.164(56)) on amendment of regulation 11.1.1: entered into force on 1 December 2008;
- 2011 amendments (resolution MEPC.200(62)) on special area provisions and the designation of the Baltic Sea as a special area: entered into force on 1 January 2013;
- 2012 amendments (resolution MEPC.216(63)) on regional arrangements for port reception facilities: entered into force on 1 August 2013;
- 2014 amendments (resolution MEPC.246(66)) to make the use of the III Code mandatory: entered into force on 1 January 2016; and
- 2015 amendments (resolution MEPC.265(68)) to make the use of environment-related provisions of the Polar Code mandatory: entered into force on 1 January 2017.

Annex V – Regulations for the prevention of pollution by garbage from ships

Annex V entered into force on 31 December 1988. Subsequent amendments have been adopted by the MEPC and have entered into force as summarized below:

- 1989 amendments (resolution MEPC.36(28)) on designation of the North Sea as a special area and amendment of regulation 6, Exceptions: entered into force on 18 February 1991;
- 1990 amendments (resolution MEPC.42(30)) on designation of the Antarctic area as a special area: entered into force on 17 March 1992;
- 1991 amendments (resolution MEPC.48(31)) on designation of the Wider Caribbean area as a special area: entered into force on 4 April 1993;
- 1994 amendments (resolution 3 adopted on 2 November 1994 by the Conference of Parties to MARPOL 73/78) on port State control on operational requirements: entered into force on 3 March 1996;
- 1995 amendments (resolution MEPC.65(37)) on amendment of regulation 2 and the addition of a new regulation 9 of Annex V: entered into force on 1 July 1997;
- 2000 amendments (resolution MEPC.89(45)) on amendments to regulations 1, 3, 5 and 9 and to the Record of Garbage Discharge: entered into force on 1 March 2002;
- 2004 amendments (resolution MEPC.116(51) on amendments to the appendix to Annex V: entered into force on 1 August 2005;
- 2011 amendments (resolution MEPC.201(62)) on the revised Annex V: entered into force on 1 January 2013;
- 2012 amendments (resolution MEPC.216(63)) on regional arrangements for port reception facilities: entered into force on 1 August 2013;
- 2014 amendments (resolution MEPC.246(66)) to make the use of the III Code mandatory: entered into force on 1 January 2016; and
- 2015 amendments (resolution MEPC.265(68)) to make the use of environment-related provisions of the Polar Code mandatory: entered into force on 1 January 2017.

Annex VI – Regulations for the prevention of air pollution from ships

Annex VI is appended to the Protocol of 1997 to amend the International Convention for the Prevention of Pollution from Ships, 1973, as modified by the Protocol of 1978 relating thereto, which was adopted by the International Conference of Parties to the MARPOL Convention in September 1997. Annex VI entered into force on 19 May 2005. Subsequent amendments have been adopted by the MEPC as summarized below:

- 2005 amendments (resolution MEPC.132(53)) on introducing the Harmonized System of Survey and Certification to the Annex and on designation of the North Sea as a new SO_x emission control area (SECA): entered into force on 22 November 2006;
- 2008 amendments (resolution MEPC.176(58)) on the revised Annex VI: entered into force on 1 July 2010;
- 2010 amendments (resolution MEPC.190(60)) on the North American emission control area: entered into force on 1 August 2011;
- 2010 amendments (resolution MEPC.194(61)) on revised form of supplement to the IAPP Certificate: entered into force on 1 February 2012;
- 2011 amendments (resolution MEPC.202(62)) to regulations 13 and 14 of Annex VI: entered into force on 1 January 2013; and resolution MEPC.203(62) on regulations on energy efficiency for ships: entered into force on 1 January 2013;

- 2012 amendments (resolution MEPC.217(63)) on regional arrangements for port reception facilities: entered into force on 1 August 2013;
- 2014 amendments (resolution MEPC.247(66)) to make the use of the III Code mandatory: entered into force on 1 January 2016;
- 2014 amendments (resolution MEPC.251(66)) on amendments to regulations 2, 13, 19, 20 and 21 and the Supplement to the IAPP Certificate under MARPOL Annex VI and certification of dual-fuel engines under the NO_X Technical Code 2008: entered into force on 1 September 2015; and
- 2014 amendments (resolution MEPC.258(67)) to regulations 2 and 13 and the Supplement to the IAPP Certificate: entered into force on 1 March 2016.

International Convention for the Prevention of Pollution from Ships, 1973

International Convention for the Prevention of Pollution from Ships, 1973

THE PARTIES TO THE CONVENTION,

BEING CONSCIOUS of the need to preserve the human environment in general and the marine environment in particular,

RECOGNIZING that deliberate, negligent or accidental release of oil and other harmful substances from ships constitutes a serious source of pollution,

RECOGNIZING ALSO the importance of the International Convention for the Prevention of Pollution of the Sea by Oil, 1954, as being the first multilateral instrument to be concluded with the prime objective of protecting the environment, and appreciating the significant contribution which that Convention has made in preserving the seas and coastal environment from pollution,

DESIRING to achieve the complete elimination of intentional pollution of the marine environment by oil and other harmful substances and the minimization of accidental discharge of such substances,

CONSIDERING that this object may best be achieved by establishing rules not limited to oil pollution having a universal purport,

HAVE AGREED as follows:

Article 1
General obligations under the Convention

(1) The Parties to the Convention undertake to give effect to the provisions of the present Convention and those Annexes thereto by which they are bound, in order to prevent the pollution of the marine environment by the discharge of harmful substances or effluents containing such substances in contravention of the Convention.

(2) Unless expressly provided otherwise, a reference to the present Convention constitutes at the same time a reference to its Protocols and to the Annexes.

Article 2
Definitions

For the purposes of the present Convention, unless expressly provided otherwise:

(1) *Regulation* means the regulations contained in the Annexes to the present Convention.

(2) *Harmful substance* means any substance which, if introduced into the sea, is liable to create hazards to human health, to harm living resources and marine life, to damage amenities or to interfere with other legitimate uses of the sea, and includes any substance subject to control by the present Convention.

(3) **(a)** *Discharge*, in relation to harmful substances or effluents containing such substances, means any release howsoever caused from a ship and includes any escape, disposal, spilling, leaking, pumping, emitting or emptying;

(b) *Discharge* does not include:

(i) dumping within the meaning of the Convention on the Prevention of Marine Pollution by Dumping of Wastes and Other Matter, done at London on 13 November 1972; or

(ii) release of harmful substances directly arising from the exploration, exploitation and associated offshore processing of sea-bed mineral resources; or

(iii) release of harmful substances for purposes of legitimate scientific research into pollution abatement or control.

(4) *Ship* means a vessel of any type whatsoever operating in the marine environment and includes hydrofoil boats, air-cushion vehicles, submersibles, floating craft and fixed or floating platforms.

(5) *Administration* means the Government of the State under whose authority the ship is operating. With respect to a ship entitled to fly a flag of any State, the Administration is the Government of that State. With respect to fixed or floating platforms engaged in exploration and exploitation of the sea-bed and subsoil thereof adjacent to the coast over which the coastal State exercises sovereign rights for the purposes of exploration and exploitation of their natural resources, the Administration is the Government of the coastal State concerned.

(6) *Incident* means an event involving the actual or probable discharge into the sea of a harmful substance, or effluents containing such a substance.

(7) *Organization* means the Inter-Governmental Maritime Consultative Organization.*

Article 3
Application

(1) The present Convention shall apply to:

(a) ships entitled to fly the flag of a Party to the Convention; and

(b) ships not entitled to fly the flag of a Party but which operate under the authority of a Party.

(2) Nothing in the present article shall be construed as derogating from or extending the sovereign rights of the Parties under international law over the sea-bed and subsoil thereof adjacent to their coasts for the purposes of exploration and exploitation of their natural resources.

(3) The present Convention shall not apply to any warship, naval auxiliary or other ship owned or operated by a State and used, for the time being, only on government non-commercial service. However, each Party shall ensure by the adoption of appropriate measures not impairing the operations or operational capabilities of such ships owned or operated by it, that such ships act in a manner consistent, so far as is reasonable and practicable, with the present Convention.

Article 4
Violation

(1) Any violation of the requirements of the present Convention shall be prohibited and sanctions shall be established therefor under the law of the Administration of the ship concerned wherever the violation occurs. If the Administration is informed of such a violation and is satisfied that sufficient evidence is available to enable proceedings to be brought in respect of the alleged violation, it shall cause such proceedings to be taken as soon as possible, in accordance with its law.

* The name of the Organization was changed to "International Maritime Organization" by virtue of amendments to the Organization's Convention which entered into force on 22 May 1982.

(2) Any violation of the requirements of the present Convention within the jurisdiction of any Party to the Convention shall be prohibited and sanctions shall be established therefor under the law of that Party. Whenever such a violation occurs, that Party shall either:

- **(a)** cause proceedings to be taken in accordance with its law; or
- **(b)** furnish to the Administration of the ship such information and evidence as may be in its possession that a violation has occurred.

(3) Where information or evidence with respect to any violation of the present Convention by a ship is furnished to the Administration of that ship, the Administration shall promptly inform the Party which has furnished the information or evidence, and the Organization, of the action taken.

(4) The penalties specified under the law of a Party pursuant to the present article shall be adequate in severity to discourage violations of the present Convention and shall be equally severe irrespective of where the violations occur.

Article 5

Certificates and special rules on inspection of ships

(1) Subject to the provisions of paragraph (2) of the present article a certificate issued under the authority of a Party to the Convention in accordance with the provisions of the regulations shall be accepted by the other Parties and regarded for all purposes covered by the present Convention as having the same validity as a certificate issued by them.

(2) A ship required to hold a certificate in accordance with the provisions of the regulations is subject, while in the ports or offshore terminals under the jurisdiction of a Party, to inspection by officers duly authorized by that Party. Any such inspection shall be limited to verifying that there is on board a valid certificate, unless there are clear grounds for believing that the condition of the ship or its equipment does not correspond substantially with the particulars of that certificate. In that case, or if the ship does not carry a valid certificate, the Party carrying out the inspection shall take such steps as will ensure that the ship shall not sail until it can proceed to sea without presenting an unreasonable threat of harm to the marine environment. That Party may, however, grant such a ship permission to leave the port or offshore terminal for the purpose of proceeding to the nearest appropriate repair yard available.

(3) If a Party denies a foreign ship entry to the ports or offshore terminals under its jurisdiction or takes any action against such a ship for the reason that the ship does not comply with the provisions of the present Convention, the Party shall immediately inform the consul or diplomatic representative of the Party whose flag the ship is entitled to fly, or if this is not possible, the Administration of the ship concerned. Before denying entry or taking such action the Party may request consultation with the Administration of the ship concerned. Information shall also be given to the Administration when a ship does not carry a valid certificate in accordance with the provisions of the regulations.

(4) With respect to the ship of non-Parties to the Convention, Parties shall apply the requirements of the present Convention as may be necessary to ensure that no more favourable treatment is given to such ships.

Article 6

Detection of violations and enforcement of the Convention

(1) Parties to the Convention shall cooperate in the detection of violations and the enforcement of the provisions of the present Convention, using all appropriate and practicable measures of detection and environmental monitoring, adequate procedures for reporting and accumulation of evidence.

(2) A ship to which the present Convention applies may, in any port or offshore terminal of a Party, be subject to inspection by officers appointed or authorized by that Party for the purpose of verifying whether the ship has discharged any harmful substances in violation of the provisions of the regulations. If an inspection indicates a violation of the Convention, a report shall be forwarded to the Administration for any appropriate action.

(3) Any Party shall furnish to the Administration evidence, if any, that the ship has discharged harmful substances or effluents containing such substances in violation of the provisions of the regulations. If it is practicable to do so, the competent authority of the former Party shall notify the master of the ship of the alleged violation.

(4) Upon receiving such evidence, the Administration so informed shall investigate the matter, and may request the other Party to furnish further or better evidence of the alleged contravention. If the Administration is satisfied that sufficient evidence is available to enable proceedings to be brought in respect of the alleged violation, it shall cause such proceedings to be taken in accordance with its law as soon as possible. The Administration shall promptly inform the Party which has reported the alleged violation, as well as the Organization, of the action taken.

(5) A Party may also inspect a ship to which the present Convention applies when it enters the ports or offshore terminals under its jurisdiction, if a request for an investigation is received from any Party together with sufficient evidence that the ship has discharged harmful substances or effluents containing such substances in any place. The report of such investigation shall be sent to the Party requesting it and to the Administration so that the appropriate action may be taken under the present Convention.

Article 7

Undue delay to ships

(1) All possible efforts shall be made to avoid a ship being unduly detained or delayed under articles 4, 5 or 6 of the present Convention.

(2) When a ship is unduly detained or delayed under articles 4, 5 or 6 of the present Convention, it shall be entitled to compensation for any loss or damage suffered.

Article 8

Reports on incidents involving harmful substances

(1) A report of an incident shall be made without delay to the fullest extent possible in accordance with the provisions of Protocol I to the present Convention.

(2) Each Party to the Convention shall:

- **(a)** make all arrangements necessary for an appropriate officer or agency to receive and process all reports on incidents; and
- **(b)** notify the Organization with complete details of such arrangements for circulation to other Parties and Member States of the Organization.

(3) Whenever a Party receives a report under the provisions of the present article, that Party shall relay the report without delay to:

- **(a)** the Administration of the ship involved; and
- **(b)** any other State which may be affected.

(4) Each Party to the Convention undertakes to issue instructions to its maritime inspection vessels and aircraft and to other appropriate services, to report to its authorities any incident referred to in Protocol I to the present Convention. That Party shall, if it considers it appropriate, report accordingly to the Organization and to any other Party concerned.

Article 9

Other treaties and interpretation

(1) Upon its entry into force, the present Convention supersedes the International Convention for the Prevention of Pollution of the Sea by Oil, 1954, as amended, as between Parties to that Convention.

(2) Nothing in the present Convention shall prejudice the codification and development of the law of the sea by the United Nations Conference on the Law of the Sea convened pursuant to resolution 2750 C(XXV) of the General Assembly of the United Nations nor the present or future claims and legal views of any State concerning the law of the sea and the nature and extent of coastal and flag State jurisdiction.

(3) The term "jurisdiction" in the present Convention shall be construed in the light of international law in force at the time of application or interpretation of the present Convention.

Article 10
Settlement of disputes

Any dispute between two or more Parties to the Convention concerning the interpretation or application of the present Convention shall, if settlement by negotiation between the Parties involved has not been possible, and if these Parties do not otherwise agree, be submitted upon request of any of them to arbitration as set out in Protocol II to the present Convention.

Article 11
Communication of information

(1) The Parties to the Convention undertake to communicate to the Organization:

- **(a)** the text of laws, orders, decrees and regulations and other instruments which have been promulgated on the various matters within the scope of the present Convention;
- **(b)** a list of non-governmental agencies which are authorized to act on their behalf in matters relating to the design, construction and equipment of ships carrying harmful substances in accordance with the provisions of the regulations;*
- **(c)** a sufficient number of specimens of their certificates issued under the provisions of the regulations;
- **(d)** a list of reception facilities including their location, capacity and available facilities and other characteristics;
- **(e)** official reports or summaries of official reports in so far as they show the results of the application of the present Convention; and
- **(f)** an annual statistical report, in a form standardized by the Organization, of penalties actually imposed for infringement of the present Convention.

(2) The Organization shall notify Parties of the receipt of any communications under the present article and circulate to all Parties any information communicated to it under subparagraphs (1)(b) to (f) of the present article.

Article 12
Casualties to ships

(1) Each Administration undertakes to conduct an investigation of any casualty occurring to any of its ships subject to the provisions of the regulations if such casualty has produced a major deleterious effect upon the marine environment.

(2) Each Party to the Convention undertakes to supply the Organization with information concerning the findings of such investigation, when it judges that such information may assist in determining what changes in the present Convention might be desirable.

* The text of this subparagraph is replaced by that contained in article III of the 1978 Protocol.

Article 13
Signature, ratification, acceptance, approval and accession

(1) The present Convention shall remain open for signature at the Headquarters of the Organization from 15 January 1974 until 31 December 1974 and shall thereafter remain open for accession. States may become Parties to the present Convention by:

(a) signature without reservation as to ratification, acceptance or approval; or

(b) signature subject to ratification, acceptance or approval, followed by ratification, acceptance or approval; or

(c) accession.

(2) Ratification, acceptance, approval or accession shall be effected by the deposit of an instrument to that effect with the Secretary-General of the Organization.

(3) The Secretary-General of the Organization shall inform all States which have signed the present Convention or acceded to it of any signature or of the deposit of any new instrument of ratification, acceptance, approval or accession and the date of its deposit.

Article 14
Optional Annexes

(1) A State may at the time of signing, ratifying, accepting, approving or acceding to the present Convention declare that it does not accept any one or all of Annexes III, IV and V (hereinafter referred to as "Optional Annexes") of the present Convention. Subject to the above, Parties to the Convention shall be bound by any Annex in its entirety.

(2) A State which has declared that it is not bound by an Optional Annex may at any time accept such Annex by depositing with the Organization an instrument of the kind referred to in article 13(2).

(3) A State which makes a declaration under paragraph (1) of the present article in respect of an Optional Annex and which has not subsequently accepted that Annex in accordance with paragraph (2) of the present article shall not be under any obligation nor entitled to claim any privileges under the present Convention in respect of matters related to such Annex and all references to Parties in the present Convention shall not include that State in so far as matters related to such Annex are concerned.

(4) The Organization shall inform the States which have signed or acceded to the present Convention of any declaration under the present article as well as the receipt of any instrument deposited in accordance with the provisions of paragraph (2) of the present article.

Article 15
Entry in force

(1) The present Convention shall enter into force 12 months after the date on which not less than 15 States, the combined merchant fleets of which constitute not less than 50 per cent of the gross tonnage of the world's merchant shipping, have become parties to it in accordance with article 13.

(2) An Optional Annex shall enter into force 12 months after the date on which the conditions stipulated in paragraph (1) of the present article have been satisfied in relation to that Annex.

(3) The Organization shall inform the States which have signed the present Convention or acceded to it of the date on which it enters into force and of the date on which an Optional Annex enters into force in accordance with paragraph (2) of the present article.

(4) For States which have deposited an instrument of ratification, acceptance, approval or accession in respect of the present Convention or any Optional Annex after the requirements for entry into force thereof have been met but prior to the date of entry into force, the ratification, acceptance, approval or accession shall take effect on the date of entry into force of the Convention or such Annex or three months after the date of deposit of the instrument whichever is the later date.

(5) For States which have deposited an instrument of ratification, acceptance, approval or accession after the date on which the Convention or an Optional Annex entered into force, the Convention or the Optional Annex shall become effective three months after the date of deposit of the instrument.

(6) After the date on which all the conditions required under article 16 to bring an amendment to the present Convention or an Optional Annex into force have been fulfilled, any instrument of ratification, acceptance, approval or accession deposited shall apply to the Convention or Annex as amended.

Article 16
Amendments

(1) The present Convention may be amended by any of the procedures specified in the following paragraphs.

(2) Amendments after consideration by the Organization:

(a) any amendment proposed by a Party to the Convention shall be submitted to the Organization and circulated by its Secretary-General to all Members of the Organization and all Parties at least six months prior to its consideration;

(b) any amendment proposed and circulated as above shall be submitted to an appropriate body by the Organization for consideration;

(c) Parties to the Convention, whether or not Members of the Organization, shall be entitled to participate in the proceedings of the appropriate body;

(d) amendments shall be adopted by a two-thirds majority of only the Parties to the Convention present and voting;

(e) if adopted in accordance with subparagraph (d) above, amendments shall be communicated by the Secretary-General of the Organization to all the Parties to the Convention for acceptance;

(f) an amendment shall be deemed to have been accepted in the following circumstances:

(i) an amendment to an article of the Convention shall be deemed to have been accepted on the date on which it is accepted by two thirds of the Parties, the combined merchant fleets of which constitute not less than 50 per cent of the gross tonnage of the world's merchant fleet;

(ii) an amendment to an Annex to the Convention shall be deemed to have been accepted in accordance with the procedure specified in subparagraph (f)(iii) unless the appropriate body, at the time of its adoption, determines that the amendment shall be deemed to have been accepted on the date on which it is accepted by two thirds of the Parties, the combined merchant fleets of which constitute not less than 50 per cent of the gross tonnage of the world's merchant fleet. Nevertheless, at any time before the entry into force of an amendment to an Annex to the Convention, a Party may notify the Secretary-General of the Organization that its express approval will be necessary before the amendment enters into force for it. The latter shall bring such notification and the date of its receipt to the notice of Parties;

(iii) an amendment to an appendix to an Annex to the Convention shall be deemed to have been accepted at the end of a period to be determined by the appropriate body at the time of its adoption, which period shall be not less than ten months, unless within that period an objection is communicated to the Organization by not less than one third of the Parties or by the Parties the combined merchant fleets of which constitute not less than 50 per cent of the gross tonnage of the world's merchant fleet whichever condition is fulfilled;

(iv) an amendment to Protocol I to the Convention shall be subject to the same procedures as for the amendments to the Annexes to the Convention, as provided for in subparagraphs (f)(ii) or (f)(iii) above;

(v) an amendment to Protocol II to the Convention shall be subject to the same procedures as for the amendments to an article of the Convention, as provided for in subparagraph (f)(i) above;

(g) the amendment shall enter into force under the following conditions:

(i) in the case of an amendment to an article of the Convention, to Protocol II, or to Protocol I or to an Annex to the Convention not under the procedure specified in subparagraph (f)(iii), the amendment accepted in conformity with the foregoing provisions shall enter into force six months after the date of its acceptance with respect to the Parties which have declared that they have accepted it;

(ii) in the case of an amendment to Protocol I, to an appendix to an Annex or to an Annex to the Convention under the procedure specified in subparagraph (f)(iii), the amendment deemed to have been accepted in accordance with the foregoing conditions shall enter into force six months after its acceptance for all the Parties with the exception of those which, before that date, have made a declaration that they do not accept it or a declaration under subparagraph (f)(ii), that their express approval is necessary.

(3) Amendment by a Conference:

(a) Upon the request of a Party, concurred in by at least one third of the Parties, the Organization shall convene a Conference of Parties to the Convention to consider amendments to the present Convention.

(b) Every amendment adopted by such a Conference by a two-thirds majority of those present and voting of the Parties shall be communicated by the Secretary-General of the Organization to all Contracting Parties for their acceptance.

(c) Unless the Conference decides otherwise, the amendment shall be deemed to have been accepted and to have entered into force in accordance with the procedures specified for that purpose in paragraph (2)(f) and (g) above.

(4) **(a)** In the case of an amendment to an Optional Annex, a reference in the present article to a "Party to the Convention" shall be deemed to mean a reference to a Party bound by that Annex.

(b) Any Party which has declined to accept an amendment to an Annex shall be treated as a non-Party only for the purpose of application of that amendment.

(5) The adoption and entry into force of a new Annex shall be subject to the same procedures as for the adoption and entry into force of an amendment to an article of the Convention.

(6) Unless expressly provided otherwise, any amendment to the present Convention made under this article, which relates to the structure of a ship, shall apply only to ships for which the building contract is placed, or in the absence of a building contract, the keel of which is laid, on or after the date on which the amendment comes into force.

(7) Any amendment to a Protocol or to an Annex shall relate to the substance of that Protocol or Annex and shall be consistent with the articles of the present Convention.

(8) The Secretary-General of the Organization shall inform all Parties of any amendments which enter into force under the present article, together with the date on which each such amendment enters into force.

(9) Any declaration of acceptance or of objection to an amendment under the present article shall be notified in writing to the Secretary-General of the Organization. The latter shall bring such notification and the date of its receipt to the notice of the Parties to the Convention.

Article 17
Promotion of technical cooperation

The Parties to the Convention shall promote, in consultation with the Organization and other international bodies, with assistance and coordination by the Executive Director of the United Nations Environment Programme, support for those Parties which request technical assistance for:

(a) the training of scientific and technical personnel;

(b) the supply of necessary equipment and facilities for reception and monitoring;

(c) the facilitation of other measures and arrangements to prevent or mitigate pollution of the marine environment by ships; and

(d) the encouragement of research;

preferably within the countries concerned, so furthering the aims and purposes of the present Convention.

Article 18
Denunciation

(1) The present Convention or any Optional Annex may be denounced by any Parties to the Convention at any time after the expiry of five years from the date on which the Convention or such Annex enters into force for that Party.

(2) Denunciation shall be effected by notification in writing to the Secretary-General of the Organization who shall inform all the other Parties of any such notification received and of the date of its receipt as well as the date on which such denunciation takes effect.

(3) A denunciation shall take effect 12 months after receipt of the notification of denunciation by the Secretary-General of the Organization or after the expiry of any other longer period which may be indicated in the notification.

Article 19
Deposit and registration

(1) The present Convention shall be deposited with the Secretary-General of the Organization who shall transmit certified true copies thereof to all States which have signed the present Convention or acceded to it.

(2) As soon as the present Convention enters into force, the text shall be transmitted by the Secretary-General of the Organization to the Secretary-General of the United Nations for registration and publication, in accordance with Article 102 of the Charter of the United Nations.

Article 20
Languages

The present Convention is established in a single copy in the English, French, Russian and Spanish languages, each text being equally authentic. Official translations in the Arabic, German, Italian and Japanese languages shall be prepared and deposited with the signed original.

IN WITNESS WHEREOF the undersigned* being duly authorized by their respective governments for that purpose have signed the present convention.

DONE AT LONDON this second day of November, one thousand nine hundred and seventy-three.

* Signatures omitted.

Protocol I

Provisions concerning reports on incidents involving harmful substances

Protocol I

Provisions concerning reports on incidents involving harmful substances
(in accordance with article 8 of the Convention)

Article I
Duty to report

(1) The master or other person having charge of any ship involved in an incident referred to in article II of this Protocol shall report the particulars of such incident without delay and to the fullest extent possible in accordance with the provisions of this Protocol.

(2) In the event of the ship referred to in paragraph (1) of this article being abandoned, or in the event of a report from such a ship being incomplete or unobtainable, the owner, charterer, manager or operator of the ship, or their agent shall, to the fullest extent possible, assume the obligations placed upon the master under the provisions of this Protocol.

Article II
When to make reports

(1) The report shall be made when an incident involves:

(a) a discharge above the permitted level or probable discharge of oil or of noxious liquid substances for whatever reason including those for the purpose of securing the safety of the ship or for saving life at sea; or

(b) a discharge or probable discharge of harmful substances in packaged form, including those in freight containers, portable tanks, road and rail vehicles and shipborne barges; or

(c) damage, failure or breakdown of a ship of 15 metres in length or above which:

(i) affects the safety of the ship; including but not limited to collision, grounding, fire, explosion, structural failure, flooding and cargo shifting; or

(ii) results in impairment of the safety of navigation; including but not limited to, failure or breakdown of steering gear, propulsion plant, electrical generating system, and essential shipborne navigational aids; or

(d) a discharge during the operation of the ship of oil or noxious liquid substances in excess of the quantity or instantaneous rate permitted under the present Convention.

(2) For the purposes of this Protocol:

(a) *Oil* referred to in subparagraph (1)(a) of this article means oil as defined in regulation 1.1 of Annex I of the Convention.

(b) *Noxious liquid substances* referred to in subparagraph (1)(a) of this article means noxious liquid substances as defined in regulation 1.10 of Annex II of the Convention.

(c) *Harmful substances* in packaged form referred to in subparagraph (1)(b) of this article means substances which are identified as marine pollutants in the International Maritime Dangerous Goods Code (IMDG Code).

Article III
Contents of report

Reports shall in any case include:

(a) identity of ships involved;

(b) time, type and location of incident;

(c) quantity and type of harmful substance involved;

(d) assistance and salvage measures.

Article IV
Supplementary report

Any person who is obliged under the provisions of this Protocol to send a report shall, when possible:

(a) supplement the initial report, as necessary, and provide information concerning further developments; and

(b) comply as fully as possible with requests from affected States for additional information.

Article V
Reporting procedures

(1) Reports shall be made by the fastest telecommunications channels available with the highest possible priority to the nearest coastal State.

(2) In order to implement the provisions of this Protocol, Parties to the present Convention shall issue, or cause to be issued, regulations or instructions on the procedures to be followed in reporting incidents involving harmful substances, based on guidelines developed by the Organization.*

* Refer to the General Principles for Ship Reporting Systems and Ship Reporting Requirements, including Guidelines for Reporting Incidents Involving Dangerous Goods, Harmful Substances and/or Marine Pollutants (resolution A.851(20), as amended by resolution MEPC.138(53)).

Protocol II

Arbitration

Protocol II

Arbitration
(in accordance with article 10 of the Convention)

Article I

Arbitration procedure, unless the Parties to the dispute decide otherwise, shall be in accordance with the rules set out in this Protocol.

Article II

(1) An Arbitration Tribunal shall be established upon the request of one Party to the Convention addressed to another in application of article 10 of the present Convention. The request for arbitration shall consist of a statement of the case together with any supporting documents.

(2) The requesting Party shall inform the Secretary-General of the Organization of the fact that it has applied for the establishment of a Tribunal, of the names of the Parties to the dispute, and of the articles of the Convention or Regulations over which there is in its opinion disagreement concerning their interpretation or application. The Secretary-General shall transmit this information to all Parties.

Article III

The Tribunal shall consist of three members: one Arbitrator nominated by each Party to the dispute and a third Arbitrator who shall be nominated by agreement between the two first named, and shall act as its Chairman.

Article IV

(1) If, at the end of a period of 60 days from the nomination of the second Arbitrator, the Chairman of the Tribunal shall not have been nominated, the Secretary-General of the Organization upon request of either Party shall within a further period of 60 days proceed to such nomination, selecting him from a list of qualified persons previously drawn up by the Council of the Organization.

(2) If, within a period of 60 days from the date of the receipt of the request, one of the Parties shall not have nominated the member of the Tribunal for whose designation it is responsible, the other Party may directly inform the Secretary-General of the Organization who shall nominate the Chairman of the Tribunal within a period of 60 days, selecting him from the list prescribed in paragraph (1) of the present article.

(3) The Chairman of the Tribunal shall, upon nomination, request the Party which has not provided an Arbitrator, to do so in the same manner and under the same conditions. If the Party does not make the required nomination, the Chairman of the Tribunal shall request the Secretary-General of the Organization to make the nomination in the form and conditions prescribed in the preceding paragraph.

(4) The Chairman of the Tribunal, if nominated under the provisions of the present article, shall not be or have been a national of one of the Parties concerned, except with the consent of the other Party.

(5) In the case of the decease or default of an Arbitrator for whose nomination one of the Parties is responsible, the said Party shall nominate a replacement within a period of 60 days from the date of decease or default. Should the said Party not make the nomination, the arbitration shall proceed under the remaining Arbitrators. In case of the decease or default of the Chairman of the Tribunal, a replacement shall be nominated in accordance with the provisions of article III above, or in the absence of agreement between the members of the Tribunal within a period of 60 days of the decease or default, according to the provisions of the present article.

Article V

The Tribunal may hear and determine counter-claims arising directly out of the subject matter of the dispute.

Article VI

Each Party shall be responsible for the remuneration of its Arbitrator and connected costs and for the costs entailed by the preparation of its own case. The remuneration of the Chairman of the Tribunal and of all general expenses incurred by the Arbitration shall be borne equally by the Parties. The Tribunal shall keep a record of all its expenses and shall furnish a final statement thereof.

Article VII

Any Party to the Convention which has an interest of a legal nature and which may be affected by the decision in the case may, after giving written notice to the Parties which have originally initiated the procedure, join in the arbitration procedure with the consent of the Tribunal.

Article VIII

Any Arbitration Tribunal established under the provisions of the present Protocol shall decide its own rules of procedure.

Article IX

(1) Decisions of the Tribunal both as to its procedure and its place of meeting and as to any question laid before it, shall be taken by majority votes of its members; the absence or abstention of one of the members of the Tribunal for whose nomination the Parties were responsible, shall not constitute an impediment to the Tribunal reaching a decision. In cases of equal voting, the vote of the Chairman shall be decisive.

(2) The Parties shall facilitate the work of the Tribunal and in particular, in accordance with their legislation, and using all means at their disposal:

- **(a)** provide the Tribunal with the necessary documents and information;
- **(b)** enable the Tribunal to enter their territory, to hear witnesses or experts, and to visit the scene.

(3) Absence or default of one Party shall not constitute an impediment to the procedure.

Article X

(1) The Tribunal shall render its award within a period of five months from the time it is established unless it decides, in the case of necessity, to extend the time limit for a further period not exceeding three months. The award of the Tribunal shall be accompanied by a statement of reasons. It shall be final and without appeal and shall be communicated to the Secretary-General of the Organization. The Parties shall immediately comply with the award.

(2) Any controversy which may arise between the Parties as regards interpretation or execution of the award may be submitted by either Party for judgment to the Tribunal which made the award, or, if it is not available to another Tribunal constituted for this purpose, in the same manner as the original Tribunal.

Protocol of 1978 relating to the International Convention for the Prevention of Pollution from Ships, 1973

Protocol of 1978 relating to the International Convention for the Prevention of Pollution from Ships, 1973

THE PARTIES TO THE PRESENT PROTOCOL,

RECOGNIZING the significant contribution which can be made by the International Convention for the Prevention of Pollution from Ships, 1973, to the protection of the marine environment from pollution from ships,

RECOGNIZING ALSO the need to improve further the prevention and control of marine pollution from ships, particularly oil tankers,

RECOGNIZING FURTHER the need for implementing the regulations for the prevention of pollution by oil contained in Annex I of that Convention as early and as widely as possible,

ACKNOWLEDGING HOWEVER the need to defer the application of Annex II of that Convention until certain technical problems have been satisfactorily resolved,

CONSIDERING that these objectives may best be achieved by the conclusion of a Protocol relating to the International Convention for the Prevention of Pollution from Ships, 1973,

HAVE AGREED as follows:

Article I
General obligations

1 The Parties to the present Protocol undertake to give effect to the provisions of:

- **(a)** the present Protocol and the Annex hereto which shall constitute an integral part of the present Protocol; and
- **(b)** the International Convention for the Prevention of Pollution from Ships, 1973 (hereinafter referred to as "the Convention"), subject to the modifications and additions set out in the present Protocol.

2 The provisions of the Convention and the present Protocol shall be read and interpreted together as one single instrument.

3 Every reference to the present Protocol constitutes at the same time a reference to the Annex hereto.

Article II
Implementation of Annex II of the Convention

1 Notwithstanding the provisions of article 14(1) of the Convention, the Parties to the present Protocol agree that they shall not be bound by the provisions of Annex II of the Convention for a period of three years from the date of entry into force of the present Protocol or for such longer period as may be decided by a two-thirds majority of the Parties to the present Protocol in the Marine Environment Protection Committee (hereinafter referred to as "the Committee") of the Inter-Governmental Maritime Consultative Organization (hereinafter referred to as "the Organization").*

* The name of the Organization was changed to "International Maritime Organization" by virtue of amendments to the Organization's Convention which entered into force on 22 May 1982.

2 During the period specified in paragraph 1 of this article, the Parties to the present Protocol shall not be under any obligations nor entitled to claim any privileges under the Convention in respect of matters relating to Annex II of the Convention and all reference to Parties in the Convention shall not include the Parties to the present Protocol in so far as matters relating to that Annex are concerned.

Article III
Communication of information

The text of article 11(1)(b) of the Convention is replaced by the following:

> "a list of nominated surveyors or recognized organizations which are authorized to act on their behalf in the administration of matters relating to the design, construction, equipment and operation of ships carrying harmful substances in accordance with the provisions of the regulations for circulation to the Parties for information of their officers. The Administration shall therefore notify the Organization of the specific responsibilities and conditions of the authority delegated to nominated surveyors or recognized organizations;".

Article IV
Signature, ratification, acceptance, approval and accession

1 The present Protocol shall be open for signature at the Headquarters of the Organization from 1 June 1978 to 31 May 1979 and shall thereafter remain open for accession. States may become Parties to the present Protocol by:

- **(a)** signature without reservation as to ratification, acceptance or approval; or
- **(b)** signature, subject to ratification, acceptance or approval, followed by ratification, acceptance or approval; or
- **(c)** accession.

2 Ratification, acceptance, approval or accession shall be effected by the deposit of an instrument to that effect with the Secretary-General of the Organization.

Article V
Entry into force

1 The present Protocol shall enter into force 12 months after the date on which not less than 15 States, the combined merchant fleets of which constitute not less than 50 per cent of the gross tonnage of the world's merchant shipping, have become Parties to it in accordance with article IV of the present Protocol.

2 Any instrument of ratification, acceptance, approval or accession deposited after the date on which the present Protocol enters into force shall take effect three months after the date of deposit.

3 After the date on which an amendment to the present Protocol is deemed to have been accepted in accordance with article 16 of the Convention, any instrument of ratification, acceptance, approval or accession deposited shall apply to the present Protocol as amended.

Article VI
Amendments

The procedures set out in article 16 of the Convention in respect of amendments to the articles, an Annex and an appendix to an Annex of the Convention shall apply respectively to amendments to the articles, the Annex and an appendix to the Annex of the present Protocol.

Article VII
Denunciation

1 The present Protocol may be denounced by any Party to the present Protocol at any time after the expiry of five years from the date on which the Protocol enters into force for that Party.

2 Denunciation shall be effected by the deposit of an instrument of denunciation with the Secretary-General of the Organization.

3 A denunciation shall take effect 12 months after receipt of the notification by the Secretary-General of the Organization or after the expiry of any other longer period which may be indicated in the notification.

Article VIII
Depositary

1 The present Protocol shall be deposited with the Secretary-General of the Organization (hereinafter referred to as "the Depositary").

2 The Depositary shall:

(a) inform all States which have signed the present Protocol or acceded thereto of:

(i) each new signature or deposit of an instrument of ratification, acceptance, approval or accession, together with the date thereof;

(ii) the date of entry into force of the present Protocol;

(iii) the deposit of any instrument of denunciation of the present Protocol together with the date on which it was received and the date on which the denunciation takes effect;

(iv) any decision made in accordance with article II(1) of the present Protocol;

(b) transmit certified true copies of the present Protocol to all States which have signed the present Protocol or acceded thereto.

3 As soon as the present Protocol enters into force, a certified true copy thereof shall be transmitted by the Depositary to the Secretariat of the United Nations for registration and publication in accordance with Article 102 of the Charter of the United Nations.

Article IX
Languages

The present Protocol is established in a single original in the English, French, Russian and Spanish languages, each text being equally authentic. Official translations in the Arabic, German, Italian and Japanese languages shall be prepared and deposited with the signed original.

IN WITNESS WHEREOF the undersigned[*] being duly authorized by their respective governments for that purpose have signed the present protocol.

DONE AT LONDON this seventeenth day of February one thousand nine hundred and seventy-eight.

[*] Signatures omitted.

Protocol of 1997 to amend the International Convention for the Prevention of Pollution from Ships, 1973, as modified by the Protocol of 1978 relating thereto

Protocol of 1997 to amend the International Convention for the Prevention of Pollution from Ships, 1973, as modified by the Protocol of 1978 relating thereto

THE PARTIES TO THE PRESENT PROTOCOL,

BEING parties to the Protocol of 1978 relating to the International Convention for the Prevention of Pollution from Ships, 1973,

RECOGNIZING the need to prevent and control air pollution from ships,

RECALLING principle 15 of the Rio Declaration on Environment and Development which calls for the application of a precautionary approach,

CONSIDERING that this objective could best be achieved by the conclusion of a Protocol of 1997 to amend the International Convention for the Prevention of Pollution from Ships, 1973, as modified by the Protocol of 1978 relating thereto,

HAVE AGREED as follows:

Article 1
Instrument to be amended

The instrument which the present Protocol amends is the International Convention for the Prevention of Pollution from Ships, 1973, as modified by the Protocol of 1978 relating thereto (hereinafter referred to as the "Convention").

Article 2
Addition of Annex VI to the Convention

Annex VI entitled Regulations for the prevention of air pollution from ships, the text of which is set out in the annex to the present Protocol, is added.

Article 3
General obligations

1 The Convention and the present Protocol shall, as between the Parties to the present Protocol, be read and interpreted together as one single instrument.

2 Every reference to the present Protocol constitutes at the same time a reference to the annex hereto.

Article 4
Amendment procedure

In applying article 16 of the Convention to an amendment to Annex VI and its appendices, the reference to "a Party to the Convention" shall be deemed to mean the reference to a Party bound by that Annex.

FINAL CLAUSES

Article 5
Signature, ratification, acceptance, approval and accession

1 The present Protocol shall be open for signature at the Headquarters of the International Maritime Organization (hereinafter referred to as the "Organization") from 1 January 1998 until 31 December 1998 and shall thereafter remain open for accession. Only Contracting States to the Protocol of 1978 relating to the International Convention for the Prevention of Pollution from Ships, 1973 (hereinafter referred to as the "1978 Protocol") may become Parties to the present Protocol by:

(a) signature without reservation as to ratification, acceptance or approval; or

(b) signature, subject to ratification, acceptance or approval, followed by ratification, acceptance or approval; or

(c) accession.

2 Ratification, acceptance, approval or accession shall be effected by the deposit of an instrument to that effect with the Secretary-General of the Organization (hereinafter referred to as the "Secretary-General").

Article 6
Entry into force

1 The present Protocol shall enter into force twelve months after the date on which not less than fifteen States, the combined merchant fleets of which constitute not less than 50 per cent of the gross tonnage of the world's merchant shipping, have become Parties to it in accordance with article 5 of the present Protocol.

2 Any instrument of ratification, acceptance, approval or accession deposited after the date on which the present Protocol enters into force shall take effect three months after the date of deposit.

3 After the date on which an amendment to the present Protocol is deemed to have been accepted in accordance with article 16 of the Convention, any instrument of ratification, acceptance, approval or accession deposited shall apply to the present Protocol as amended.

Article 7
Denunciation

1 The present Protocol may be denounced by any Party to the present Protocol at any time after the expiry of five years from the date on which the Protocol enters into force for that Party.

2 Denunciation shall be effected by the deposit of an instrument of denunciation with the Secretary-General.

3 A denunciation shall take effect twelve months after receipt of the notification by the Secretary-General or after the expiry of any other longer period which may be indicated in the notification.

4 A denunciation of the 1978 Protocol in accordance with article VII thereof shall be deemed to include a denunciation of the present Protocol in accordance with this article. Such denunciation shall take effect on the date on which denunciation of the 1978 Protocol takes effect in accordance with article VII of that Protocol.

Article 8
Depositary

1 The present Protocol shall be deposited with the Secretary-General (hereinafter referred to as the "Depositary").

2 The Depositary shall:

(a) inform all States which have signed the present Protocol or acceded thereto of:

(i) each new signature or deposit of an instrument of ratification, acceptance, approval or accession, together with the date thereof;

(ii) the date of entry into force of the present Protocol; and

(iii) the deposit of any instrument of denunciation of the present Protocol, together with the date on which it was received and the date on which the denunciation takes effect; and

(b) transmit certified true copies of the present Protocol to all States which have signed the present Protocol or acceded thereto.

3 As soon as the present Protocol enters into force, a certified true copy thereof shall be transmitted by the Depositary to the Secretariat of the United Nations for registration and publication in accordance with Article 102 of the Charter of the United Nations.

Article 9
Languages

THE PRESENT PROTOCOL is established in a single copy in the Arabic, Chinese, English, French, Russian and Spanish languages, each text being equally authentic.

IN WITNESS WHEREOF the undersigned, being duly authorized by their respective governments for that purpose, have signed* the present protocol.

DONE AT LONDON this twenty-sixth day of September, one thousand nine hundred and ninety-seven.

* Signatures omitted.

MARPOL Annex I

Regulations for the prevention of pollution by oil

MARPOL Annex I

Regulations for the prevention of pollution by oil

Chapter 1 – General

Regulation 1
Definitions

For the purposes of this Annex:

1 *Oil* means petroleum in any form including crude oil, fuel oil, sludge, oil refuse and refined products (other than those petrochemicals which are subject to the provisions of Annex II of the present Convention) and, without limiting the generality of the foregoing, includes the substances listed in appendix I to this Annex.

SEE INTERPRETATION 1

2 *Crude oil* means any liquid hydrocarbon mixture occurring naturally in the earth whether or not treated to render it suitable for transportation and includes:

.1 crude oil from which certain distillate fractions may have been removed; and

.2 crude oil to which certain distillate fractions may have been added.

3 *Oily mixture* means a mixture with any oil content.

4 *Oil fuel* means any oil used as fuel in connection with the propulsion and auxiliary machinery of the ship in which such oil is carried.

5 *Oil tanker* means a ship constructed or adapted primarily to carry oil in bulk in its cargo spaces and includes combination carriers, any "NLS tanker" as defined in Annex II of the present Convention and any gas carrier as defined in regulation 3.20 of chapter II-1 of SOLAS 74 (as amended), when carrying a cargo or part cargo of oil in bulk.

SEE INTERPRETATION 2

6 *Crude oil tanker* means an oil tanker engaged in the trade of carrying crude oil.

7 *Product carrier* means an oil tanker engaged in the trade of carrying oil other than crude oil.

8 *Combination carrier* means a ship designed to carry either oil or solid cargoes in bulk.

9 *Major conversion*:

SEE INTERPRETATION 3

.1 means a conversion of a ship:

.1.1 which substantially alters the dimensions or carrying capacity of the ship; or

.1.2 which changes the type of the ship; or

.1.3 the intent of which in the opinion of the Administration is substantially to prolong its life; or

.1.4 which otherwise so alters the ship that, if it were a new ship, it would become subject to relevant provisions of the present Convention not applicable to it as an existing ship.

.2 Notwithstanding the provisions of this definition:

.2.1 conversion of an oil tanker of 20,000 tonnes deadweight and above delivered on or before 1 June 1982, as defined in regulation 1.28.3, to meet the requirements of regulation 18 of this Annex shall not be deemed to constitute a major conversion for the purpose of this Annex; and

.2.2 conversion of an oil tanker delivered before 6 July 1996, as defined in regulation 1.28.5, to meet the requirements of regulation 19 or 20 of this Annex shall not be deemed to constitute a major conversion for the purpose of this Annex.

10 *Nearest land.* The term "from the nearest land" means from the baseline from which the territorial sea of the territory in question is established in accordance with international law, except that, for the purposes of the present Convention "from the nearest land" off the north-eastern coast of Australia shall mean from a line drawn from a point on the coast of Australia in:

latitude 11°00′ S, longitude 142°08′ E
to a point in latitude 10°35′ S, longitude 141°55′ E,
thence to a point latitude 10°00′ S, longitude 142°00′ E,
thence to a point latitude 09°10′ S, longitude 143°52′ E,
thence to a point latitude 09°00′ S, longitude 144°30′ E,
thence to a point latitude 10°41′ S, longitude 145°00′ E,
thence to a point latitude 13°00′ S, longitude 145°00′ E,
thence to a point latitude 15°00′ S, longitude 146°00′ E,
thence to a point latitude 17°30′ S, longitude 147°00′ E,
thence to a point latitude 21°00′ S, longitude 152°55′ E,
thence to a point latitude 24°30′ S, longitude 154°00′ E,
thence to a point on the coast of Australia
in latitude 24°42′ S, longitude 153°15′ E.

11 *Special area* means a sea area where for recognized technical reasons in relation to its oceanographical and ecological condition and to the particular character of its traffic the adoption of special mandatory methods for the prevention of sea pollution by oil is required.

For the purposes of this Annex, the special areas are defined as follows:

.1 *the Mediterranean Sea area* means the Mediterranean Sea proper including the gulfs and seas therein with the boundary between the Mediterranean and the Black Sea constituted by the 41° N parallel and bounded to the west by the Straits of Gibraltar at the meridian of 005°36′ W;

.2 *the Baltic Sea area* means the Baltic Sea proper with the Gulf of Bothnia, the Gulf of Finland and the entrance to the Baltic Sea bounded by the parallel of the Skaw in the Skagerrak at 57°44′.8 N;

.3 *the Black Sea area* means the Black Sea proper with the boundary between the Mediterranean Sea and the Black Sea constituted by the parallel 41° N;

.4 *the Red Sea area* means the Red Sea proper including the Gulfs of Suez and Aqaba bounded at the south by the rhumb line between Ras si Ane (12°28′.5 N, 043°19′.6 E) and Husn Murad (12°40′.4 N, 043°30′.2 E);

.5 *the Gulfs area* means the sea area located north-west of the rhumb line between Ras al Hadd (22°30′ N, 059°48′ E) and Ras al Fasteh (25°04′ N, 061° 25′ E);

.6 *the Gulf of Aden area* means that part of the Gulf of Aden between the Red Sea and the Arabian Sea bounded to the west by the rhumb line between Ras si Ane (12°28′.5 N, 043°19′.6 E) and Husn Murad (12°40′.4 N, 043°30′.2 E) and to the east by the rhumb line between Ras Asir (11°50′ N, 051°16′.9 E) and the Ras Fartak (15°35′ N, 052°13′.8 E);

.7 *the Antarctic area* means the sea area south of latitude 60° S; and

.8 *the North West European waters* include the North Sea and its approaches, the Irish Sea and its approaches, the Celtic Sea, the English Channel and its approaches and part of the North East Atlantic immediately to the west of Ireland. The area is bounded by lines joining the following points:

48°27′ N on the French coast
48°27′ N; 006°25′ W
49°52′ N; 007°44′ W
50°30′ N; 012° W
56°30′ N; 012° W
62° N; 003° W
62° N on the Norwegian coast
57°44′.8 N on the Danish and Swedish coasts

.9 *the Oman area of the Arabian Sea* means the sea area enclosed by the following coordinates:

22°30′.00 N; 059°48′.00 E
23°47′.27 N; 060°35′.73 E
22°40′.62 N; 062°25′.29 E
21°47′.40 N; 063°22′.22 E
20°30′.37 N; 062°52′.41 E
19°45′.90 N; 062°25′.97 E
18°49′.92 N; 062°02′.94 E
17°44′.36 N; 061°05′.53 E
16°43′.71 N; 060°25′.62 E
16°03′.90 N; 059°32′.24 E
15°15′.20 N; 058°58′.52 E
14°36′.93 N; 058°10′.23 E
14°18′.93 N; 057°27′.03 E
14°11′.53 N; 056°53′.75 E
13°53′.80 N; 056°19′.24 E
13°45′.86 N; 055°54′.53 E
14°27′.38 N; 054°51′.42 E
14°40′.10 N; 054°27′.35 E
14°46′.21 N; 054°08′.56 E
15°20′.74 N; 053°38′.33 E
15°48′.69 N; 053°32′.07 E
16°23′.02 N; 053°14′.82 E
16°39′.06 N; 053°06′.52 E

.10 *the Southern South African waters* means the sea area enclosed by the following coordinates:

31°14′ S; 017°50′ E
31°30′ S; 017°12′ E
32°00′ S; 017°06′ E
32°32′ S; 016°52′ E
34°06′ S; 017°24′ E
36°58′ S; 020°54′ E
36°00′ S; 022°30′ E
35°14′ S; 022°54′ E
34°30′ S; 026°00′ E
33°48′ S; 027°25′ E
33°27′ S; 027°12′ E

12 *Instantaneous rate of discharge of oil content* means the rate of discharge of oil in litres per hour at any instant divided by the speed of the ship in knots at the same instant.

13 *Tank* means an enclosed space which is formed by the permanent structure of a ship and which is designed for the carriage of liquid in bulk.

14 *Wing tank* means any tank adjacent to the side shell plating.

15 *Centre tank* means any tank inboard of a longitudinal bulkhead.

16 *Slop tank* means a tank specifically designated for the collection of tank drainings, tank washings and other oily mixtures.

17 *Clean ballast* means the ballast in a tank which, since oil was last carried therein, has been so cleaned that effluent therefrom if it were discharged from a ship which is stationary into clean calm water on a clear day would not produce visible traces of oil on the surface of the water or on adjoining shorelines or cause a sludge or emulsion to be deposited beneath the surface of the water or upon adjoining shorelines. If the ballast is discharged through an oil discharge monitoring and control system approved by the Administration, evidence based on such a system to the effect that the oil content of the effluent did not exceed 15 ppm shall be determinative that the ballast was clean, notwithstanding the presence of visible traces.

18 *Segregated ballast* means the ballast water introduced into a tank which is completely separated from the cargo oil and oil fuel system and which is permanently allocated to the carriage of ballast or to the carriage of ballast or cargoes other than oil or noxious liquid substances as variously defined in the Annexes of the present Convention.

SEE INTERPRETATION 4

19 *Length* (*L*) means 96% of the total length on a waterline at 85% of the least moulded depth measured from the top of the keel, or the length from the foreside of the stem to the axis of the rudder stock on that waterline, if that be greater. In ships designed with a rake of keel the waterline on which this length is measured shall be parallel to the designed waterline. The length (*L*) shall be measured in metres.

20 *Forward and after perpendiculars* shall be taken at the forward and after ends of the length (*L*). The forward perpendicular shall coincide with the foreside of the stem on the waterline on which the length is measured.

21 *Amidships* is at the middle of the length (*L*).

22 *Breadth* (*B*) means the maximum breadth of the ship, measured amidships to the moulded line of the frame in a ship with a metal shell and to the outer surface of the hull in a ship with a shell of any other material. The breadth (*B*) shall be measured in metres.

23 *Deadweight* (*DW*) means the difference in tonnes between the displacement of a ship in water of a relative density of 1.025 at the load waterline corresponding to the assigned summer freeboard and the lightweight of the ship.

24 *Lightweight* means the displacement of a ship in tonnes without cargo, fuel, lubricating oil, ballast water, fresh water and feed water in tanks, consumable stores, and passengers and crew and their effects.

SEE INTERPRETATION 5

25 *Permeability* of a space means the ratio of the volume within that space which is assumed to be occupied by water to the total volume of that space.

26 *Volumes and areas* in a ship shall be calculated in all cases to moulded lines.

27 *Anniversary date* means the day and the month of each year, which will correspond to the date of expiry of the International Oil Pollution Prevention Certificate.

28.1 *Ship delivered on or before 31 December 1979* means a ship:

.1 for which the building contract is placed on or before 31 December 1975; or

.2 in the absence of a building contract, the keel of which is laid or which is at a similar stage of construction on or before 30 June 1976; or

.3 the delivery of which is on or before 31 December 1979; or

.4 which has undergone a major conversion:

.4.1 for which the contract is placed on or before 31 December 1975; or

.4.2 in the absence of a contract, the construction work of which is begun on or before 30 June 1976; or

.4.3 which is completed on or before 31 December 1979.

SEE INTERPRETATIONS 6 AND 7

28.2 *Ship delivered after 31 December 1979* means a ship:

.1 for which the building contract is placed after 31 December 1975; or

.2 in the absence of a building contract, the keel of which is laid or which is at a similar stage of construction after 30 June 1976; or

.3 the delivery of which is after 31 December 1979; or

.4 which has undergone a major conversion:

.4.1 for which the contract is placed after 31 December 1975; or

.4.2 in the absence of a contract, the construction work of which is begun after 30 June 1976; or

.4.3 which is completed after 31 December 1979.

SEE INTERPRETATIONS 7 AND 8

28.3 *Oil tanker delivered on or before 1 June 1982* means an oil tanker:

.1 for which the building contract is placed on or before 1 June 1979; or

.2 in the absence of a building contract, the keel of which is laid or which is at a similar stage of construction on or before 1 January 1980; or

.3 the delivery of which is on or before 1 June 1982; or

.4 which has undergone a major conversion:

.4.1 for which the contract is placed on or before 1 June 1979; or

.4.2 in the absence of a contract, the construction work of which is begun on or before 1 January 1980; or

.4.3 which is completed on or before 1 June 1982

28.4 *Oil tanker delivered after 1 June 1982* means an oil tanker:

.1 for which the building contract is placed after 1 June 1979; or

.2 in the absence of a building contract, the keel of which is laid or which is at a similar stage of construction after 1 January 1980; or

.3 the delivery of which is after 1 June 1982; or

.4 which has undergone a major conversion:

.4.1 for which the contract is placed after 1 June 1979; or

.4.2 in the absence of a contract, the construction work of which is begun after 1 January 1980; or

.4.3 which is completed after 1 June 1982.

SEE INTERPRETATIONS 7 AND 8

28.5 *Oil tanker delivered before 6 July 1996* means an oil tanker:

.1 for which the building contract is placed before 6 July 1993; or

.2 in the absence of a building contract, the keel of which is laid or which is at a similar stage of construction before 6 January 1994; or

.3 the delivery of which is before 6 July 1996; or

.4 which has undergone a major conversion:

.4.1 for which the contract is placed before 6 July 1993; or

.4.2 in the absence of a contract, the construction work of which is begun before 6 January 1994; or

.4.3 which is completed before 6 July 1996.

28.6 *Oil tanker delivered on or after 6 July 1996* means an oil tanker:

.1 for which the building contract is placed on or after 6 July 1993; or

.2 in the absence of a building contract, the keel of which is laid or which is at a similar stage of construction on or after 6 January 1994; or

.3 the delivery of which is on or after 6 July 1996; or

.4 which has undergone a major conversion:

.4.1 for which the contract is placed on or after 6 July 1993; or

.4.2 in the absence of a contract, the construction work of which is begun on or after 6 January 1994; or

.4.3 which is completed on or after 6 July 1996.

SEE INTERPRETATIONS 7 AND 8

28.7 *Oil tanker delivered on or after 1 February 2002* means an oil tanker:

.1 for which the building contract is placed on or after 1 February 1999; or

.2 in the absence of a building contract, the keel of which is laid or which is at a similar stage of construction on or after 1 August 1999; or

.3 the delivery of which is on or after 1 February 2002; or

.4 which has undergone a major conversion:

.4.1 for which the contract is placed on or after 1 February 1999; or

.4.2 in the absence of a contract, the construction work of which is begun on or after 1 August 1999; or

.4.3 which is completed on or after 1 February 2002.

SEE INTERPRETATIONS 7 AND 8

28.8 *Oil tanker delivered on or after 1 January 2010* means an oil tanker:

.1 for which the building contract is placed on or after 1 January 2007; or

.2 in the absence of a building contract, the keel of which is laid or which is at a similar stage of construction on or after 1 July 2007; or

.3 the delivery of which is on or after 1 January 2010; or

.4 which has undergone a major conversion:

.4.1 for which the contract is placed on or after 1 January 2007; or

.4.2 in the absence of a contract, the construction work of which is begun on or after 1 July 2007; or

.4.3 which is completed on or after 1 January 2010.

SEE INTERPRETATIONS 7 AND 8

28.9 *Ship delivered on or after 1 August 2010* means a ship:

.1 for which the building contract is placed on or after 1 August 2007; or

.2 in the absence of a building contract, the keels of which are laid or which are at a similar stage of construction on or after 1 February 2008; or

.3 the delivery of which is on or after 1 August 2010; or

.4 which have undergone a major conversion:*

.4.1 for which the contract is placed after 1 August 2007; or

.4.2 in the absence of contract, the construction work of which is begun after 1 February 2008; or

.4.3 which is completed after 1 August 2010.

SEE INTERPRETATIONS 7 AND 8

29 *Parts per million (ppm)* means parts of oil per million parts of water by volume.

30 *Constructed* means a ship the keel of which is laid or which is at a similar stage of construction.

SEE INTERPRETATION 7

31 *Oil residue (sludge)* means the residual waste oil products generated during the normal operation of a ship such as those resulting from the purification of fuel or lubricating oil for main or auxiliary machinery, separated waste oil from oil filtering equipment, waste oil collected in drip trays, and waste hydraulic and lubricating oils.

32 *Oil residue (sludge) tank* means a tank which holds oil residue (sludge) from which sludge may be disposed directly through the standard discharge connection or any other approved means of disposal.

33 *Oily bilge water* means water which may be contaminated by oil resulting from things such as leakage or maintenance work in machinery spaces. Any liquid entering the bilge system including bilge wells, bilge piping, tank top or bilge holding tanks is considered oily bilge water.

34 *Oily bilge water holding tank* means a tank collecting oily bilge water prior to its discharge, transfer or disposal.

35 *Audit* means a systematic, independent and documented process for obtaining audit evidence and evaluating it objectively to determine the extent to which audit criteria are fulfilled.

36 *Audit Scheme* means the IMO Member State Audit Scheme established by the Organization and taking into account the guidelines developed by the Organization.†

37 *Code for Implementation* means the IMO Instruments Implementation Code (III Code) adopted by the Organization by resolution A.1070(28).

38 *Audit Standard* means the Code for Implementation.

* MEPC 59 agreed (MEPC 59/24, paragraph 6.18) that the clarification of the requirements of MARPOL Annex I regulation 12A is also applicable to major conversions as defined in regulation 1.28.9.

† Refer to the Framework and Procedures for the IMO Member State Audit Scheme (resolution A.1067(28)).

Regulation 2
Application

1 Unless expressly provided otherwise, the provisions of this Annex shall apply to all ships.

2 In ships other than oil tankers fitted with cargo spaces which are constructed and utilized to carry oil in bulk of an aggregate capacity of 200 m^3 or more, the requirements of regulations 16, 26.4, 29, 30, 31, 32, 34 and 36 of this Annex for oil tankers shall also apply to the construction and operation of those spaces, except that where such aggregate capacity is less than 1,000 m^3 the requirements of regulation 34.6 of this Annex may apply in lieu of regulations 29, 31 and 32.

3 Where a cargo subject to the provisions of Annex II of the present Convention is carried in a cargo space of an oil tanker, the appropriate requirements of Annex II of the present Convention shall also apply.

4 The requirements of regulations 29, 31 and 32 of this Annex shall not apply to oil tankers carrying asphalt or other products subject to the provisions of this Annex, which through their physical properties inhibit effective product/water separation and monitoring, for which the control of discharge under regulation 34 of this Annex shall be effected by the retention of residues on board with discharge of all contaminated washings to reception facilities.

SEE INTERPRETATION 9

5 Subject to the provisions of paragraph 6 of this regulation, regulations 18.6 to 18.8 of this Annex shall not apply to an oil tanker delivered on or before 1 June 1982, as defined in regulation 1.28.3, solely engaged in specific trades between:

.1 ports or terminals within a State Party to the present Convention; or

.2 ports or terminals of States Parties to the present Convention, where:

.2.1 the voyage is entirely within a Special Area; or

.2.2 the voyage is entirely within other limits designated by the Organization.

6 The provisions of paragraph 5 of this regulation shall only apply when the ports or terminals where cargo is loaded on such voyages are provided with reception facilities adequate for the reception and treatment of all the ballast and tank washing water from oil tankers using them and all the following conditions are complied with:

.1 subject to the exceptions provided for in regulation 4 of this Annex, all ballast water, including clean ballast water, and tank washing residues are retained on board and transferred to the reception facilities and the appropriate entry in the Oil Record Book Part II referred to in regulation 36 of this Annex is endorsed by the competent Port State Authority;

.2 agreement has been reached between the Administration and the Governments of the Port States referred to in paragraphs 5.1 or 5.2 of this regulation concerning the use of an oil tanker delivered on or before 1 June 1982, as defined in regulation 1.28.3, for a specific trade;

.3 the adequacy of the reception facilities in accordance with the relevant provisions of this Annex at the ports or terminals referred to above, for the purpose of this regulation, is approved by the Governments of the States Parties to the present Convention within which such ports or terminals are situated; and

.4 the International Oil Pollution Prevention Certificate is endorsed to the effect that the oil tanker is solely engaged in such specific trade.

Regulation 3

Exemptions and waivers

1 Any ship such as hydrofoil, air-cushion vehicle, near-surface craft and submarine craft etc., whose constructional features are such as to render the application of any of the provisions of chapters 3 and 4 of this Annex or section 1.2 of part II-A of the Polar Code relating to construction and equipment unreasonable or impracticable may be exempted by the Administration from such provisions, provided that the construction and equipment of that ship provides equivalent protection against pollution by oil, having regard to the service for which it is intended.

2 Particulars of any such exemption granted by the Administration shall be indicated in the Certificate referred to in regulation 7 of this Annex.

3 The Administration which allows any such exemption shall, as soon as possible, but not more than 90 days thereafter, communicate to the Organization particulars of same and the reasons therefor, which the Organization shall circulate to the Parties to the present Convention for their information and appropriate action, if any.

4 The Administration may waive the requirements of regulations 29, 31 and 32 of this Annex, for any oil tanker which engages exclusively on voyages both of 72 h or less in duration and within 50 nautical miles from the nearest land, provided that the oil tanker is engaged exclusively in trades between ports or terminals within a State Party to the present Convention. Any such waiver shall be subject to the requirement that the oil tanker shall retain on board all oily mixtures for subsequent discharge to reception facilities and to the determination by the Administration that facilities available to receive such oily mixtures are adequate.

SEE INTERPRETATIONS 10, 11 AND 12

5 The Administration may waive the requirements of regulations 31 and 32 of this Annex for oil tankers other than those referred to in paragraph 4 of this regulation in cases where:

.1 the tanker is an oil tanker delivered on or before 1 June 1982, as defined in regulation 1.28.3, of 40,000 tonnes deadweight or above, as referred to in regulation 2.5 of this Annex, solely engaged in specific trades, and the conditions specified in regulation 2.6 of this Annex are complied with; or

.2 the tanker is engaged exclusively in one or more of the following categories of voyages:

.2.1 voyages within special areas; or

.2.2 voyages within Arctic waters; or

.2.3 voyages within 50 nautical miles from the nearest land outside special areas or Arctic waters where the tanker is engaged in:

.2.3.1 trades between ports or terminals of a State Party to the present Convention; or

.2.3.2 restricted voyages as determined by the Administration, and of 72 h or less in duration;

SEE INTERPRETATION 11

provided that all of the following conditions are complied with:

.2.4 all oily mixtures are retained on board for subsequent discharge to reception facilities;

SEE INTERPRETATION 12

.2.5 for voyages specified in paragraph 5.2.3 of this regulation, the Administration has determined that adequate reception facilities are available to receive such oily mixtures in those oil loading ports or terminals the tanker calls at;

.2.6 the International Oil Pollution Prevention Certificate, when required, is endorsed to the effect that the ship is exclusively engaged in one or more of the categories of voyages specified in paragraphs 5.2.1 and 5.2.3.2 of this regulation; and

.2.7 the quantity, time and port of discharge are recorded in the Oil Record Book.

SEE INTERPRETATION 10

6 The Administration may waive the requirements of regulation 28(6) for the following oil tankers if loaded in accordance with the conditions approved by the Administration taking into account the guidelines developed by the Organization:*

.1 oil tankers which are on a dedicated service, with a limited number of permutations of loading such that all anticipated conditions have been approved in the stability information provided to the master in accordance with regulation 28(5);

.2 oil tankers where stability verification is made remotely by a means approved by the Administration;

.3 oil tankers which are loaded within an approved range of loading conditions; or

.4 oil tankers constructed before 1 January 2016 provided with approved limiting KG/GM curves covering all applicable intact and damage stability requirements.

Regulation 4
Exceptions

Regulations 15 and 34 of this Annex and paragraph 1.1.1 of part II-A of the Polar Code shall not apply to:

.1 the discharge into the sea of oil or oily mixture necessary for the purpose of securing the safety of a ship or saving life at sea; or

.2 the discharge into the sea of oil or oily mixture resulting from damage to a ship or its equipment:

.2.1 provided that all reasonable precautions have been taken after the occurrence of the damage or discovery of the discharge for the purpose of preventing or minimizing the discharge; and

.2.2 except if the owner or the master acted either with intent to cause damage, or recklessly and with knowledge that damage would probably result; or

.3 the discharge into the sea of substances containing oil, approved by the Administration, when being used for the purpose of combating specific pollution incidents in order to minimize the damage from pollution. Any such discharge shall be subject to the approval of any Government in whose jurisdiction it is contemplated the discharge will occur.

Regulation 5
Equivalents

SEE INTERPRETATION 13

1 The Administration may allow any fitting, material, appliance or apparatus to be fitted in a ship as an alternative to that required by this Annex if such fitting, material, appliance or apparatus is at least as effective as that required by this Annex. This authority of the Administration shall not extend to substitution of operational methods to effect the control of discharge of oil as equivalent to those design and construction features which are prescribed by regulations in this Annex.

2 The Administration which allows a fitting, material, appliance or apparatus to be fitted in a ship as an alternative to that required by this Annex shall communicate particulars thereof to the Organization for circulation to the Parties to the Convention for their information and appropriate action, if any.

* Refer to operational guidance provided in part 2 of the Guidelines for verification of damage stability requirements for tankers (MSC.1/Circ.1461).

Chapter 2 – Surveys and certification

Regulation 6
Surveys

1 Every oil tanker of 150 gross tonnage and above, and every other ship of 400 gross tonnage and above shall be subject to the surveys specified below:

.1 an initial survey before the ship is put in service or before the Certificate required under regulation 7 of this Annex is issued for the first time, which shall include a complete survey of its structure, equipment, systems, fittings, arrangements and material in so far as the ship is covered by this Annex. This survey shall be such as to ensure that the structure, equipment, systems, fittings, arrangements and material fully comply with the applicable requirements of this Annex;

.2 a renewal survey at intervals specified by the Administration, but not exceeding five years, except where regulation 10.2.2, 10.5, 10.6 or 10.7 of this Annex is applicable. The renewal survey shall be such as to ensure that the structure, equipment, systems, fittings, arrangements and material fully comply with applicable requirements of this Annex;

.3 an intermediate survey within three months before or after the second anniversary date or within three months before or after the third anniversary date of the Certificate which shall take the place of one of the annual surveys specified in paragraph 1.4 of this regulation. The intermediate survey shall be such as to ensure that the equipment and associated pump and piping systems, including oil discharge monitoring and control systems, crude oil washing systems, oily-water separating equipment and oil filtering systems, fully comply with the applicable requirements of this Annex and are in good working order. Such intermediate surveys shall be endorsed on the Certificate issued under regulation 7 or 8 of this Annex;

SEE INTERPRETATION 14

.4 an annual survey within three months before or after each anniversary date of the Certificate, including a general inspection of the structure, equipment, systems, fittings, arrangements and material referred to in paragraph 1.1 of this regulation to ensure that they have been maintained in accordance with paragraphs 4.1 and 4.2 of this regulation and that they remain satisfactory for the service for which the ship is intended. Such annual surveys shall be endorsed on the Certificate issued under regulation 7 or 8 of this Annex; and

SEE INTERPRETATION 14

.5 an additional survey either general or partial, according to the circumstances, shall be made after a repair resulting from investigations prescribed in paragraph 4.3 of this regulation, or whenever any important repairs or renewals are made. The survey shall be such as to ensure that the necessary repairs or renewals have been effectively made, that the material and workmanship of such repairs or renewals are in all respects satisfactory and that the ship complies in all respects with the requirements of this Annex.

2 The Administration shall establish appropriate measures for ships which are not subject to the provisions of paragraph 1 of this regulation in order to ensure that the applicable provisions of this Annex are complied with.

3.1 Surveys of ships as regards the enforcement of the provisions of this Annex shall be carried out by officers of the Administration. The Administration may, however, entrust the surveys either to surveyors nominated for the purpose or to organizations recognized by it. Such organizations, including classification societies, shall be authorized by the Administration in accordance with the provisions of the present Convention and with the Code for recognized organizations (RO Code), consisting of part 1 and part 2 (the provisions of which shall be treated as mandatory) and part 3 (the provisions of which shall be treated as recommendatory), as adopted by the Organization by resolution MEPC.237(65), as may be amended by the Organization, provided that:

.1 amendments to part 1 and part 2 of the RO Code are adopted, brought into force and take effect in accordance with the provisions of article 16 of the present Convention concerning the amendment procedures applicable to this annex;

.2 amendments to part 3 of the RO Code are adopted by the Marine Environment Protection Committee in accordance with its Rules of Procedure; and

.3 any amendments referred to in .1 and .2 adopted by the Maritime Safety Committee and the Marine Environment Protection Committee are identical and come into force or take effect at the same time, as appropriate.

3.2 An Administration nominating surveyors or recognizing organizations to conduct surveys as set forth in paragraph 3.1 of this regulation shall, as a minimum, empower any nominated surveyor or recognized organization to:

.1 require repairs to a ship; and

.2 carry out surveys, if requested by the appropriate authorities of a port State.

The Administration shall notify the Organization of the specific responsibilities and conditions of the authority delegated to the nominated surveyors or recognized organizations, for circulation to Parties to the present Convention for the information of their officers.

3.3 When a nominated surveyor or recognized organization determines that the condition of the ship or its equipment does not correspond substantially with the particulars of the Certificate or is such that the ship is not fit to proceed to sea without presenting an unreasonable threat of harm to the marine environment, such surveyor or organization shall immediately ensure that corrective action is taken and shall in due course notify the Administration. If such corrective action is not taken the Certificate shall be withdrawn and the Administration shall be notified immediately; and if the ship is in a port of another Party, the appropriate authorities of the port State shall also be notified immediately. When an officer of the Administration, a nominated surveyor or a recognized organization has notified the appropriate authorities of the port State, the Government of the port State concerned shall give such officer, surveyor or organization any necessary assistance to carry out their obligations under this regulation. When applicable, the Government of the port State concerned shall take such steps as will ensure that the ship shall not sail until it can proceed to sea or leave the port for the purpose of proceeding to the nearest appropriate repair yard available without presenting an unreasonable threat of harm to the marine environment.

3.4 In every case, the Administration concerned shall fully guarantee the completeness and efficiency of the survey and shall undertake to ensure the necessary arrangements to satisfy this obligation.

4.1 The condition of the ship and its equipment shall be maintained to conform with the provisions of the present Convention to ensure that the ship in all respects will remain fit to proceed to sea without presenting an unreasonable threat of harm to the marine environment.

4.2 After any survey of the ship under paragraph 1 of this regulation has been completed, no change shall be made in the structure, equipment, systems, fittings, arrangements or material covered by the survey, without the sanction of the Administration, except the direct replacement of such equipment and fittings.

4.3 Whenever an accident occurs to a ship or a defect is discovered which substantially affects the integrity of the ship or the efficiency or completeness of its equipment covered by this Annex the master or owner of the ship shall report at the earliest opportunity to the Administration, the recognized organization or the nominated surveyor responsible for issuing the relevant Certificate, who shall cause investigations to be initiated to determine whether a survey as required by paragraph 1 of this regulation is necessary. If the ship is in a port of another Party, the master or owner shall also report immediately to the appropriate authorities of the port State and the nominated surveyor or recognized organization shall ascertain that such report has been made.

Regulation 7
Issue or endorsement of certificate

SEE INTERPRETATION 15

1 An International Oil Pollution Prevention Certificate shall be issued, after an initial or renewal survey in accordance with the provisions of regulation 6 of this Annex, to any oil tanker of 150 gross tonnage and above and any other ships of 400 gross tonnage and above which are engaged in voyages to ports or offshore terminals under the jurisdiction of other Parties to the present Convention.

2 Such certificate shall be issued or endorsed as appropriate either by the Administration or by any persons or organization duly authorized by it. In every case the Administration assumes full responsibility for the certificate.

Regulation 8
Issue or endorsement of certificate by another Government

1 The Government of a Party to the present Convention may, at the request of the Administration, cause a ship to be surveyed and, if satisfied that the provisions of this Annex are complied with, shall issue or authorize the issue of an International Oil Pollution Prevention Certificate to the ship and, where appropriate, endorse or authorize the endorsement of that certificate on the ship in accordance with this Annex.

2 A copy of the certificate and a copy of the survey report shall be transmitted as soon as possible to the requesting Administration.

3 A certificate so issued shall contain a statement to the effect that it has been issued at the request of the Administration and it shall have the same force and receive the same recognition as the certificate issued under regulation 7 of this Annex.

4 No International Oil Pollution Prevention Certificate shall be issued to a ship which is entitled to fly the flag of a State which is not a Party.

Regulation 9
Form of certificate

SEE INTERPRETATION 16

The International Oil Pollution Prevention Certificate shall be drawn up in the form corresponding to the model given in appendix II to this Annex and shall be in at least English, French or Spanish. If an official language of the issuing country is also used, this shall prevail in case of a dispute or discrepancy.

Regulation 10
Duration and validity of certificate

SEE INTERPRETATION 17

1 An International Oil Pollution Prevention Certificate shall be issued for a period specified by the Administration, which shall not exceed five years.

2.1 Notwithstanding the requirements of paragraph 1 of this regulation, when the renewal survey is completed within three months before the expiry date of the existing certificate, the new certificate shall be valid from the date of completion of the renewal survey to a date not exceeding five years from the date of expiry of the existing certificate.

2.2 When the renewal survey is completed after the expiry date of the existing certificate, the new certificate shall be valid from the date of completion of the renewal survey to a date not exceeding five years from the date of expiry of the existing certificate.

2.3 When the renewal survey is completed more than three months before the expiry date of the existing certificate, the new certificate shall be valid from the date of completion of the renewal survey to a date not exceeding five years from the date of completion of the renewal survey.

3 If a certificate is issued for a period of less than five years, the Administration may extend the validity of the certificate beyond the expiry date to the maximum period specified in paragraph 1 of this regulation, provided that the surveys referred to in regulations 6.1.3 and 6.1.4 of this Annex applicable when a certificate is issued for a period of five years are carried out as appropriate.

4 If a renewal survey has been completed and a new certificate cannot be issued or placed on board the ship before the expiry date of the existing certificate, the person or organization authorized by the Administration may endorse the existing certificate and such a certificate shall be accepted as valid for a further period which shall not exceed five months from the expiry date.

5 If a ship at the time when a certificate expires is not in a port in which it is to be surveyed, the Administration may extend the period of validity of the certificate but this extension shall be granted only for the purpose of allowing the ship to complete its voyage to the port in which it is to be surveyed, and then only in cases where it appears proper and reasonable to do so. No certificate shall be extended for a period longer than three months, and a ship to which an extension is granted shall not, on its arrival in the port in which it is to be surveyed, be entitled by virtue of such extension to leave that port without having a new certificate. When the renewal survey is completed, the new certificate shall be valid to a date not exceeding five years from the date of expiry of the existing certificate before the extension was granted.

6 A certificate issued to a ship engaged on short voyages which has not been extended under the foregoing provisions of this regulation may be extended by the Administration for a period of grace of up to one month from the date of expiry stated on it. When the renewal survey is completed, the new certificate shall be valid to a date not exceeding five years from the date of expiry of the existing certificate before the extension was granted.

7 In special circumstances, as determined by the Administration, a new certificate need not be dated from the date of expiry of the existing certificate as required by paragraphs 2.2, 5 or 6 of this regulation. In these special circumstances, the new certificate shall be valid to a date not exceeding five years from the date of completion of the renewal survey.

8 If an annual or intermediate survey is completed before the period specified in regulation 6 of this Annex, then:

.1 the anniversary date shown on the certificate shall be amended by endorsement to a date which shall not be more than three months later than the date on which the survey was completed;

.2 the subsequent annual or intermediate survey required by regulation 6.1 of this Annex shall be completed at the intervals prescribed by that regulation using the new anniversary date; and

.3 the expiry date may remain unchanged provided one or more annual or intermediate surveys, as appropriate, are carried out so that the maximum intervals between the surveys prescribed by regulation 6.1 of this Annex are not exceeded.

9 A certificate issued under regulation 7 or 8 of this Annex shall cease to be valid in any of the following cases:

.1 if the relevant surveys are not completed within the periods specified under regulation 6.1 of this Annex;

.2 if the certificate is not endorsed in accordance with regulation 6.1.3 or 6.1.4 of this Annex; or

.3 upon transfer of the ship to the flag of another State. A new certificate shall only be issued when the Government issuing the new certificate is fully satisfied that the ship is in compliance with the requirements of regulations 6.4.1 and 6.4.2 of this Annex. In the case of a transfer between Parties, if requested within three months after the transfer has taken place, the Government of the Party whose flag the ship was formerly entitled to fly shall, as soon as possible, transmit to the Administration copies of the certificate carried by the ship before the transfer and, if available, copies of the relevant survey reports.

Regulation 11

Port State control on operational requirements[*]

1 A ship when in a port or an offshore terminal of another Party is subject to inspection by officers duly authorized by such Party concerning operational requirements under this Annex, where there are clear grounds for believing that the master or crew are not familiar with essential shipboard procedures relating to the prevention of pollution by oil.

2 In the circumstances given in paragraph 1 of this regulation, the Party shall take such steps as will ensure that the ship shall not sail until the situation has been brought to order in accordance with the requirements of this Annex.

3 Procedures relating to the port State control prescribed in article 5 of the present Convention shall apply to this regulation.

4 Nothing in this regulation shall be construed to limit the rights and obligations of a Party carrying out control over operational requirements specifically provided for in the present Convention.

[*] Refer to Procedures for port State control, 2011 (resolution A.1052(27)).

Chapter 3 – Requirements for machinery spaces of all ships

Part A – Construction

Regulation 12
Tanks for oil residues (sludge)

1 Unless indicated otherwise, this regulation applies to every ship of 400 gross tonnage and above except that paragraph 3.5 of this regulation need only be applied as far as is reasonable and practicable to ships delivered on or before 31 December 1979, as defined in regulation 1.28.1.

2 Oil residue (sludge) may be disposed of directly from the oil residue (sludge) tank(s) to reception facilities through the standard discharge connection referred to in regulation 13, or to any other approved means of disposal of oil residue (sludge), such as an incinerator, auxiliary boiler suitable for burning oil residues (sludge) or other acceptable means which shall be annotated in item 3.2 of the Supplement to IOPP Certificate Form A or B.

3 Oil residue (sludge) tank(s) shall be provided and:

.1 shall be of adequate capacity, having regard to the type of machinery and length of voyage, to receive the oil residues (sludge) which cannot be dealt with otherwise in accordance with the requirements of this Annex;

SEE INTERPRETATION 18

.2 shall be provided with a designated pump that is capable of taking suction from the oil residue (sludge) tank(s) for disposal of oil residue (sludge) by means as described in regulation 12.2;

SEE INTERPRETATION 19

.3 shall have no discharge connections to the bilge system, oily bilge water holding tank(s), tank top or oily water separators, except that:

- **.1** the tank(s) may be fitted with drains, with manually operated self-closing valves and arrangements for subsequent visual monitoring of the settled water that lead to an oily bilge water holding tank or bilge well, or an alternative arrangement provided such arrangement does not connect directly to the bilge discharge piping system; and
- **.2** the sludge tank discharge piping and bilge-water piping may be connected to a common piping leading to the standard discharge connection referred to in regulation 13; the connection of both systems to the possible common piping leading to the standard discharge connection referred to in regulation 13 shall not allow for the transfer of sludge to the bilge system;

SEE INTERPRETATION 20

.4 shall not be arranged with any piping that has direct connection overboard, other than the standard discharge connection referred to in regulation 13; and

SEE INTERPRETATION 21

.5 shall be designed and constructed so as to facilitate their cleaning and the discharge of residues to reception facilities.

SEE INTERPRETATION 22

4 Ships constructed before 1 January 2017 shall be arranged to comply with paragraph 3.3 of this regulation not later than the first renewal survey carried out on or after 1 January 2017.

Regulation 12A*

Oil fuel tank protection

1 This regulation shall apply to all ships with an aggregate oil fuel capacity of 600 m^3 and above which are delivered on or after 1 August 2010, as defined in regulation 1.28.9 of this Annex.

2 The application of this regulation in determining the location of tanks used to carry oil fuel does not govern over the provisions of regulation 19 of this Annex.

3 For the purpose of this regulation, the following definitions shall apply:

.1 *Oil fuel* means any oil used as fuel oil in connection with the propulsion and auxiliary machinery of the ship in which such oil is carried.

.2 *Load line draught (d_S)* is the vertical distance, in metres, from the moulded baseline at mid-length to the waterline corresponding to the summer freeboard draught to be assigned to the ship.

.3 *Light ship draught* is the moulded draught amidships corresponding to the lightweight.

.4 *Partial load line draught (d_P)* is the light ship draught plus 60% of the difference between the light ship draught and the load line draught (d_S). The partial load line draught (d_P) shall be measured in metres.

.5 *Waterline (d_B)* is the vertical distance, in metres, from the moulded baseline at mid-length to the waterline corresponding to 30% of the depth D_S.

.6 *Breadth (B_S)* is the greatest moulded breadth of the ship, in metres, at or below the deepest load line draught d_S.

.7 *Breadth (B_B)* is the greatest moulded breadth of the ship, in metres, at or below the waterline d_B.

.8 *Depth (D_S)* is the moulded depth, in metres, measured at mid-length to the upper deck at side. For the purpose of the application, "upper deck" means the highest deck to which the watertight transverse bulkheads except aft peak bulkheads extend.

.9 *Length (L)* means 96% of the total length on a waterline at 85% of the least moulded depth measured from the top of the keel, or the length from the foreside of the stem to the axis of the rudder stock on that waterline, if that be greater. In ships designed with a rake of keel the waterline on which this length is measured shall be parallel to the designed waterline. The length (*L*) shall be measured in metres.

.10 *Breadth (B)* means the maximum breadth of the ship, in metres, measured amidships to the moulded line of the frame in a ship with a metal shell and to the outer surface of the hull in a ship with a shell of any other material.

.11 *Oil fuel tank* means a tank in which oil fuel is carried, but excludes those tanks which would not contain oil fuel in normal operation, such as overflow tanks.

.12 *Small oil fuel tank* is an oil fuel tank with a maximum individual capacity not greater than 30 m^3.

.13 *C* is the ship's total volume of oil fuel, including that of the small oil fuel tanks, in cubic metres, at 98% tank filling.

.14 *Oil fuel capacity* means the volume of a tank in cubic metres, at 98% filling.

4 The provisions of this regulation shall apply to all oil fuel tanks except small oil fuel tanks, as defined in 3.12, provided that the aggregate capacity of such excluded tanks is not greater than 600 m^3.

5 Individual oil fuel tanks shall not have a capacity of over 2,500 m^3.

6 For ships, other than self-elevating drilling units, having an aggregate oil fuel capacity of 600 m^3 and above, oil fuel tanks shall be located above the moulded line of the bottom shell plating nowhere less than the distance *h* as specified below:

$h = \frac{B}{20}$ (m) or

$h = 2.0$ m, whichever is the lesser.

The minimum value of $h = 0.76$ m.

* MEPC 58 decided (MEPC 58/23, paragraph 6.10) that, with regard to conversions from single hull oil tankers to bulk/ore carriers, regulation 12A should be applied to the entire bulk/ore carrier, i.e. all new and existing fuel oil tanks.

In the turn of the bilge area and at locations without a clearly defined turn of the bilge, the oil fuel tank boundary line shall run parallel to the line of the midship flat bottom as shown in figure 1.

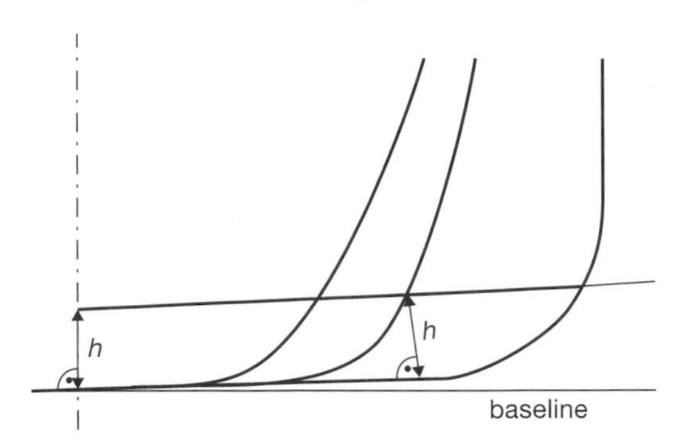

Figure 1 – *Oil fuel tank boundary lines*

7 For ships having an aggregate oil fuel capacity of 600 m^3 or more but less than 5,000 m^3, oil fuel tanks shall be located inboard of the moulded line of the side shell plating, nowhere less than the distance *w* which, as shown in figure 2, is measured at any cross-section at right angles to the side shell, as specified below:

$$w = 0.4 + \frac{2.4C}{20,000} \text{ (m)}$$

The minimum value of $w = 1.0$ m; however, for individual tanks with an oil fuel capacity of less than 500 m^3 the minimum value is 0.76 m.

8 For ships having an aggregate oil fuel capacity of 5,000 m^3 and over, oil fuel tanks shall be located inboard of the moulded line of the side shell plating, nowhere less than the distance *w* which, as shown in figure 2, is measured at any cross-section at right angles to the side shell, as specified below:

$$w = 0.5 + \frac{C}{20,000} \text{ (m) or}$$

$w = 2.0$ m, whichever is the lesser.

The minimum value of $w = 1.0$ m.

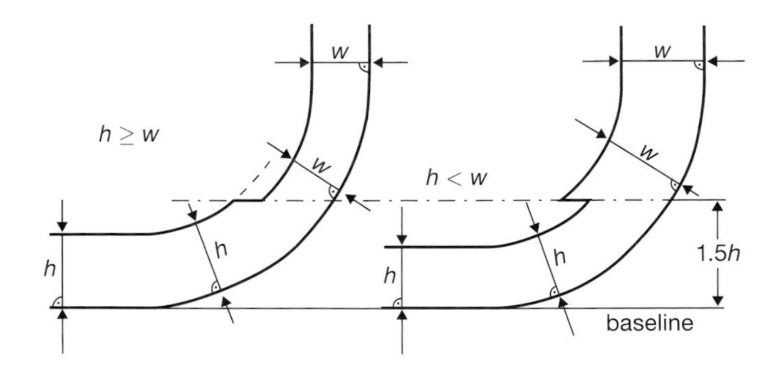

Figure 2 – *Oil fuel tank boundary lines*

9 Lines of oil fuel piping located at a distance from the ship's bottom of less than *h*, as defined in paragraph 6, or from the ship's side less than *w*, as defined in paragraphs 7 and 8, shall be fitted with valves or similar closing devices within or immediately adjacent to the oil fuel tank. These valves shall be capable of being brought into operation from a readily accessible enclosed space the location of which is accessible from the navigation bridge or propulsion machinery control position without traversing exposed freeboard or superstructure decks. The valves shall close in case of remote control system failure (fail in a closed position) and shall be kept closed at sea at any time when the tank contains oil fuel except that they may be opened during oil fuel transfer operations.

10 Suction wells in oil fuel tanks may protrude into the double bottom below the boundary line defined by the distance *h* provided that such wells are as small as practicable and the distance between the well bottom and the bottom shell plating is not less than 0.5*h*.

11 Alternatively to paragraphs 6 and either 7 or 8, ships shall comply with the accidental oil fuel outflow performance standard specified below:

.1 The level of protection against oil fuel pollution in the event of collision or grounding shall be assessed on the basis of the mean oil outflow parameter as follows:

$O_M \leq 0.0157 - 1.14E - 6C$ for $600\ m^3 \leq C < 5{,}000\ m^3$

$O_M \leq 0.010$ for $C \geq 5{,}000\ m^3$

where:

O_M = mean oil outflow parameter;

C = total oil fuel volume.

.2 The following general assumption shall apply when calculating the mean oil outflow parameter:

.2.1 the ship shall be assumed loaded to the partial load line draught (d_P) without trim or heel;

.2.2 all oil fuel tanks shall be assumed loaded to 98% of their volumetric capacity;

.2.3 the nominal density of the oil fuel (ρ_n) shall generally be taken as 1,000 kg/m^3. If the density of the oil fuel is specifically restricted to a lesser value, the lesser value may be applied; and

.2.4 for the purpose of these outflow calculations, the permeability of each oil fuel tank shall be taken as 0.99, unless proven otherwise.

.3 The following assumptions shall be used when combining the oil outflow parameters:

.3.1 The mean oil outflow shall be calculated independently for side damage and for bottom damage and then combined into a non-dimensional oil outflow parameter O_M, as follows:

$$O_M = \frac{0.4O_{MS} + 0.6O_{MB}}{C}$$

where:

O_{MS} = mean outflow for side damage, in m^3

O_{MB} = mean outflow for bottom damage, in m^3

C = total oil fuel volume.

.3.2 For bottom damage, independent calculations for mean outflow shall be done for 0 m and minus 2.5 m tide conditions, and then combined as follows:

$$O_{MB} = 0.7O_{MB(0)} + 0.3O_{MB(2.5)}$$

where:

$O_{MB(0)}$ = mean outflow for 0 m tide condition, and

$O_{MB(2.5)}$ = mean outflow for minus 2.5 m tide condition, in m^3.

.4 The mean outflow for side damage O_{MS} shall be calculated as follows:

$$O_{MS} = \sum_{i}^{n} P_{S(i)} O_{S(i)} \ (m^3)$$

where:

i = each oil fuel tank under consideration;

n = total number of oil fuel tanks;

$P_{S(i)}$ = the probability of penetrating oil fuel tank i from side damage, calculated in accordance with paragraph 11.6 of this regulation;

$O_{S(i)}$ = the outflow, in m^3, from side damage to oil fuel tank i, which is assumed equal to the total volume in oil fuel tank i at 98% filling.

.5 The mean outflow for bottom damage shall be calculated for each tidal condition as follows:

.5.1 $O_{MB(0)} = \sum_{i}^{n} P_{B(i)} O_{B(i)} C_{DB(i)}$ (m^3)

where:

i = each oil fuel tank under consideration;

n = total number of oil fuel tanks;

$P_{B(i)}$ = the probability of penetrating oil fuel tank i from bottom damage, calculated in accordance with paragraph 11.7 of this regulation;

$O_{B(i)}$ = the outflow from oil fuel tank i, in m^3, calculated in accordance with paragraph 11.5.3 of this regulation; and

$C_{DB(i)}$ = factor to account for oil capture as defined in paragraph 11.5.4.

.5.2 $O_{MB(2.5)} = \sum_{i}^{n} P_{B(i)} O_{B(i)} C_{DB(i)}$ (m^3)

where:

i, n, $P_{B(i)}$ and $C_{DB(i)}$ = as defined in subparagraph .5.1 above

$O_{B(i)}$ = the outflow from oil fuel tank i, in m^3, after tidal change.

.5.3 The oil outflow $O_{B(i)}$ for each oil fuel tank shall be calculated based on pressure balance principles, in accordance with the following assumptions:

.5.3.1 The ship shall be assumed stranded with zero trim and heel, with the stranded draught prior to tidal change equal to the partial load line draught d_P.

.5.3.2 The oil fuel level after damage shall be calculated as follows:

$$h_F = \frac{(d_P + t_c - Z_l)\rho_s}{\rho_n}$$

where:

h_F = the height of the oil fuel surface above Z_l, in metres;

t_c = the tidal change, in metres. Reductions in tide shall be expressed as negative values;

Z_l = the height of the lowest point in the oil fuel tank above the baseline, in metres;

ρ_s = density of seawater, to be taken as 1,025 kg/m^3; and

ρ_n = nominal density of the oil fuel, as defined in 11.2.3.

.5.3.3 The oil outflow $O_{B(i)}$ for any tank bounding the bottom shell plating shall be taken not less than the following formula, but no more than the tank capacity:

$$O_{B(i)} = H_W \cdot A$$

where:

H_W = 1.0 m, when $Y_B = 0$

H_W = $\frac{B_B}{50}$ but not greater than 0.4 m, when Y_B is greater than $\frac{B_B}{5}$ or 11.5 m, whichever is less

H_W is to be measured upwards from the midship flat bottom line. In the turn of the bilge area and at locations without a clearly defined turn of the bilge, H_W is to be measured from a line parallel to the midship flat bottom, as shown for distance h in figure 1.

For Y_B values outboard $\frac{B_B}{5}$ or 11.5 m, whichever is less, H_W is to be linearly interpolated.

Y_B = the minimum value of Y_B over the length of the oil fuel tank, where at any given location, Y_B is the transverse distance between the side shell at waterline d_B and the tank at or below waterline d_B.

A = the maximum horizontal projected area of the oil fuel tank up to the level of H_W from the bottom of the tank.

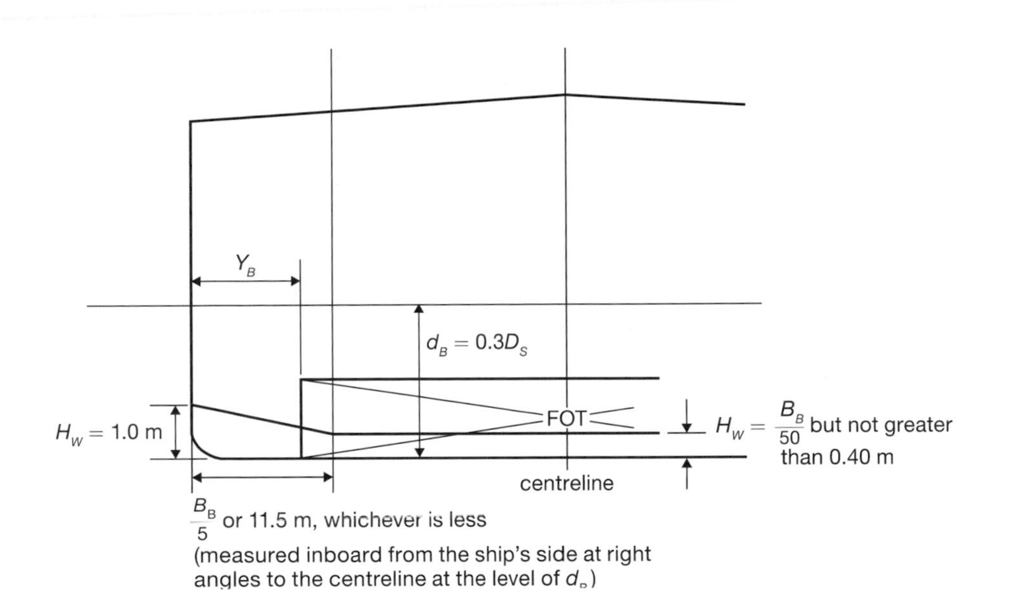

Figure 3 – *Dimensions for calculation of the minimum oil outflow*

.5.4 In the case of bottom damage, a portion from the outflow from an oil fuel tank may be captured by non-oil compartments. This effect is approximated by application of the factor $C_{DB(i)}$ for each tank, which shall be taken as follows:

$C_{DB(i)} = 0.6$ for oil fuel tanks bounded from below by non-oil compartments;

$C_{DB(i)} = 1$ otherwise.

.6 The probability P_S of breaching a compartment from side damage shall be calculated as follows:

.6.1 $P_S = P_{SL} \cdot P_{SV} \cdot P_{ST}$

where:

$P_{SL} = (1 - P_{Sf} - P_{Sa})$ = probability the damage will extend into the longitudinal zone bounded by X_a and X_f;

$P_{SV} = (1 - P_{Su} - P_{Sl})$ = probability the damage will extend into the vertical zone bounded by Z_l and Z_u;

$P_{ST} = (1 - P_{Sy})$ = probability the damage will extend transversely beyond the boundary defined by y;

.6.2 P_{Sa}, P_{Sf}, P_{Su} and P_{Sl} shall be determined by linear interpolation from the tables of probabilities for side damage provided in 11.6.3, and P_{Sy} shall be calculated from the formulas provided in 11.6.3, where:

P_{Sa} = the probability the damage will lie entirely aft of location $\frac{X_a}{L}$;

P_{Sf} = the probability the damage will lie entirely forward of location $\frac{X_f}{L}$;

P_{Sl} = probability the damage will lie entirely below the tank;

P_{Su} = probability the damage will lie entirely above the tank; and

P_{Sy} = probability the damage will lie entirely outboard the tank.

Compartment boundaries X_a, X_f, Z_l, Z_u and y shall be developed as follows:

X_a = the longitudinal distance from aft terminal of L to the aftmost point on the compartment being considered, in metres;

X_f = the longitudinal distance from aft terminal of L to the foremost point on the compartment being considered, in metres;

Z_l = the vertical distance from the moulded baseline to the lowest point on the compartment being considered, in metres. Where Z_l is greater than D_S, Z_l shall be taken as D_S;

Z_u = the vertical distance from the moulded baseline to the highest point on the compartment being considered, in metres. Where Z_u is greater than D_S, Z_u shall be taken as D_S; and

y = the minimum horizontal distance measured at right angles to the centreline between the compartment under consideration and the side shell, in metres.*

In way of the turn of the bilge, y need not to be considered below a distance h above baseline, where h is lesser of $\frac{B}{10}$, 3 m or the top of the tank.

.6.3 Tables of probabilities for side damage

$\frac{X_a}{L}$	P_{Sa}	$\frac{X_f}{L}$	P_{Sf}	$\frac{Z_l}{D_S}$	P_{Sl}	$\frac{Z_u}{D_S}$	P_{Su}
0.00	0.000	0.00	0.967	0.00	0.000	0.00	0.968
0.05	0.023	0.05	0.917	0.05	0.000	0.05	0.952
0.10	0.068	0.10	0.867	0.10	0.001	0.10	0.931
0.15	0.117	0.15	0.817	0.15	0.003	0.15	0.905
0.20	0.167	0.20	0.767	0.20	0.007	0.20	0.873
0.25	0.217	0.25	0.717	0.25	0.013	0.25	0.836
0.30	0.267	0.30	0.667	0.30	0.021	0.30	0.789
0.35	0.317	0.35	0.617	0.35	0.034	0.35	0.733
0.40	0.367	0.40	0.567	0.40	0.055	0.40	0.670
0.45	0.417	0.45	0.517	0.45	0.085	0.45	0.599
0.50	0.467	0.50	0.467	0.50	0.123	0.50	0.525
0.55	0.517	0.55	0.417	0.55	0.172	0.55	0.452
0.60	0.567	0.60	0.367	0.60	0.226	0.60	0.383
0.65	0.617	0.65	0.317	0.65	0.285	0.65	0.317
0.70	0.667	0.70	0.267	0.70	0.347	0.70	0.255
0.75	0.717	0.75	0.217	0.75	0.413	0.75	0.197
0.80	0.767	0.80	0.167	0.80	0.482	0.80	0.143
0.85	0.817	0.85	0.117	0.85	0.553	0.85	0.092
0.90	0.867	0.90	0.068	0.90	0.626	0.90	0.046
0.95	0.917	0.95	0.023	0.95	0.700	0.95	0.013
1.00	0.967	1.00	0.000	1.00	0.775	1.00	0.000

P_{Sy} shall be calculated as follows:

$$P_{Sy} = \left(\frac{24.96 - 199.6y}{B_S}\right)\left(\frac{y}{B_S}\right) \quad \text{for } \frac{y}{B_S} \leq 0.05$$

$$P_{Sy} = 0.749 + \left(5 - 44.4\left(\frac{y}{B_S} - 0.05\right)\right)\left(\frac{y}{B_S}\right) - 0.05\Big) \quad \text{for } 0.05 < \frac{y}{B_S} < 0.1$$

$$P_{Sy} = 0.888 + 0.56\left(\frac{y}{B_S} - 0.1\right) \quad \text{for } \frac{y}{B_S} \geq 0.1$$

P_{Sy} is not to be taken greater than 1.

.7 The probability P_B of breaching a compartment from bottom damage shall be calculated as follows:

.7.1 $P_B = P_{BL} \cdot P_{BT} \cdot P_{BV}$

where:

$P_{BL} = (1 - P_{Bf} - P_{Ba})$ = probability the damage will extend into the longitudinal zone bounded by X_a and X_f;

$P_{BT} = (1 - P_{Bp} - P_{Bs})$ = probability the damage will extend into transverse zone bounded by Y_p and Y_s; and

* For symmetrical tank arrangements, damages are considered for one ship only, in which case all "y" dimensions are to be measured from that side. For asymmetrical arrangements, refer to Explanatory Notes on matters related to the accidental oil outflow performance (resolution MEPC.122(52), as amended).

$P_{BV} = (1 - P_{Bz})$ = probability the damage will extend vertically above the boundary defined by z;

.7.2 P_{Ba}, P_{Bf}, P_{Bp} and P_{Bs} shall be determined by linear interpolation from the tables of probabilities for bottom damage provided in 11.7.3, and P_{Bz} shall be calculated from the formulas provided in 11.7.3, where:

P_{Ba} = the probability the damage will lie entirely aft of location $\frac{X_a}{L}$;

P_{Bf} = the probability the damage will lie entirely forward of location $\frac{X_f}{L}$;

P_{Bp} = probability the damage will lie entirely to port of the tank;

P_{Bs} = probability the damage will lie entirely to starboard of the tank; and

P_{Bz} = probability the damage will lie entirely below the tank.

Compartment boundaries X_a, X_f, Y_p, Y_s and z shall be developed as follows:

X_a and X_f as defined in 11.6.2;

Y_p = the transverse distance from the port-most point on the compartment located at or below the waterline d_B, to a vertical plane located $\frac{B_B}{2}$ to starboard of the ship's centreline;

Y_s = the transverse distance from the starboard-most point on the compartment located at or below the waterline d_B, to a vertical plane located $\frac{B_B}{2}$ to starboard of the ship's centreline; and

z = the minimum value of z over the length of the compartment, where, at any given longitudinal location, z is the vertical distance from the lower point of the bottom shell at that longitudinal location to the lower point of the compartment at that longitudinal location.

.7.3 Tables of probabilities for bottom damage

$\frac{X_a}{L}$	P_{Ba}
0.00	0.000
0.05	0.002
0.10	0.008
0.15	0.017
0.20	0.029
0.25	0.042
0.30	0.058
0.35	0.076
0.40	0.096
0.45	0.119
0.50	0.143
0.55	0.171
0.60	0.203
0.65	0.242
0.70	0.289
0.75	0.344
0.80	0.409
0.85	0.482
0.90	0.565
0.95	0.658
1.00	0.761

$\frac{X_f}{L}$	P_{Bf}
0.00	0.969
0.05	0.953
0.10	0.936
0.15	0.916
0.20	0.894
0.25	0.870
0.30	0.842
0.35	0.810
0.40	0.775
0.45	0.734
0.50	0.687
0.55	0.630
0.60	0.563
0.65	0.489
0.70	0.413
0.75	0.333
0.80	0.252
0.85	0.170
0.90	0.089
0.95	0.026
1.00	0.000

$\frac{Y_p}{B_B}$	P_{Bp}
0.00	0.844
0.05	0.794
0.10	0.744
0.15	0.694
0.20	0.644
0.25	0.594
0.30	0.544
0.35	0.494
0.40	0.444
0.45	0.394
0.50	0.344
0.55	0.297
0.60	0.253
0.65	0.211
0.70	0.171
0.75	0.133
0.80	0.097
0.85	0.063
0.90	0.032
0.95	0.009
1.00	0.000

$\frac{Y_s}{B_B}$	P_{Bs}
0.00	0.000
0.05	0.009
0.10	0.032
0.15	0.063
0.20	0.097
0.25	0.133
0.30	0.171
0.35	0.211
0.40	0.253
0.45	0.297
0.50	0.344
0.55	0.394
0.60	0.444
0.65	0.494
0.70	0.544
0.75	0.594
0.80	0.644
0.85	0.694
0.90	0.744
0.95	0.794
1.00	0.844

P_{Bz} shall be calculated as follows:

$$P_{Bz} = \left(14.5 - \frac{67z}{D_S}\right)\left(\frac{z}{D_S}\right) \quad \text{for } \frac{z}{D_S} \leq 0.1,$$

$$P_{Bz} = 0.78 + 1.1\left(\frac{z}{D_S} - 0.1\right) \quad \text{for } \frac{z}{D_S} > 0.1.$$

P_{Bz} is not to be taken greater than 1.

.8 For the purpose of maintenance and inspection, any oil fuel tanks that do not border the outer shell plating shall be located no closer to the bottom shell plating than the minimum value of h in paragraph 6 and no closer to the side shell plating than the applicable minimum value of w in paragraph 7 or 8.

12 In approving the design and construction of ships to be built in accordance with this regulation, Administrations shall have due regard to the general safety aspects, including the need for maintenance and inspection of wing and double bottom tanks or spaces.

SEE INTERPRETATIONS 23, 24 AND 25

Regulation 13
Standard discharge connection

To enable pipes of reception facilities to be connected with the ship's discharge pipeline for residues from machinery bilges and from oil residue (sludge) tanks, both lines shall be fitted with a standard discharge connection in accordance with the following table:

Standard dimensions of flanges for discharge connections

Description	Dimension
Outside diameter	215 mm
Inner diameter	According to pipe outside diameter
Bolt circle diameter	183 mm
Slots in flange	6 holes 22 mm in diameter equidistantly placed on a bolt circle of the above diameter, slotted to the flange periphery. The slot width to be 22 mm
Flange thickness	20 mm
Bolts and nuts: quantity, diameter	6, each of 20 mm in diameter and of suitable length
The flange is designed to accept pipes up to a maximum internal diameter of 125 mm and shall be of steel or other equivalent material having a flat face. This flange, together with a gasket of oil-proof material, shall be suitable for a service pressure of 600 kPa.	

Part B – Equipment

Regulation 14
Oil filtering equipment

SEE INTERPRETATION 26

1 Except as specified in paragraph 3 of this regulation, any ship of 400 gross tonnage and above but less than 10,000 gross tonnage shall be fitted with oil filtering equipment complying with paragraph 6 of this regulation. Any such ship which may discharge into the sea ballast water retained in oil fuel tanks in accordance with regulation 16.2 shall comply with paragraph 2 of this regulation.

SEE INTERPRETATIONS 27 AND 28

2 Except as specified in paragraph 3 of this regulation, any ship of 10,000 gross tonnage and above shall be fitted with oil filtering equipment complying with paragraph 7 of this regulation.

SEE INTERPRETATION 28

3 Ships, such as hotel ships, storage vessels, etc., which are stationary except for non-cargo-carrying relocation voyages need not be provided with oil filtering equipment. Such ships shall be provided with a holding tank having a volume adequate, to the satisfaction of the Administration, for the total retention on board of the oily bilge water. All oily bilge water shall be retained on board for subsequent discharge to reception facilities.

4 The Administration shall ensure that ships of less than 400 gross tonnage are equipped, as far as practicable, to retain on board oil or oily mixtures or discharge them in accordance with the requirements of regulation 15.6 of this Annex.

5 The Administration may waive the requirements of paragraphs 1 and 2 of this regulation for:

.1 any ship engaged exclusively on voyages within special areas or Arctic waters, or

.2 any ship certified under the International Code of Safety for High-Speed Craft (or otherwise within the scope of this Code with regard to size and design) engaged on a scheduled service with a turn-around time not exceeding 24 h and covering also non-passenger/cargo-carrying relocation voyages for these ships,

.3 with regard to the provision of subparagraphs .1 and .2 above, the following conditions shall be complied with:

.3.1 the ship is fitted with a holding tank having a volume adequate, to the satisfaction of the Administration, for the total retention on board of the oily bilge water;

.3.2 all oily bilge water is retained on board for subsequent discharge to reception facilities;

.3.3 the Administration has determined that adequate reception facilities are available to receive such oily bilge water in a sufficient number of ports or terminals the ship calls at;

.3.4 the International Oil Pollution Prevention Certificate, when required, is endorsed to the effect that the ship is exclusively engaged on the voyages within special areas or Arctic waters or has been accepted as a high-speed craft for the purpose of this regulation and the service is identified; and

SEE INTERPRETATION 29

.3.5 the quantity, time, and port of the discharge are recorded in the Oil Record Book Part I.

SEE INTERPRETATION 10

6 Oil filtering equipment referred to in paragraph 1 of this regulation shall be of a design approved by the Administration and shall be such as will ensure that any oily mixture discharged into the sea after passing through the system has an oil content not exceeding 15 ppm. In considering the design of such equipment, the Administration shall have regard to the specification recommended by the Organization.*

7 Oil filtering equipment referred to in paragraph 2 of this regulation shall comply with paragraph 6 of this regulation. In addition, it shall be provided with alarm arrangements to indicate when this level cannot be maintained. The system shall also be provided with arrangements to ensure that any discharge of oily mixtures is automatically stopped when the oil content of the effluent exceeds 15 ppm. In considering the design of such equipment and approvals, the Administration shall have regard to the specification recommended by the Organization.*

* Refer to Recommendation on international performance and test specification for oily-water separating equipment and oil content meters (resolution A.393(X)), Guidelines and specifications for pollution prevention equipment for machinery space bilges of ships (resolution MEPC.60(33)), 2011 Guidelines and specifications for add-on equipment for upgrading resolution MEPC.60(33)-compliant oil filtering equipment (resolution MEPC.205(62)), or Revised guidelines and specification for pollution prevention equipment for machinery space bilges of ships (resolution MEPC.107(49), as amended by resolution MEPC.285(70)).

Part C – Control of operational discharge of oil

Regulation 15
Control of discharge of oil

SEE INTERPRETATIONS 26 AND 30

1 Subject to the provisions of regulation 4 of this Annex and paragraphs 2, 3, and 6 of this regulation, any discharge into the sea of oil or oily mixtures from ships shall be prohibited.

A Discharges outside special areas except in Arctic waters

2 Any discharge into the sea of oil or oily mixtures from ships of 400 gross tonnage and above shall be prohibited except when all the following conditions are satisfied:

.1 the ship is proceeding en route;

SEE INTERPRETATION 31

.2 the oily mixture is processed through an oil filtering equipment meeting the requirements of regulation 14 of this Annex;

.3 the oil content of the effluent without dilution does not exceed 15 ppm;

.4 the oily mixture does not originate from cargo pump-room bilges on oil tankers; and

.5 the oily mixture, in case of oil tankers, is not mixed with oil cargo residues.

B Discharges in special areas

3 Any discharge into the sea of oil or oily mixtures from ships of 400 gross tonnage and above shall be prohibited except when all of the following conditions are satisfied:

.1 the ship is proceeding en route;

.2 the oily mixture is processed through an oil filtering equipment meeting the requirements of regulation 14.7 of this Annex;

.3 the oil content of the effluent without dilution does not exceed 15 ppm;

.4 the oily mixture does not originate from cargo pump-room bilges on oil tankers; and

.5 the oily mixture, in case of oil tankers, is not mixed with oil cargo residues.

4 In respect of the Antarctic area, any discharge into the sea of oil or oily mixtures from any ship shall be prohibited.

5 Nothing in this regulation shall prohibit a ship on a voyage only part of which is in a special area from discharging outside a special area in accordance with paragraph 2 of this regulation.

C Requirements for ships of less than 400 gross tonnage in all areas except the Antarctic area and Arctic waters

6 In the case of a ship of less than 400 gross tonnage, oil and all oily mixtures shall either be retained on board for subsequent discharge to reception facilities or discharged into the sea in accordance with the following provisions:

.1 the ship is proceeding en route;

.2 the ship has in operation equipment of a design approved by the Administration that ensures that the oil content of the effluent without dilution does not exceed 15 ppm;

ANNEX I ANNEX II ANNEX III ANNEX IV ANNEX V ANNEX VI UNIFIED INTERPRETATIONS

.3 the oily mixture does not originate from cargo pump-room bilges on oil tankers; and

.4 the oily mixture, in case of oil tankers, is not mixed with oil cargo residues.

D General requirements

7 Whenever visible traces of oil are observed on or below the surface of the water in the immediate vicinity of a ship or its wake, Governments of Parties to the present Convention should, to the extent they are reasonably able to do so, promptly investigate the facts bearing on the issue of whether there has been a violation of the provisions of this regulation. The investigation should include, in particular, the wind and sea conditions, the track and speed of the ship, other possible sources of the visible traces in the vicinity, and any relevant oil discharge records.

8 No discharge into the sea shall contain chemicals or other substances in quantities or concentrations which are hazardous to the marine environment or chemicals or other substances introduced for the purpose of circumventing the conditions of discharge specified in this regulation.

9 The oil residues which cannot be discharged into the sea in compliance with this regulation shall be retained on board for subsequent discharge to reception facilities.

Regulation 16

Segregation of oil and water ballast and carriage of oil in forepeak tanks

1 Except as provided in paragraph 2 of this regulation, in ships delivered after 31 December 1979, as defined in regulation 1.28.2, of 4,000 gross tonnage and above other than oil tankers, and in oil tankers delivered after 31 December 1979, as defined in regulation 1.28.2, of 150 gross tonnage and above, no ballast water shall be carried in any oil fuel tank.

2 Where the need to carry large quantities of oil fuel render it necessary to carry ballast water which is not a clean ballast in any oil fuel tank, such ballast water shall be discharged to reception facilities or into the sea in compliance with regulation 15 of this Annex using the equipment specified in regulation 14.2 of this Annex, and an entry shall be made in the Oil Record Book to this effect.

SEE INTERPRETATION 32

3 In a ship of 400 gross tonnage and above, for which the building contract is placed after 1 January 1982 or, in the absence of a building contract, the keel of which is laid or which is at a similar stage of construction after 1 July 1982, oil shall not be carried in a forepeak tank or a tank forward of the collision bulkhead.

4 All ships other than those subject to paragraphs 1 and 3 of this regulation shall comply with the provisions of those paragraphs as far as is reasonable and practicable.

SEE INTERPRETATION 33

Regulation 17

Oil Record Book Part I – Machinery space operations

1 Every oil tanker of 150 gross tonnage and above and every ship of 400 gross tonnage and above other than an oil tanker shall be provided with an Oil Record Book Part I (Machinery space operations). The Oil Record Book, whether as a part of the ship's official logbook or otherwise, shall be in the form specified in appendix III to this Annex.

2 The Oil Record Book Part I shall be completed on each occasion, on a tank-to-tank basis if appropriate, whenever any of the following machinery space operations takes place in the ship:

.1 ballasting or cleaning of oil fuel tanks;

.2 discharge of dirty ballast or cleaning water from oil fuel tanks;

.3 collection and disposal of oil residues (sludge);

.4 discharge overboard or disposal otherwise of bilge water which has accumulated in machinery spaces; and

.5 bunkering of fuel or bulk lubricating oil.

3 In the event of such discharge of oil or oily mixture as is referred to in regulation 4 of this Annex or in the event of accidental or other exceptional discharge of oil not excepted by that regulation, a statement shall be made in the Oil Record Book Part I of the circumstances of, and the reasons for, the discharge.

4 Each operation described in paragraph 2 of this regulation shall be fully recorded without delay in the Oil Record Book Part I, so that all entries in the book appropriate to that operation are completed. Each completed operation shall be signed by the officer or officers in charge of the operations concerned and each completed page shall be signed by the master of ship. The entries in the Oil Record Book Part I, for ships holding an International Oil Pollution Prevention Certificate, shall be at least in English, French or Spanish. Where entries in an official national language of the State whose flag the ship is entitled to fly are also used, this shall prevail in case of a dispute or discrepancy.

5 Any failure of the oil filtering equipment shall be recorded in the Oil Record Book Part I.

6 The Oil Record Book Part I shall be kept in such a place as to be readily available for inspection at all reasonable times and, except in the case of unmanned ships under tow, shall be kept on board the ship. It shall be preserved for a period of three years after the last entry has been made.

7 The competent authority of the Government of a Party to the present Convention may inspect the Oil Record Book Part I on board any ship to which this Annex applies while the ship is in its port or offshore terminals and may make a copy of any entry in that book and may require the master of the ship to certify that the copy is a true copy of such entry. Any copy so made which has been certified by the master of the ship as a true copy of an entry in the ship's Oil Record Book Part I shall be made admissible in any judicial proceedings as evidence of the facts stated in the entry. The inspection of an Oil Record Book Part I and the taking of a certified copy by the competent authority under this paragraph shall be performed as expeditiously as possible without causing the ship to be unduly delayed.

Chapter 4 – Requirements for the cargo area of oil tankers

Part A – Construction

Regulation 18
Segregated ballast tanks

SEE INTERPRETATION 34

Oil tankers of 20,000 tonnes deadweight and above delivered after 1 June 1982

1 Every crude oil tanker of 20,000 tonnes deadweight and above and every product carrier of 30,000 tonnes deadweight and above delivered after 1 June 1982, as defined in regulation 1.28.4, shall be provided with segregated ballast tanks and shall comply with paragraphs 2, 3 and 4, or 5 as appropriate, of this regulation.

2 The capacity of the segregated ballast tanks shall be so determined that the ship may operate safely on ballast voyages without recourse to the use of cargo tanks for water ballast except as provided for in paragraph 3 or 4 of this regulation. In all cases, however, the capacity of segregated ballast tanks shall be at least such that, in any ballast condition at any part of the voyage, including the conditions consisting of lightweight plus segregated ballast only, the ship's draughts and trim can meet the following requirements:

.1 the moulded draught amidships (d_m) in metres (without taking into account any ship's deformation) shall not be less than:

$$d_m = 2.0 + 0.02L$$

.2 the draughts at the forward and after perpendiculars shall correspond to those determined by the draught amidships (d_m) as specified in paragraph 2.1 of this regulation, in association with the trim by the stern of not greater than $0.015L$; and

.3 in any case the draught at the after perpendicular shall not be less than that which is necessary to obtain full immersion of the propeller(s).

3 In no case shall ballast water be carried in cargo tanks, except:

.1 on those rare voyages when weather conditions are so severe that, in the opinion of the master, it is necessary to carry additional ballast water in cargo tanks for the safety of the ship; and

.2 in exceptional cases where the particular character of the operation of an oil tanker renders it necessary to carry ballast water in excess of the quantity required under paragraph 2 of this regulation, provided that such operation of the oil tanker falls under the category of exceptional cases as established by the Organization.

SEE INTERPRETATION 35

Such additional ballast water shall be processed and discharged in compliance with regulation 34 of this Annex and an entry shall be made in the Oil Record Book Part II referred to in regulation 36 of this Annex.

4 In the case of crude oil tankers, the additional ballast permitted in paragraph 3 of this regulation shall be carried in cargo tanks only if such tanks have been crude oil washed in accordance with regulation 35 of this Annex before departure from an oil unloading port or terminal.

5 Notwithstanding the provisions of paragraph 2 of this regulation, the segregated ballast conditions for oil tankers less than 150 m in length shall be to the satisfaction of the Administration.

SEE INTERPRETATION 36

Crude oil tankers of 40,000 tonnes deadweight and above delivered on or before 1 June 1982

6 Subject to the provisions of paragraph 7 of this regulation, every crude oil tanker of 40,000 tonnes deadweight and above delivered on or before 1 June 1982, as defined in regulation 1.28.3, shall be provided with segregated ballast tanks and shall comply with the requirements of paragraphs 2 and 3 of this regulation.

7 Crude oil tankers referred to in paragraph 6 of this regulation may, in lieu of being provided with segregated tanks, operate with a cargo tank cleaning procedure using crude oil washing in accordance with regulations 33 and 35 of this Annex unless the crude oil tanker is intended to carry crude oil which is not suitable for crude oil washing.

SEE INTERPRETATION 37

Product carriers of 40,000 tonnes deadweight and above delivered on or before 1 June 1982

8 Every product carrier of 40,000 tonnes deadweight and above delivered on or before 1 June 1982, as defined in regulation 1.28.3, shall be provided with segregated ballast tanks and shall comply with the requirements of paragraphs 2 and 3 of this regulation, or alternatively operate with dedicated clean ballast tanks in accordance with the following provisions:

.1 The product carrier shall have adequate tank capacity, dedicated solely to the carriage of clean ballast as defined in regulation 1.17 of this Annex, to meet the requirements of paragraphs 2 and 3 of this regulation.

.2 The arrangements and operational procedures for dedicated clean ballast tanks shall comply with the requirements established by the Administration. Such requirements shall contain at least all the provisions of the revised Specifications for Oil Tankers with Dedicated Clean Ballast Tanks adopted by the Organization by resolution A.495(XII).

.3 The product carrier shall be equipped with an oil content meter, approved by the Administration on the basis of specifications recommended by the Organization, to enable supervision of the oil content in ballast water being discharged.*

SEE INTERPRETATION 39

.4 Every product carrier operating with dedicated clean ballast tanks shall be provided with a Dedicated Clean Ballast Tank Operation Manual† detailing the system and specifying operational procedures. Such a Manual shall be to the satisfaction of the Administration and shall contain all the information set out in the Specifications referred to in subparagraph 8.2 of this regulation. If an alteration affecting the dedicated clean ballast tank system is made, the Operation Manual shall be revised accordingly.

SEE INTERPRETATIONS 37 AND 38

* For oil content meters installed on oil tankers built prior to 2 October 1986, refer to the Recommendation on international performance and test specifications for oily-water separating equipment and oil content meters (resolution A.393(X)). For oil content meters as part of discharge monitoring and control systems installed on oil tankers built on or after 2 October 1986, refer to Guidelines and specifications for oil discharge monitoring and control systems for oil tankers (resolution A.586(14)). For oil content meters as part of discharge monitoring and control systems installed on oil tankers built on or after 1 January 2005, refer to Revised guidelines and specifications for oil discharge monitoring and control systems for oil tankers (resolution MEPC.108(49), as amended by resolution MEPC.240(65)).

† Refer to resolution A.495(XII) for the standard format of the Manual.

An oil tanker qualified as a segregated ballast oil tanker

9 Any oil tanker which is not required to be provided with segregated ballast tanks in accordance with paragraphs 1, 6 or 8 of this regulation may, however, be qualified as a segregated ballast tanker, provided that it complies with the requirements of paragraphs 2 and 3 or 5, as appropriate, of this regulation.

Oil tankers delivered on or before 1 June 1982 having special ballast arrangements

10 Oil tankers delivered on or before 1 June 1982, as defined in regulation 1.28.3, having special ballast arrangements:

.1 Where an oil tanker delivered on or before 1 June 1982, as defined in regulation 1.28.3, is so constructed or operates in such a manner that it complies at all times with the draught and trim requirements set out in paragraph 2 of this regulation without recourse to the use of ballast water, it shall be deemed to comply with the segregated ballast tank requirements referred to in paragraph 6 of this regulation, provided that all of the following conditions are complied with:

.1.1 operational procedures and ballast arrangements are approved by the Administration;

.1.2 agreement is reached between the Administration and the Governments of the port States Parties to the present Convention concerned when the draught and trim requirements are achieved through an operational procedure; and

.1.3 the International Oil Pollution Prevention Certificate is endorsed to the effect that the oil tanker is operating with special ballast arrangements.

.2 In no case shall ballast water be carried in oil tanks except on those rare voyages when weather conditions are so severe that, in the opinion of the master, it is necessary to carry additional ballast water in cargo tanks for the safety of the ship. Such additional ballast water shall be processed and discharged in compliance with regulation 34 of this Annex and in accordance with the requirements of regulations 29, 31 and 32 of this Annex, and an entry shall be made in the Oil Record Book referred to in regulation 36 of this Annex.

.3 An Administration which has endorsed a Certificate in accordance with subparagraph 10.1.3 of this regulation shall communicate to the Organization the particulars thereof for circulation to the Parties to the present Convention.

Oil tankers of 70,000 tonnes deadweight and above delivered after 31 December 1979

11 Oil tankers of 70,000 tonnes deadweight and above delivered after 31 December 1979, as defined in regulation 1.28.2, shall be provided with segregated ballast tanks and shall comply with paragraphs 2, 3 and 4 or paragraph 5 as appropriate of this regulation.

Protective location of segregated ballast

12 *Protective location of segregated ballast spaces*

In every crude oil tanker of 20,000 tonnes deadweight and above and every product carrier of 30,000 tonnes deadweight and above delivered after 1 June 1982, as defined in regulation 1.28.4, except those tankers that meet regulation 19, the segregated ballast tanks required to provide the capacity to comply with the requirements of paragraph 2 of this regulation, which are located within the cargo tank length, shall be arranged in accordance with the requirements of paragraphs 13, 14 and 15 of this regulation to provide a measure of protection against oil outflow in the event of grounding or collision.

SEE INTERPRETATION 40

13 Segregated ballast tanks and spaces other than oil tanks within the cargo tanks length (L_t) shall be so arranged as to comply with the following requirement:

$$\sum PA_c + \sum PA_s \geq J[L_t(B + 2D)]$$

where:

PA_c = the side shell area in square metres for each segregated ballast tank or space other than an oil tank based on projected moulded dimensions,

PA_s = the bottom shell area in square metres for each such tank or space based on projected moulded dimensions,

L_t = length in metres between the forward and after extremities of the cargo tanks,

B = maximum breadth of the ship in metres as defined in regulation 1.22 of this Annex,

D = moulded depth in metres measured vertically from the top of the keel to the top of the freeboard deck beam at side amidships. In ships having rounded gunwales, the moulded depth shall be measured to the point of intersection of the moulded lines of the deck and side shell plating, the lines extending as though the gunwale were of angular design,

J = 0.45 for oil tankers of 20,000 tonnes deadweight, 0.30 for oil tankers of 200,000 tonnes deadweight and above, subject to the provisions of paragraph 14 of this regulation.

For intermediate values of deadweight the value of J shall be determined by linear interpolation.

Whenever symbols given in this paragraph appear in this regulation, they have the meaning as defined in this paragraph.

SEE INTERPRETATION 40

14 For tankers of 200,000 tonnes deadweight and above the value of J may be reduced as follows:

$$J_{reduced} = J - \left(a - \frac{O_c + O_s}{4O_A}\right) \quad \text{or 0.2 whichever is greater}$$

where:

a = 0.25 for oil tankers of 200,000 tonnes deadweight,

a = 0.40 for oil tankers of 300,000 tonnes deadweight,

a = 0.50 for oil tankers of 420,000 tonnes deadweight and above.

For intermediate values of deadweight the value of a shall be determined by linear interpolation.

O_c = as defined in regulation 25.1.1 of this Annex,

O_s = as defined in regulation 25.1.2 of this Annex,

O_A = the allowable oil outflow as required by regulation 26.2 of this Annex.

SEE INTERPRETATION 40

15 In the determination of PA_c and PA_s for segregated ballast tanks and spaces other than oil tanks the following shall apply:

.1 the minimum width of each wing tank or space either of which extends for the full depth of the ship's side or from the deck to the top of the double bottom shall be not less than 2 m. The width shall be measured inboard from the ship's side at right angles to the centreline. Where a lesser width is provided, the wing tank or space shall not be taken into account when calculating the protecting area PA_c; and

.2 the minimum vertical depth of each double bottom tank or space shall be $\frac{B}{15}$ or 2 m, whichever is the lesser. Where a lesser depth is provided, the bottom tank or space shall not be taken into account when calculating the protecting area PA_s.

The minimum width and depth of wing tanks and double bottom tanks shall be measured clear of the bilge area and, in the case of minimum width, shall be measured clear of any rounded gunwale area.

SEE INTERPRETATION 40

Regulation 19

Double hull and double bottom requirements for oil tankers[*] *delivered on or after 6 July 1996*

SEE INTERPRETATIONS 15, 34 AND 41

1 This regulation shall apply to oil tankers of 600 tonnes deadweight and above delivered on or after 6 July 1996, as defined in regulation 1.28.6, as follows:

2 Every oil tanker of 5,000 tonnes deadweight and above shall:

.1 in lieu of paragraphs 12 to 15 of regulation 18, as applicable, comply with the requirements of paragraph 3 of this regulation unless it is subject to the provisions of paragraphs 4 and 5 of this regulation; and

.2 comply, if applicable, with the requirements of regulation 28.7.

3 The entire cargo tank length shall be protected by ballast tanks or spaces other than tanks that carry oil as follows:

.1 *Wing tanks or spaces*

Wing tanks or spaces shall extend either for the full depth of the ship's side or from the top of the double bottom to the uppermost deck, disregarding a rounded gunwale where fitted. They shall be arranged such that the cargo tanks are located inboard of the moulded line of the side shell plating nowhere less than the distance w, which, as shown in figure 1, is measured at any cross-section at right angles to the side shell, as specified below:

$$w = 0.5 + \frac{DW}{20{,}000} \text{ (m) or}$$

$w = 2.0$ m, whichever is the lesser.

The minimum value of $w = 1.0$ m.

.2 *Double bottom tanks or spaces*

At any cross-section, the depth of each double bottom tank or space shall be such that the distance h between the bottom of the cargo tanks and the moulded line of the bottom shell plating measured at right angles to the bottom shell plating as shown in figure 1 is not less than specified below:

$$h = \frac{B}{15} \text{ (m) or}$$

$h = 2.0$ m, whichever is the lesser.

The minimum value of $h = 1.0$ m.

.3 *Turn of the bilge area or at locations without a clearly defined turn of the bilge*

When the distances h and w are different, the distance w shall have preference at levels exceeding $1.5h$ above the baseline as shown in figure 1.

SEE INTERPRETATION 42

[*] Refer to Unified Interpretations on measurement of distances (MSC-MEPC.5/Circ.5).

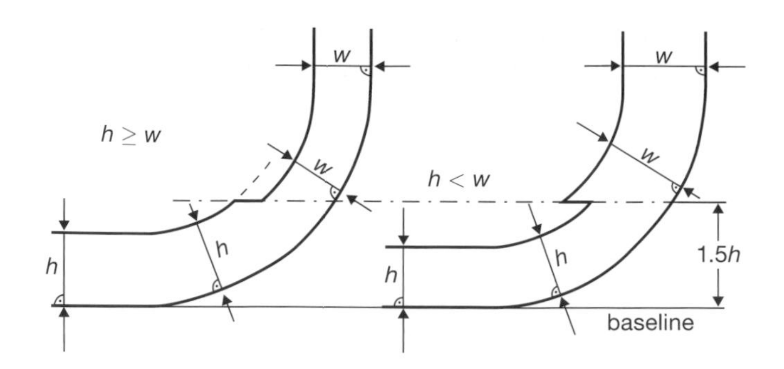

Figure 1 – *Cargo tank boundary lines*

.4 *The aggregate capacity of ballast tanks*

On crude oil tankers of 20,000 tonnes deadweight and above and product carriers of 30,000 tonnes deadweight and above, the aggregate capacity of wing tanks, double bottom tanks, forepeak tanks and after peak tanks shall not be less than the capacity of segregated ballast tanks necessary to meet the requirements of regulation 18 of this Annex. Wing tanks or spaces and double bottom tanks used to meet the requirements of regulation 18 shall be located as uniformly as practicable along the cargo tank length. Additional segregated ballast capacity provided for reducing longitudinal hull girder bending stress, trim, etc. may be located anywhere within the ship.

.5 *Suction wells in cargo tanks*

Suction wells in cargo tanks may protrude into the double bottom below the boundary line defined by the distance h provided that such wells are as small as practicable and the distance between the well bottom and bottom shell plating is not less than $0.5h$.

.6 *Ballast and cargo piping*

Ballast piping and other piping such as sounding and vent piping to ballast tanks shall not pass through cargo tanks. Cargo piping and similar piping to cargo tanks shall not pass through ballast tanks. Exemptions to this requirement may be granted for short lengths of piping, provided that they are completely welded or equivalent.

4 The following applies for double bottom tanks or spaces:

.1 Double bottom tanks or spaces as required by paragraph 3.2 of this regulation may be dispensed with, provided that the design of the tanker is such that the cargo and vapour pressure exerted on the bottom shell plating forming a single boundary between the cargo and the sea does not exceed the external hydrostatic water pressure, as expressed by the following formula:

$$f \times h_c \times \rho_c \times g + p \leq d_n \times \rho_s \times g$$

where:

h_c = height of cargo in contact with the bottom shell plating in metres
ρ_c = maximum cargo density in kg/m^3
d_n = minimum operating draught under any expected loading condition in metres
ρ_s = density of seawater in kg/m^3
p = maximum set pressure above atmospheric pressure (gauge pressure) of pressure/vacuum valve provided for the cargo tank in pascals
f = safety factor = 1.1
g = standard acceleration of gravity (9.81 m/s^2).

.2 Any horizontal partition necessary to fulfil the above requirements shall be located at a height not less than $\frac{B}{6}$ or 6 m, whichever is the lesser, but not more than $0.6D$, above the baseline where D is the moulded depth amidships.

.3 The location of wing tanks or spaces shall be as defined in paragraph 3.1 of this regulation except that, below a level 1.5h above the baseline where h is as defined in paragraph 3.2 of this regulation, the cargo tank boundary line may be vertical down to the bottom plating, as shown in figure 2.

SEE INTERPRETATION 43

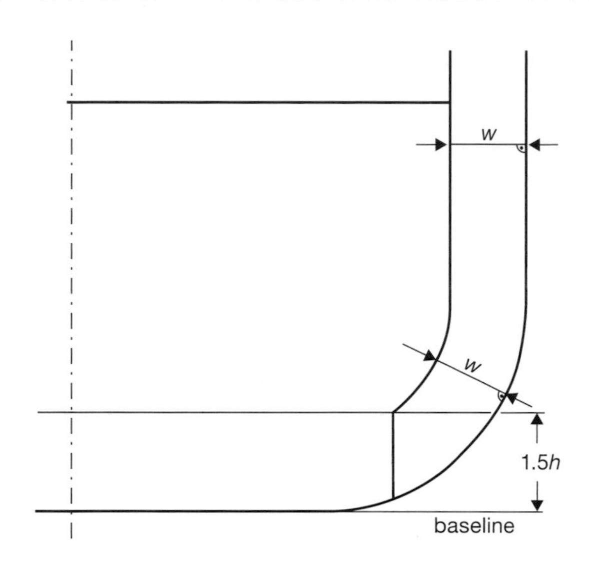

Figure 2 – *Cargo tank boundary lines*

5 Other methods of design and construction of oil tankers may also be accepted as alternatives to the requirements prescribed in paragraph 3 of this regulation, provided that such methods ensure at least the same level of protection against oil pollution in the event of collision or stranding and are approved in principle by the Marine Environment Protection Committee based on guidelines developed by the Organization.*

6 Every oil tanker of less than 5,000 tonnes deadweight shall comply with paragraphs 3 and 4 of this regulation, or shall:

.1 at least be fitted with double bottom tanks or spaces having such a depth that the distance h specified in paragraph 3.2 of this regulation complies with the following:

$$h = \frac{B}{15} \text{ (m)}$$

with a minimum value of $h = 0.76$ m;

in the turn of the bilge area and at locations without a clearly defined turn of the bilge, the cargo tank boundary line shall run parallel to the line of the midship flat bottom as shown in figure 3; and

.2 be provided with cargo tanks so arranged that the capacity of each cargo tank does not exceed 700 m^3 unless wing tanks or spaces are arranged in accordance with paragraph 3.1 of this regulation, complying with the following:

$$w = 0.4 + \frac{2.4\text{DW}}{20{,}000} \text{ (m)}$$

with a minimum value of $w = 0.76$ m.

SEE INTERPRETATION 44

* Refer to Revised Interim Guidelines for the approval of alternative methods of design and construction of oil tankers (resolution MEPC.110(49)).

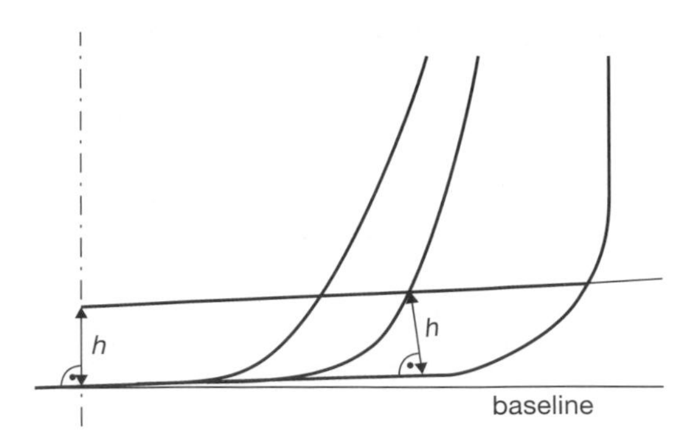

Figure 3 – *Cargo tank boundary lines*

7 Oil shall not be carried in any space extending forward of a collision bulkhead located in accordance with regulation II-1/11 of the International Convention for the Safety of Life at Sea, 1974, as amended.* An oil tanker that is not required to have a collision bulkhead in accordance with that regulation shall not carry oil in any space extending forward of the transverse plane perpendicular to the centreline that is located as if it were a collision bulkhead located in accordance with that regulation.

8 In approving the design and construction of oil tankers to be built in accordance with the provisions of this regulation, Administrations shall have due regard to the general safety aspects, including the need for the maintenance and inspections of wing and double bottom tanks or spaces.

Regulation 20

Double hull and double bottom requirements for oil tankers delivered before 6 July 1996

SEE INTERPRETATION 34

1 Unless expressly provided otherwise this regulation shall:

.1 apply to oil tankers of 5,000 tonnes deadweight and above, which are delivered before 6 July 1996, as defined in regulation 1.28.5 of this Annex; and

.2 not apply to oil tankers complying with regulation 19 and regulation 28 in respect of paragraph 28.7, which are delivered before 6 July 1996, as defined in regulation 1.28.5 of this Annex; and

.3 not apply to oil tankers covered by subparagraph 1 above which comply with regulation 19.3.1 and 19.3.2 or 19.4 or 19.5 of this Annex, except that the requirement for minimum distances between the cargo tank boundaries and the ship side and bottom plating need not be met in all respects. In that event, the side protection distances shall not be less than those specified in the International Bulk Chemical Code for type 2 cargo tank location and the bottom protection distances at centreline shall comply with regulation 18.15.2 of this Annex.

2 For the purpose of this regulation:

.1 *Heavy diesel oil* means diesel oil other than those distillates of which more than 50% by volume distils at a temperature not exceeding 340°C when tested by the method acceptable to the Organization.†

.2 *Fuel oil* means heavy distillates or residues from crude oil or blends of such materials intended for use as a fuel for the production of heat or power of a quality equivalent to the specification acceptable to the Organization.‡

* Refer to 2006 (Chapters II-1, II-2, III and XII and appendix) amendments (resolution MSC.216(82)).

† Refer to the American Society for Testing and Materials' Standard Test Method (Designation D86).

‡ Refer to the American Society for Testing and Materials' Specification for Number Four Fuel Oil (Designation D396) or heavier.

3 For the purpose of this regulation, oil tankers are divided into the following categories:

.1 *Category 1 oil tanker* means an oil tanker of 20,000 tonnes deadweight and above carrying crude oil, fuel oil, heavy diesel oil or lubricating oil as cargo, and of 30,000 tonnes deadweight and above carrying oil other than the above, which does not comply with the requirements for oil tankers delivered after 1 June 1982, as defined in regulation 1.28.4 of this Annex;

.2 *Category 2 oil tanker* means an oil tanker of 20,000 tonnes deadweight and above carrying crude oil, fuel oil, heavy diesel oil or lubricating oil as cargo, and of 30,000 tonnes deadweight and above carrying oil other than the above, which complies with the requirements for oil tankers delivered after 1 June 1982, as defined in regulation 1.28.4 of this Annex; and

SEE INTERPRETATION 45

.3 *Category 3 oil tanker* means an oil tanker of 5,000 tonnes deadweight and above but less than that specified in subparagraph 1 or 2 of this paragraph.

4 An oil tanker to which this regulation applies shall comply with the requirements of paragraphs 2 to 5, 7 and 8 of regulation 19 and regulation 28 in respect of paragraph 28.7 of this Annex not later than 5 April 2005 or the anniversary of the date of delivery of the ship on the date or in the year specified in the following table:

Category of oil tanker	Date or year
Category 1	5 April 2005 for ships delivered on 5 April 1982 or earlier
	2005 for ships delivered after 5 April 1982
Category 2 and Category 3	5 April 2005 for ships delivered on 5 April 1977 or earlier
	2005 for ships delivered after 5 April 1977 but before 1 January 1978
	2006 for ships delivered in 1978 and 1979
	2007 for ships delivered in 1980 and 1981
	2008 for ships delivered in 1982
	2009 for ships delivered in 1983
	2010 for ships delivered in 1984 or later

SEE INTERPRETATION 46

5 Notwithstanding the provisions of paragraph 4 of this regulation, in the case of a Category 2 or 3 oil tanker fitted with only double bottoms or double sides not used for the carriage of oil and extending to the entire cargo tank length or double hull spaces which are not used for the carriage of oil and extend to the entire cargo tank length, but which does not fulfil conditions for being exempted from the provisions of paragraph 1.3 of this regulation, the Administration may allow continued operation of such a ship beyond the date specified in paragraph 4 of this regulation, provided that:

.1 the ship was in service on 1 July 2001;

.2 the Administration is satisfied by verification of the official records that the ship complied with the conditions specified above;

.3 the conditions of the ship specified above remain unchanged; and

.4 such continued operation does not go beyond the date on which the ship reaches 25 years after the date of its delivery.

6 A Category 2 or 3 oil tanker of 15 years and over after the date of its delivery shall comply with the Condition Assessment Scheme adopted by the Marine Environment Protection Committee by resolution MEPC.94(46), as amended, provided that such amendments shall be adopted, brought into force and take effect in accordance with the provisions of article 16 of the present Convention relating to amendment procedures applicable to an appendix to an Annex.

SEE INTERPRETATION 47

7 The Administration may allow continued operation of a Category 2 or 3 oil tanker beyond the date specified in paragraph 4 of this regulation, if satisfactory results of the Condition Assessment Scheme warrant that, in the opinion of the Administration, the ship is fit to continue such operation, provided that the operation shall not go beyond the anniversary of the date of delivery of the ship in 2015 or the date on which the ship reaches 25 years after the date of its delivery, whichever is the earlier date.

8.1 The Administration of a Party to the present Convention which allows the application of paragraph 5 of this regulation, or allows, suspends, withdraws or declines the application of paragraph 7 of this regulation, to a ship entitled to fly its flag shall forthwith communicate to the Organization for circulation to the Parties to the present Convention particulars thereof, for their information and appropriate action, if any.

8.2 A Party to the present Convention shall be entitled to deny entry into the ports or offshore terminals under its jurisdiction of oil tankers operating in accordance with the provisions of:

.1 paragraph 5 of this regulation beyond the anniversary of the date of delivery of the ship in 2015; or

.2 paragraph 7 of this regulation.

In such cases, that Party shall communicate to the Organization for circulation to the Parties to the present Convention particulars thereof for their information.

Regulation 21
Prevention of oil pollution from oil tankers carrying heavy grade oil as cargo

1 This regulation shall:

.1 apply to oil tankers of 600 tonnes deadweight and above carrying heavy grade oil as cargo regardless of the date of delivery; and

.2 not apply to oil tankers covered by subparagraph 1 above which comply with regulations 19.3.1 and 19.3.2 or 19.4 or 19.5 of this Annex, except that the requirement for minimum distances between the cargo tank boundaries and the ship side and bottom plating need not be met in all respects. In that event, the side protection distances shall not be less than those specified in the International Bulk Chemical Code for type 2 cargo tank location and the bottom protection distances at centreline shall comply with regulation 18.15.2 of this Annex.

2 For the purpose of this regulation *heavy grade oil* means any of the following:

.1 crude oils having a density at 15°C higher than 900 kg/m^3;

.2 oils, other than crude oils, having either a density at 15°C higher than 900 kg/m^3 or a kinematic viscosity at 50°C higher than 180 mm^2/s; or

.3 bitumen, tar and their emulsions.

3 An oil tanker to which this regulation applies shall comply with the provisions of paragraphs 4 to 8 of this regulation in addition to complying with the applicable provisions of regulation 20.

4 Subject to the provisions of paragraphs 5, 6 and 7 of this regulation, an oil tanker to which this regulation applies shall:

.1 if 5,000 tonnes deadweight and above, comply with the requirements of regulation 19 of this Annex not later than 5 April 2005; or

.2 if 600 tonnes deadweight and above but less than 5,000 tonnes deadweight, be fitted with both double bottom tanks or spaces complying with the provisions of regulation 19.6.1 of this Annex, and wing tanks or spaces arranged in accordance with regulation 19.3.1 and complying with the requirement for distance *w* as referred to in regulation 19.6.2, not later than the anniversary of the date of delivery of the ship in the year 2008.

5 In the case of an oil tanker of 5,000 tonnes deadweight and above, carrying heavy grade oil as cargo fitted with only double bottoms or double sides not used for the carriage of oil and extending to the entire cargo tank length or double hull spaces which are not used for the carriage of oil and extend to the entire cargo tank length, but which does not fulfil conditions for being exempted from the provisions of paragraph 1.2 of this regulation, the Administration may allow continued operation of such a ship beyond the date specified in paragraph 4 of this regulation, provided that:

.1 the ship was in service on 4 December 2003;

.2 the Administration is satisfied by verification of the official records that the ship complied with the conditions specified above;

.3 the conditions of the ship specified above remain unchanged; and

.4 such continued operation does not go beyond the date on which the ship reaches 25 years after the date of its delivery.

6.1 The Administration may allow continued operation of an oil tanker of 5,000 tonnes deadweight and above, carrying crude oil having a density at 15°C higher than 900 kg/m^3 but lower than 945 kg/m^3, beyond the date specified in paragraph 4.1 of this regulation, if satisfactory results of the Condition Assessment Scheme referred to in regulation 20.6 warrant that, in the opinion of the Administration, the ship is fit to continue such operation, having regard to the size, age, operational area and structural conditions of the ship and provided that the operation shall not go beyond the date on which the ship reaches 25 years after the date of its delivery.

SEE INTERPRETATION 48

6.2 The Administration may allow continued operation of an oil tanker of 600 tonnes deadweight and above but less than 5,000 tonnes deadweight, carrying heavy grade oil as cargo, beyond the date specified in paragraph 4.2 of this regulation, if, in the opinion of the Administration, the ship is fit to continue such operation, having regard to the size, age, operational area and structural conditions of the ship, provided that the operation shall not go beyond the date on which the ship reaches 25 years after the date of its delivery.

7 The Administration of a Party to the present Convention may exempt an oil tanker of 600 tonnes deadweight and above carrying heavy grade oil as cargo from the provisions of this regulation if the oil tanker:

.1 either is engaged in voyages exclusively within an area under its jurisdiction, or operates as a floating storage unit of heavy grade oil located within an area under its jurisdiction; or

.2 either is engaged in voyages exclusively within an area under the jurisdiction of another Party, or operates as a floating storage unit of heavy grade oil located within an area under the jurisdiction of another Party, provided that the Party within whose jurisdiction the oil tanker will be operating agrees to the operation of the oil tanker within an area under its jurisdiction.

8.1 The Administration of a Party to the present Convention which allows, suspends, withdraws or declines the application of paragraph 5, 6 or 7 of this regulation to a ship entitled to fly its flag shall forthwith communicate to the Organization for circulation to the Parties to the present Convention particulars thereof, for their information and appropriate action, if any.

8.2 Subject to the provisions of international law, a Party to the present Convention shall be entitled to deny entry of oil tankers operating in accordance with the provisions of paragraph 5 or 6 of this regulation into the ports or offshore terminals under its jurisdiction, or deny ship-to-ship transfer of heavy grade oil in areas under its jurisdiction except when this is necessary for the purpose of securing the safety of a ship or saving life at sea. In such cases, that Party shall communicate to the Organization for circulation to the Parties to the present Convention particulars thereof for their information.

Regulation 22
Pump-room bottom protection

1 This regulation applies to oil tankers of 5,000 tonnes deadweight and above constructed on or after 1 January 2007.

2 The pump-room shall be provided with a double bottom such that at any cross-section the depth of each double bottom tank or space shall be such that the distance h between the bottom of the pump-room and the ship's baseline measured at right angles to the ship's baseline is not less than specified below:

$$h = \frac{B}{15} \text{ (m) or}$$

$$h = 2 \text{ m, whichever is the lesser.}$$

The minimum value of $h = 1$ m.

3 In case of pump-rooms whose bottom plate is located above the baseline by at least the minimum height required in paragraph 2 above (e.g. gondola stern designs), there will be no need for a double bottom construction in way of the pump-room.

4 Ballast pumps shall be provided with suitable arrangements to ensure efficient suction from double bottom tanks.

5 Notwithstanding the provisions of paragraphs 2 and 3 above, where the flooding of the pump-room would not render the ballast or cargo pumping system inoperative, a double bottom need not be fitted.

SEE INTERPRETATION 49

Regulation 23
Accidental oil outflow performance

1 This regulation shall apply to oil tankers delivered on or after 1 January 2010, as defined in regulation 1.28.8.

2 For the purpose of this regulation, the following definitions shall apply:

.1 *Load line draught* (d_S) is the vertical distance, in metres, from the moulded baseline at mid-length to the waterline corresponding to the summer freeboard to be assigned to the ship. Calculations pertaining to this regulation should be based on draught d_S, notwithstanding assigned draughts that may exceed d_S, such as the tropical load line.

.2 *Waterline* (d_B) is the vertical distance, in metres, from the moulded baseline at mid-length to the waterline corresponding to 30% of the depth D_S.

.3 *Breadth* (B_S) is the greatest moulded breadth of the ship, in metres, at or below the deepest load line draught d_S.

.4 *Breadth* (B_B) is the greatest moulded breadth of the ship, in metres, at or below the waterline d_B.

.5 *Depth* (D_S) is the moulded depth, in metres, measured at mid-length to the upper deck at side.

.6 *Length* (L) and *deadweight* (DW) are as defined in regulations 1.19 and 1.23, respectively.

3 To provide adequate protection against oil pollution in the event of collision or stranding, the following shall be complied with:

.1 for oil tankers of 5,000 tonnes deadweight (DWT) and above, the mean oil outflow parameter shall be as follows:

$$O_M \leq 0.015 \quad \text{for } C \leq 200{,}000 \text{ m}^3$$

$$O_M \leq 0.012 + \frac{0.003}{200{,}000}(400{,}000 - C) \quad \text{for } 200{,}000 \text{ m}^3 < C < 400{,}000 \text{ m}^3$$

$$O_M \leq 0.012 \quad \text{for } C \geq 400{,}000 \text{ m}^3$$

for combination carriers between 5,000 tonnes deadweight (DWT) and 200,000 m^3 capacity, the mean oil outflow parameter may be applied, provided calculations are submitted to the satisfaction of the Administration, demonstrating that, after accounting for its increased structural strength, the combination carrier has at least equivalent oil outflow performance to a standard double hull tanker of the same size having a $O_M \leq 0.015$.

$$O_M \leq 0.021 \quad \text{for } C \leq 100{,}000 \text{ m}^3$$

$$O_M \leq 0.015 + \left(\frac{0.006}{100{,}000}\right)(200{,}000 - C)$$

$$\text{for } 100{,}000 \text{ m}^3 < C \leq 200{,}000 \text{ m}^3$$

where:

O_M = mean oil outflow parameter

C = total volume of cargo oil, in m^3, at 98% tank filling.

.2 for oil tankers of less than 5,000 tonnes deadweight (DWT), the length of each cargo tank shall not exceed 10 m or one of the following values, whichever is the greater:

.2.1 where no longitudinal bulkhead is provided inside the cargo tanks:

$$\left(0.5\frac{b_i}{B} + 0.1\right)L \quad \text{but not to exceed } 0.2L$$

.2.2 where a centreline longitudinal bulkhead is provided inside the cargo tanks:

$$\left(0.25\frac{b_i}{B} + 0.15\right)L$$

.2.3 where two or more longitudinal bulkheads are provided inside the cargo tanks:

.2.3.1 for wing cargo tanks: $0.2L$

.2.3.2 for centre cargo tanks:

.2.3.2.1 if $\frac{b_i}{B} \geq 0.2L$: $0.2L$

.2.3.2.2 if $\frac{b_i}{B} < 0.2L$:

.2.3.2.2.1 where no centreline longitudinal bulkhead is provided:

$$\left(0.5\frac{b_i}{B} + 0.1\right)L$$

.2.3.2.2.2 where a centreline longitudinal bulkhead is provided:

$$\left(0.25\frac{b_i}{B} + 0.15\right)L$$

b_i is the minimum distance from the ship's side to the outer longitudinal bulkhead of the tank in question measured inboard at right angles to the centreline at the level corresponding to the assigned summer freeboard.

4 The following general assumptions shall apply when calculating the mean oil outflow parameter:

.1 The cargo block length extends between the forward and aft extremities of all tanks arranged for the carriage of cargo oil, including slop tanks.

.2 Where this regulation refers to cargo tanks, it shall be understood to include all cargo tanks, slop tanks and fuel tanks located within the cargo block length.

.3 The ship shall be assumed loaded to the load line draught d_S without trim or heel.

.4 All cargo oil tanks shall be assumed loaded to 98% of their volumetric capacity. The nominal density of the cargo oil (ρ_n) shall be calculated as follows:

$$\rho_n = \frac{1{,}000(DWT)}{C} \text{ (kg/m}^3\text{)}$$

.5 For the purposes of these outflow calculations, the permeability of each space within the cargo block, including cargo tanks, ballast tanks and other non-oil spaces, shall be taken as 0.99, unless proven otherwise.

.6 Suction wells may be neglected in the determination of tank location provided that such wells are as small as practicable and the distance between the well bottom and bottom shell plating is not less than 0.5*h*, where *h* is the height as defined in regulation 19.3.2.

5 The following assumptions shall be used when combining the oil outflow parameters:

.1 The mean oil outflow shall be calculated independently for side damage and for bottom damage and then combined into the non-dimensional oil outflow parameter O_M, as follows:

$$O_M = \frac{0.4O_{MS} + 0.6O_{MB}}{C}$$

where:

O_{MS} = mean outflow for side damage, in m^3; and

O_{MB} = mean outflow for bottom damage, in m^3.

.2 For bottom damage, independent calculations for mean outflow shall be done for 0 m and minus 2.5 m tide conditions, and then combined as follows:

$$O_{MB} = 0.7O_{MB(0)} + 0.3O_{MB(2.5)}$$

where:

$O_{MB(0)}$ = mean outflow for 0 m tide condition; and

$O_{MB(2.5)}$ = mean outflow for minus 2.5 m tide condition, in m^3.

6 The mean outflow for side damage O_{MS} shall be calculated as follows:

$$O_{MS} = C_3 \sum_i^n P_{S(i)} O_{S(i)} \qquad \text{(m}^3\text{)}$$

where:

i represents each cargo tank under consideration;

n = total number of cargo tanks;

$P_{S(i)}$ = the probability of penetrating cargo tank *i* from side damage, calculated in accordance with paragraph 8.1 of this regulation;

$O_{S(i)}$ = the outflow, in m^3, from side damage to cargo tank *i*, which is assumed equal to the total volume in cargo tank *i* at 98% filling, unless it is proven through the application of the Guidelines referred to in regulation 19.5 that any significant cargo volume will be retained; and

C_3 = 0.77 for ships having two longitudinal bulkheads inside the cargo tanks, provided these bulkheads are continuous over the cargo block and $P_{S(i)}$ is developed in accordance with this regulation. C_3 equals 1.0 for all other ships or when $P_{S(i)}$ is developed in accordance with paragraph 10 of this regulation.

7 The mean outflow for bottom damage shall be calculated for each tidal condition as follows:

.1 $O_{MB(0)} = \sum_{i}^{n} P_{B(i)} O_{B(i)} C_{DB(i)}$ (m^3)

where:

i represents each cargo tank under consideration;

n = the total number of cargo tanks;

$P_{B(i)}$ = the probability of penetrating cargo tank i from bottom damage, calculated in accordance with paragraph 9.1 of this regulation;

$O_{B(i)}$ = the outflow from cargo tank i, in m^3, calculated in accordance with paragraph 7.3 of this regulation; and

$C_{DB(i)}$ = factor to account for oil capture as defined in paragraph 7.4 of this regulation

.2 $O_{MB(2.5)} = \sum_{i}^{n} P_{B(i)} O_{B(i)} C_{DB(i)}$ (m^3)

where:

i, n, $P_{B(i)}$ and $C_{DB(i)}$ = as defined in subparagraph .1 above;

$O_{B(i)}$ = the outflow from cargo tank i, in m^3, after tidal change.

.3 The oil outflow $O_{B(i)}$ for each cargo oil tank shall be calculated based on pressure-balance principles, in accordance with the following assumptions:

.3.1 The ship shall be assumed stranded with zero trim and heel, with the stranded draught prior to tidal change equal to the load line draught d_S.

.3.2 The cargo level after damage shall be calculated as follows:

$$h_c = \frac{(d_S + t_c - Z_l)(\rho_s) - \frac{1{,}000p}{g}}{\rho_n}$$

where:

h_c = the height of the cargo oil above Z_l, in metres;

t_c = the tidal change, in metres. Reductions in tide shall be expressed as negative values;

Z_l = the height of the lowest point in the cargo tank above baseline, in metres;

ρ_s = density of seawater, to be taken as 1,025 kg/m^3;

p = if an inert gas system is fitted, the normal overpressure, in kilopascals, to be taken as not less than 5 kPa; if an inert gas system is not fitted, the overpressure may be taken as 0;

SEE INTERPRETATION 50

g = the acceleration of gravity, to be taken as 9.81 m/s^2; and

ρ_n = nominal density of cargo oil, calculated in accordance with paragraph 4.4 of this regulation.

.3.3 For cargo tanks bounded by the bottom shell, unless proven otherwise, oil outflow $O_{B(i)}$ shall be taken not less than 1% of the total volume of cargo oil loaded in cargo tank i, to account for initial exchange losses and dynamic effects due to current and waves.

.4 In the case of bottom damage, a portion from the outflow from a cargo tank may be captured by non-oil compartments. This effect is approximated by application of the factor $C_{DB(i)}$ for each tank, which shall be taken as follows:

$C_{DB(i)}$ = 0.6 for cargo tanks bounded from below by non-oil compartments;

$C_{DB(i)}$ = 1.0 for cargo tanks bounded by the bottom shell.

8 The probability P_S of breaching a compartment from side damage shall be calculated as follows:

.1 $P_S = P_{SL} \cdot P_{SV} \cdot P_{ST}$

where:

$P_{SL} = 1 - P_{Sf} - P_{Sa}$ = probability the damage will extend into the longitudinal zone bounded by X_a and X_f;

$P_{SV} = 1 - P_{Su} - P_{Sl}$ = probability the damage will extend into the vertical zone bounded by Z_l and Z_u; and

$P_{ST} = 1 - P_{Sy}$ = probability the damage will extend transversely beyond the boundary defined by y.

.2 P_{Sa}, P_{Sf}, P_{Sl}, P_{Su} and P_{Sy} shall be determined by linear interpolation from the tables of probabilities for side damage provided in paragraph 8.3 of this regulation, where:

P_{Sa} = the probability the damage will lie entirely aft of location $\frac{X_a}{L}$;

P_{Sf} = the probability the damage will lie entirely forward of location $\frac{X_f}{L}$;

P_{Sl} = the probability the damage will lie entirely below the tank;

P_{Su} = the probability the damage will lie entirely above the tank; and

P_{Sy} = the probability the damage will lie entirely outboard of the tank.

Compartment boundaries X_a, X_f, Z_l, Z_u and y shall be developed as follows:

X_a = the longitudinal distance from the aft terminal of L to the aftmost point on the compartment being considered, in metres;

X_f = the longitudinal distance from the aft terminal of L to the foremost point on the compartment being considered, in metres;

Z_l = the vertical distance from the moulded baseline to the lowest point on the compartment being considered, in metres;

Z_u = the vertical distance from the moulded baseline to the highest point on the compartment being considered, in metres. Z_u is not to be taken greater than D_S; and

y = the minimum horizontal distance measured at right angles to the centreline between the compartment under consideration and the side shell, in metres;*

* For symmetrical tank arrangements, damages are considered for one side of the ship only, in which case all "y" dimensions are to be measured from that same side. For asymmetrical arrangements, refer to Explanatory Notes on matters related to the accidental oil outflow performance (resolution MEPC.122(52), as amended).

.3 Tables of probabilities for side damage

$\frac{X_a}{L}$	P_{Sa}
0.00	0.000
0.05	0.023
0.10	0.068
0.15	0.117
0.20	0.167
0.25	0.217
0.30	0.267
0.35	0.317
0.40	0.367
0.45	0.417
0.50	0.467
0.55	0.517
0.60	0.567
0.65	0.617
0.70	0.667
0.75	0.717
0.80	0.767
0.85	0.817
0.90	0.867
0.95	0.917
1.00	0.967

$\frac{X_f}{L}$	P_{Sf}
0.00	0.967
0.05	0.917
0.10	0.867
0.15	0.817
0.20	0.767
0.25	0.717
0.30	0.667
0.35	0.617
0.40	0.567
0.45	0.517
0.50	0.467
0.55	0.417
0.60	0.367
0.65	0.317
0.70	0.267
0.75	0.217
0.80	0.167
0.85	0.117
0.90	0.068
0.95	0.023
1.00	0.000

$\frac{Z_l}{D_S}$	P_{Sl}
0.00	0.000
0.05	0.000
0.10	0.001
0.15	0.003
0.20	0.007
0.25	0.013
0.30	0.021
0.35	0.034
0.40	0.055
0.45	0.085
0.50	0.123
0.55	0.172
0.60	0.226
0.65	0.285
0.70	0.347
0.75	0.413
0.80	0.482
0.85	0.553
0.90	0.626
0.95	0.700
1.00	0.775

$\frac{Z_u}{D_S}$	P_{Su}
0.00	0.968
0.05	0.952
0.10	0.931
0.15	0.905
0.20	0.873
0.25	0.836
0.30	0.789
0.35	0.733
0.40	0.670
0.45	0.599
0.50	0.525
0.55	0.452
0.60	0.383
0.65	0.317
0.70	0.255
0.75	0.197
0.80	0.143
0.85	0.092
0.90	0.046
0.95	0.013
1.00	0.000

P_{Sy} shall be calculated as follows:

$$P_{Sy} = \left(24.96 - \frac{199.6y}{B_S}\right)\left(\frac{y}{B_S}\right) \quad \text{for } \frac{y}{B_S} \leq 0.05$$

$$P_{Sy} = 0.749 + \left(5 - 44.4\left(\frac{y}{B_S} - 0.05\right)\right)\left(\frac{y}{B_S} - 0.05\right) \text{ for } 0.05 < \frac{y}{B_S} < 0.1$$

$$P_{Sy} = 0.888 + 0.56\left(\frac{y}{B_S} - 0.1\right) \quad \text{for } \frac{y}{B_S} \geq 0.1$$

P_{Sy} shall not be taken greater than 1.

9 The probability P_B of breaching a compartment from bottom damage shall be calculated as follows:

.1 $P_B = P_{BL}\, P_{BT}\, P_{BV}$

where:

$P_{BL} = 1 - P_{Bf} - P_{Ba}$ = probability the damage will extend into the longitudinal zone bounded by X_a and X_f;

$P_{BT} = 1 - P_{Bp} - P_{Bs}$ = probability the damage will extend into the transverse zone bounded by Y_p and Y_s; and

$P_{BV} = 1 - P_{Bz}$ = probability the damage will extend vertically above the boundary defined by z.

.2 P_{Ba}, P_{Bf}, P_{Bp}, P_{Bs}, and P_{Bz} shall be determined by linear interpolation from the tables of probabilities for bottom damage provided in paragraph 9.3 of this regulation, where:

P_{Ba} = the probability the damage will lie entirely aft of location $\frac{X_a}{L}$;

P_{Bf} = the probability the damage will lie entirely forward of location X_f/L;

P_{Bp} = the probability the damage will lie entirely to port of the tank;

P_{Bs} = the probability the damage will lie entirely to starboard of the tank; and

P_{Bz} = the probability the damage will lie entirely below the tank.

Compartment boundaries X_a, X_f, Y_p, Y_s, and z shall be developed as follows:

X_a and X_f are as defined in paragraph 8.2 of this regulation;

Y_p = the transverse distance from the port-most point on the compartment located at or below the waterline d_B, to a vertical plane located $B_B/2$ to starboard of the ship's centreline, in metres;

Y_s = the transverse distance from the starboard-most point on the compartment located at or below the waterline d_B, to a vertical plane located $B_B/2$ to starboard of the ship's centreline, in metres; and

z = the minimum value of z over the length of the compartment, where, at any given longitudinal location, z is the vertical distance from the lower point of the bottom shell at that longitudinal location to the lower point of the compartment at that longitudinal location, in metres.

.3 Tables of probabilities for bottom damage

$\frac{X_a}{L}$	P_{Ba}
0.00	0.000
0.05	0.002
0.10	0.008
0.15	0.017
0.20	0.029
0.25	0.042
0.30	0.058
0.35	0.076
0.40	0.096
0.45	0.119
0.50	0.143
0.55	0.171
0.60	0.203
0.65	0.242
0.70	0.289
0.75	0.344
0.80	0.409
0.85	0.482
0.90	0.565
0.95	0.658
1.00	0.761

$\frac{X_f}{L}$	P_{Bf}
0.00	0.969
0.05	0.953
0.10	0.936
0.15	0.916
0.20	0.894
0.25	0.870
0.30	0.842
0.35	0.810
0.40	0.775
0.45	0.734
0.50	0.687
0.55	0.630
0.60	0.563
0.65	0.489
0.70	0.413
0.75	0.333
0.80	0.252
0.85	0.170
0.90	0.089
0.95	0.026
1.00	0.000

$\frac{Y_p}{B_B}$	P_{Bp}
0.00	0.844
0.05	0.794
0.10	0.744
0.15	0.694
0.20	0.644
0.25	0.594
0.30	0.544
0.35	0.494
0.40	0.444
0.45	0.394
0.50	0.344
0.55	0.297
0.60	0.253
0.65	0.211
0.70	0.171
0.75	0.133
0.80	0.097
0.85	0.063
0.90	0.032
0.95	0.009
1.00	0.000

$\frac{Y_s}{B_B}$	P_{Bs}
0.00	0.000
0.05	0.009
0.10	0.032
0.15	0.063
0.20	0.097
0.25	0.133
0.30	0.171
0.35	0.211
0.40	0.253
0.45	0.297
0.50	0.344
0.55	0.394
0.60	0.444
0.65	0.494
0.70	0.544
0.75	0.594
0.80	0.644
0.85	0.694
0.90	0.744
0.95	0.794
1.00	0.844

P_{Bz} shall be calculated as follows:

$$P_{Bz} = \left(14.5 - \frac{67z}{D_S}\right)\left(\frac{z}{D_S}\right) \quad \text{for } \frac{z}{D_S} \le 0.1,$$

$$P_{Bz} = 0.78 + 1.1\left(\frac{z}{D_S} - 0.1\right) \quad \text{for } \frac{z}{D_S} > 0.1.$$

P_{Bz} shall not be taken greater than 1.

10 This regulation uses a simplified probabilistic approach where a summation is carried out over the contributions to the mean outflow from each cargo tank. For certain designs, such as those characterized by the occurrence of steps/recesses in bulkheads/decks and for sloping bulkheads and/or a pronounced hull curvature, more rigorous calculations may be appropriate. In such cases one of the following calculation procedures may be applied:

.1 The probabilities referred to in 8 and 9 above may be calculated with more precision through application of hypothetical sub-compartments.*

.2 The probabilities referred to in 8 and 9 above may be calculated through direct application of the probability density functions contained in the Guidelines referred to in regulation 19.5.

.3 The oil outflow performance may be evaluated in accordance with the method described in the Guidelines referred to in regulation 19.5.

11 The following provisions regarding piping arrangements shall apply:

.1 Lines of piping that run through cargo tanks in a position less than $0.30B_S$ from the ship's side or less than $0.30D_S$ from the ship's bottom shall be fitted with valves or similar closing devices at the point at which they open into any cargo tank. These valves shall be kept closed at sea at any time when the tanks contain cargo oil, except that they may be opened only for cargo transfer needed for essential cargo operations.

.2 Credit for reducing oil outflow through the use of an emergency rapid cargo transfer system or other system arranged to mitigate oil outflow in the event of an accident may be taken into account only after the effectiveness and safety aspects of the system are approved by the Organization. Submittal for approval shall be made in accordance with the provisions of the Guidelines referred to in regulation 19.5.

Regulation 24
Damage assumptions

1 For the purpose of calculating hypothetical oil outflow from oil tankers in accordance with regulations 25 and 26, three dimensions of the extent of damage of a parallelepiped on the side and bottom of the ship are assumed as follows. In the case of bottom damages two conditions are set forth to be applied individually to the stated portions of the oil tanker.

.1 Side damage:

.1.1 Longitudinal extent (l_c):	$\frac{1}{3}L^{\frac{2}{3}}$ or 14.5 m, whichever is less
.1.2 Transverse extent (t_c) (inboard from the ship's side at right angles to the centreline at the level corresponding to the assigned summer freeboard):	$\frac{B}{5}$ or 11.5 m, whichever is less
.1.3 Vertical extent (v_c):	From the baseline upwards without limit

* Refer to Explanatory Notes on matters related to the accidental oil outflow performance (resolution MEPC.122(52), as amended).

.2 Bottom damage:

	For 0.3L from the forward perpendicular of the ship	*Any other part of the ship*
.2.1 Longitudinal extent (l_s):	$\frac{L}{10}$	$\frac{L}{10}$ or 5 m, whichever is less
.2.2 Transverse extent (t_s):	$\frac{B}{6}$ or 10 m, whichever is less, but not less than 5 m	5 m
.2.3 Vertical extent from the baseline (v_s):	$\frac{B}{15}$ or 6 m, whichever is less	

SEE INTERPRETATION 51

2 Wherever the symbols given in this regulation appear in this chapter, they have the meaning as defined in this regulation.

Regulation 25

Hypothetical outflow of oil

SEE INTERPRETATION 52

1 The hypothetical outflow of oil in the case of side damage (O_c) and bottom damage (O_s) shall be calculated by the following formulae with respect to compartments breached by damage to all conceivable locations along the length of the ship to the extent as defined in regulation 24 of this Annex.

.1 For side damages:

$$O_c = \sum W_i + \sum K_i C_i \qquad \text{(I)}$$

.2 For bottom damages:

$$O_s = \frac{1}{3}\left(\sum Z_i W_i + \sum Z_i C_i\right) \qquad \text{(II)}$$

where:

W_i = volume of a wing tank, in cubic metres, assumed to be breached by the damage as specified in regulation 24 of this Annex; W_i for a segregated ballast tank may be taken equal to zero.

C_i = volume of a centre tank, in cubic metres, assumed to be breached by the damage as specified in regulation 24 of this Annex; C_i for a segregated ballast tank may be taken equal to zero.

$K_i = 1 - \frac{b_i}{t_c}$; when b_i is equal to or greater than t_c, K_i shall be taken equal to zero.

$Z_i = 1 - \frac{h_i}{v_s}$; when h_i is equal to or greater than v_s, Z_i shall be taken equal to zero.

b_i = width of wing tank under consideration, in metres, measured inboard from the ship's side at right angles to the centreline at the level corresponding to the assigned summer freeboard.

h_i = minimum depth of the double bottom under consideration, in metres; where no double bottom is fitted, h_i shall be taken equal to zero.

Whenever symbols given in this paragraph appear in this chapter, they have the meaning as defined in this regulation.

SEE INTERPRETATION 53

2 If a void space or segregated ballast tank of a length less than l_c as defined in regulation 24 of this Annex is located between wing oil tanks, O_c in formula (I) may be calculated on the basis of volume W_i being the actual volume of one such tank (where they are of equal capacity) or the smaller of the two tanks (if they differ in capacity) adjacent to such space, multiplied by S_i as defined below and taking for all other wing tanks involved in such collision the value of the actual full volume.

$$S_i = 1 - \frac{l_i}{l_c}$$

where l_i = length, in metres, of void space or segregated ballast tank under consideration.

3.1 Credit shall only be given in respect of double bottom tanks which are either empty or carrying clean water when cargo is carried in the tanks above.

3..2 Where the double bottom does not extend for the full length and width of the tank involved, the double bottom is considered non-existent and the volume of the tanks above the area of the bottom damage shall be included in formula (II) even if the tank is not considered breached because of the installation of such a partial double bottom.

3.3 Suction wells may be neglected in the determination of the value h_i provided such wells are not excessive in area and extend below the tank for a minimum distance and in no case more than half the height of the double bottom. If the depth of such a well exceeds half the height of the double bottom, h_i shall be taken equal to the double bottom height minus the well height.

Piping serving such wells if installed within the double bottom shall be fitted with valves or other closing arrangements located at the point of connection to the tank served to prevent oil outflow in the event of damage to the piping. Such piping shall be installed as high from the bottom shell as possible. These valves shall be kept closed at sea at any time when the tank contains oil cargo, except that they may be opened only for cargo transfer needed for the purpose of trimming of the ship.

SEE INTERPRETATION 54

4 In the case where bottom damage simultaneously involves four centre tanks, the value of O_s may be calculated according to the formula:

$$O_s = \frac{1}{4}\left(\sum Z_i W_i + \sum Z_i C_i\right) \quad \text{(III)}$$

5 An Administration may credit as reducing oil outflow in case of bottom damage, an installed cargo transfer system having an emergency high suction in each cargo oil tank, capable of transferring from a breached tank or tanks to segregated ballast tanks or to available cargo tankage if it can be assured that such tanks will have sufficient ullage. Credit for such a system would be governed by ability to transfer in two hours of operation oil equal to one half of the largest of the breached tanks involved and by availability of equivalent receiving capacity in ballast or cargo tanks. The credit shall be confined to permitting calculation of O_s according to formula (III). The pipes for such suctions shall be installed at least at a height not less than the vertical extent of the bottom damage v_s. The Administration shall supply the Organization with the information concerning the arrangements accepted by it, for circulation to other Parties to the Convention.

6 This regulation does not apply to oil tankers delivered on or after 1 January 2010, as defined in regulation 1.28.8.

Regulation 26

Limitations of size and arrangement of cargo tanks

1 Except as provided in paragraph 7 below:

.1 every oil tanker of 150 gross tonnage and above delivered after 31 December 1979, as defined in regulation 1.28.2, and

.2 every oil tanker of 150 gross tonnage and above delivered on or before 31 December 1979, as defined in regulation 1.28.1, which falls into either of the following categories:

.2.1 a tanker, the delivery of which is after 1 January 1977, or

.2.2 a tanker to which both the following conditions apply:

.2.2.1 delivery is not later than 1 January 1977; and

.2.2.2 the building contract is placed after 1 January 1974, or in cases where no building contract has previously been placed, the keel is laid or the tanker is at a similar stage of construction after 30 June 1974

shall comply with the provisions of this regulation.

2 Cargo tanks of oil tankers shall be of such size and arrangements that the hypothetical outflow O_c or O_s calculated in accordance with the provisions of regulation 25 of this Annex anywhere in the length of the ship does not exceed 30,000 m^3 or $400\sqrt[3]{DW}$, whichever is the greater, but subject to a maximum of 40,000 m^3.

3 The volume of any one wing cargo oil tank of an oil tanker shall not exceed 75% of the limits of the hypothetical oil outflow referred to in paragraph 2 of this regulation. The volume of any one centre cargo oil tank shall not exceed 50,000 m^3. However, in segregated ballast oil tankers as defined in regulation 18 of this Annex, the permitted volume of a wing cargo oil tank situated between two segregated ballast tanks, each exceeding l_c in length, may be increased to the maximum limit of hypothetical oil outflow provided that the width of the wing tanks exceeds t_c.

4 The length of each cargo tank shall not exceed 10 m or one of the following values, whichever is the greater:

.1 where no longitudinal bulkhead is provided inside the cargo tanks:

$$\left(0.5\frac{b_i}{B} + 0.1\right)L \text{ but not to exceed } 0.2L$$

.2 where a centreline longitudinal bulkhead is provided inside the cargo tanks:

$$\left(0.25\frac{b_i}{B} + 0.15\right)L$$

.3 where two or more longitudinal bulkheads are provided inside the cargo tanks:

.3.1 for wing cargo tanks: $0.2L$

.3.2 for centre cargo tanks:

.3.2.1 if $\frac{b_i}{B}$ is equal to or greater than one fifth: $0.2L$

.3.2.2 if $\frac{b_i}{B}$ is less than one fifth:

.3.2.2.1 where no centreline longitudinal bulkhead is provided:

$$\left(0.5\frac{b_i}{B} + 0.1\right)L$$

.3.2.2.2 where a centreline longitudinal bulkhead is provided:

$$\left(0.25\frac{b_i}{B} + 0.15\right)L$$

b_i is the minimum distance from the ship's side to the outer longitudinal bulkhead of the tank in question measured inboard at right angles to the centreline at the level corresponding to the assigned summer freeboard.

5 In order not to exceed the volume limits established by paragraphs 2, 3 and 4 of this regulation and irrespective of the accepted type of cargo transfer system installed, when such system interconnects two or more cargo tanks, valves or other similar closing devices shall be provided for separating the tanks from each other. These valves or devices shall be closed when the tanker is at sea.

6 Lines of piping which run through cargo tanks in a position less than t_c from the ship's side or less than v_c from the ship's bottom shall be fitted with valves or similar closing devices at the point at which they open into any cargo tank. These valves shall be kept closed at sea at any time when the tanks contain cargo oil, except that they may be opened only for cargo transfer needed for the purpose of trimming of the ship.

7 This regulation does not apply to oil tankers delivered on or after 1 January 2010, as defined in regulation 1.28.8.

Regulation 27
Intact stability

SEE INTERPRETATION 55

1 Every oil tanker of 5,000 tonnes deadweight and above delivered on or after 1 February 2002, as defined in regulation 1.28.7, shall comply with the intact stability criteria specified in paragraphs 1.1 and 1.2 of this regulation, as appropriate, for any operating draught under the worst possible conditions of cargo and ballast loading, consistent with good operational practice, including intermediate stages of liquid transfer operations. Under all conditions the ballast tanks shall be assumed slack.

.1 In port, the initial metacentric height GM_o, corrected for the free surface measured at 0° heel, shall be not less than 0.15 m;

.2 At sea, the following criteria shall be applicable:

.2.1 the area under the righting lever curve (GZ curve) shall be not less than 0.055 m·rad up to $\theta = 30°$ angle of heel and not less than 0.09 m·rad up to $\theta = 40°$ or other angle of flooding θ_f* if this angle is less than 40°. Additionally, the area under the righting lever curve (GZ curve) between the angles of heel of 30° and 40° or between 30° and θ_f, if this angle is less than 40°, shall be not less than 0.03 m·rad;

.2.2 the righting lever GZ shall be at least 0.20 m at an angle of heel equal to or greater than 30°;

.2.3 the maximum righting arm shall occur at an angle of heel preferably exceeding 30° but not less than 25°; and

.2.4 the initial metacentric height GM_o, corrected for free surface measured at 0° heel, shall be not less than 0.15 m.

2 The requirements of paragraph 1 of this regulation shall be met through design measures. For combination carriers simple supplementary operational procedures may be allowed.

3 Simple supplementary operational procedures for liquid transfer operations referred to in paragraph 2 of this regulation shall mean written procedures made available to the master which:

.1 are approved by the Administration;

* θ_f is the angle of heel at which openings in the hull superstructures or deckhouses which cannot be closed weathertight immerse. In applying this criterion, small openings through which progressive flooding cannot take place need not be considered as open.

.2 indicate those cargo and ballast tanks which may, under any specific condition of liquid transfer and possible range of cargo densities, be slack and still allow the stability criteria to be met. The slack tanks may vary during the liquid transfer operations and be of any combination provided they satisfy the criteria;

.3 will be readily understandable to the officer-in-charge of liquid transfer operations;

.4 provide for planned sequences of cargo/ballast transfer operations;

.5 allow comparisons of attained and required stability using stability performance criteria in graphical or tabular form;

.6 require no extensive mathematical calculations by the officer-in-charge;

.7 provide for corrective actions to be taken by the officer-in-charge in case of departure from recommended values and in case of emergency situations; and

.8 are prominently displayed in the approved trim and stability booklet and at the cargo/ballast transfer control station and in any computer software by which stability calculations are performed.

Regulation 28

Subdivision and damage stability

1 Every oil tanker delivered after 31 December 1979, as defined in regulation 1.28.2, of 150 gross tonnage and above, shall comply with the subdivision and damage stability criteria as specified in paragraph 3 of this regulation, after the assumed side or bottom damage as specified in paragraph 2 of this regulation, for any operating draught reflecting actual partial or full load conditions consistent with trim and strength of the ship as well as relative densities of the cargo. Such damage shall be applied to all conceivable locations along the length of the ship as follows:

.1 in tankers of more than 225 m in length, anywhere in the ship's length;

.2 in tankers of more than 150 m, but not exceeding 225 m in length, anywhere in the ship's length except involving either after or forward bulkhead bounding the machinery space located aft. The machinery space shall be treated as a single floodable compartment; and

.3 in tankers not exceeding 150 m in length, anywhere in the ship's length between adjacent transverse bulkheads with the exception of the machinery space. For tankers of 100 m or less in length where all requirements of paragraph 3 of this regulation cannot be fulfilled without materially impairing the operational qualities of the ship, Administrations may allow relaxations from these requirements.

Ballast conditions where the tanker is not carrying oil in cargo tanks, excluding any oil residues, shall not be considered.

SEE INTERPRETATION 56

2 The following provisions regarding the extent and the character of the assumed damage shall apply:

.1 Side damage:

.1.1 Longitudinal extent:	$\frac{1}{3}(L^{\frac{2}{3}})$ or 14.5 m, whichever is less
.1.2 Transverse extent (inboard from the ship's side at right angles to the centreline at the level of the summer load line):	$\frac{B}{5}$ or 11.5 m, whichever is less
.1.3 Vertical extent:	From the moulded line of the bottom shell plating at centreline, upwards without limit

.2 Bottom damage.

	For 0.3L from the forward perpendicular of the ship	*Any other part of the ship*
.2.1 Longitudinal extent:	$\frac{1}{3}(L^{\frac{2}{3}})$ or 14.5 m, whichever is less	$\frac{1}{3}(L^{\frac{2}{3}})$ or 5 m, whichever is less
.2.2 Transverse extent:	$\frac{B}{6}$ or 10 m, whichever is less	$\frac{B}{6}$ or 5 m, whichever is less
.2.3 Vertical extent:	$\frac{B}{15}$ or 6 m, whichever is less, measured from the moulded line of the bottom shell plating at centreline	$\frac{B}{15}$ or 6 m, whichever is less, measured from the moulded line of the bottom shell plating at centreline

.3 If any damage of a lesser extent than the maximum extent of damage specified in subparagraphs 2.1 and 2.2 of this paragraph would result in a more severe condition, such damage shall be considered.

.4 Where the damage involving transverse bulkheads is envisaged as specified in subparagraphs 1.1 and 1.2 of this regulation, transverse watertight bulkheads shall be spaced at least at a distance equal to the longitudinal extent of assumed damage specified in subparagraph 2.1 of this paragraph in order to be considered effective. Where transverse bulkheads are spaced at a lesser distance, one or more of these bulkheads within such extent of damage shall be assumed as non-existent for the purpose of determining flooded compartments.

.5 Where the damage between adjacent transverse watertight bulkheads is envisaged as specified in subparagraph 1.3 of this regulation, no main transverse bulkhead or a transverse bulkhead bounding side tanks or double bottom tanks shall be assumed damaged, unless:

.5.1 the spacing of the adjacent bulkheads is less than the longitudinal extent of assumed damage specified in subparagraph 2.1 of this paragraph; or

.5.2 there is a step or recess in a transverse bulkhead of more than 3.05 m in length, located within the extent of penetration of assumed damage. The step formed by the after peak bulkhead and after peak top shall not be regarded as a step for the purpose of this regulation.

.6 If pipes, ducts or tunnels are situated within the assumed extent of damage, arrangements shall be made so that progressive flooding cannot thereby extend to compartments other than those assumed to be floodable for each case of damage.

SEE INTERPRETATION 57

3 Oil tankers shall be regarded as complying with the damage stability criteria if the following requirements are met:

.1 The final waterline, taking into account sinkage, heel and trim, shall be below the lower edge of any opening through which progressive flooding may take place. Such openings shall include air-pipes and those which are closed by means of weathertight doors or hatch covers and may exclude those openings closed by means of watertight manhole covers and flush scuttles, small watertight cargo tank hatch covers which maintain the high integrity of the deck, remotely operated watertight sliding doors, and sidescuttles of the non-opening type.

.2 In the final stage of flooding, the angle of heel due to unsymmetrical flooding shall not exceed 25°, provided that this angle may be increased up to 30° if no deck edge immersion occurs.

.3 The stability in the final stage of flooding shall be investigated and may be regarded as sufficient if the righting lever curve has at least a range of 20° beyond the position of equilibrium in association with a maximum residual righting lever of at least 0.1 m within the 20° range; the area under the curve within this range shall not be less than 0.0175 m·rad. Unprotected openings shall not be immersed within this range unless the space concerned is assumed to be flooded. Within this range, the immersion of any of the openings listed in subparagraph 3.1 of this paragraph and other openings capable of being closed weathertight may be permitted.

.4 The Administration shall be satisfied that the stability is sufficient during intermediate stages of flooding.

.5 Equalization arrangements requiring mechanical aids such as valves or cross-levelling pipes, if fitted, shall not be considered for the purpose of reducing an angle of heel or attaining the minimum range of residual stability to meet the requirements of subparagraphs 3.1, 3.2 and 3.3 of this paragraph and sufficient residual stability shall be maintained during all stages where equalization is used. Spaces which are linked by ducts of a large cross-sectional area may be considered to be common.

SEE INTERPRETATION 58

4 The requirements of paragraph 1 of this regulation shall be confirmed by calculations which take into consideration the design characteristics of the ship, the arrangements, configuration and contents of the damaged compartments; and the distribution, relative densities and the free surface effect of liquids. The calculations shall be based on the following:

.1 Account shall be taken of any empty or partially filled tank, the relative density of cargoes carried, as well as any outflow of liquids from damaged compartments.

.2 The permeabilities assumed for spaces flooded as a result of damage shall be as follows:

Spaces	*Permeabilities*
Appropriated to stores	0.60
Occupied by accommodation	0.95
Occupied by machinery	0.85
Voids	0.95
Intended for consumable liquids	0 to 0.95*
Intended for other liquids	0 to 0.95*

.3 The buoyancy of any superstructure directly above the side damage shall be disregarded. The unflooded parts of superstructures beyond the extent of damage, however, may be taken into consideration provided that they are separated from the damaged space by watertight bulkheads and the requirements of subparagraph .3.1 of this regulation in respect of these intact spaces are complied with. Hinged watertight doors may be acceptable in watertight bulkheads in the superstructure.

.4 The free surface effect shall be calculated at an angle of heel of 5° for each individual compartment. The Administration may require or allow the free surface corrections to be calculated at an angle of heel greater than 5° for partially filled tanks.

.5 In calculating the effect of free surfaces of consumable liquids it shall be assumed that, for each type of liquid, at least one transverse pair or a single centreline tank has a free surface and the tank or combination of tanks to be taken into account shall be those where the effect of free surface is the greatest.

* The permeability of partially filled compartments shall be consistent with the amount of liquid carried in the compartment. Whenever damage penetrates a tank containing liquids, it shall be assumed that the contents are completely lost from that compartment and replaced by salt water up to the level of the final plane of equilibrium.

5 The master of every oil tanker to which this regulation applies and the person in charge of a non-self-propelled oil tanker to which this regulation applies shall be supplied in an approved form with:

.1 information relative to loading and distribution of cargo necessary to ensure compliance with the provisions of this regulation; and

.2 data on the ability of the ship to comply with damage stability criteria as determined by this regulation, including the effect of relaxations that may have been allowed under subparagraph 1.3 of this regulation.

6 All oil tankers shall be fitted with a stability instrument, capable of verifying compliance with intact and damage stability requirements approved by the Administration having regard to the performance standards recommended by the Organization:*

.1 oil tankers constructed before 1 January 2016 shall comply with this regulation at the first scheduled renewal survey of the ship on or after 1 January 2016 but not later than 1 January 2021;

.2 notwithstanding the requirements of subparagraph .1, a stability instrument fitted on an oil tanker constructed before 1 January 2016 need not be replaced provided it is capable of verifying compliance with intact and damage stability, to the satisfaction of the Administration; and

.3 for the purposes of control under regulation 11, the Administration shall issue a document of approval for the stability instrument.

7 For oil tankers of 20,000 tonnes deadweight and above delivered on or after 6 July 1996, as defined in regulation 1.28.6, the damage assumptions prescribed in paragraph 2.2 of this regulation shall be supplemented by the following assumed bottom raking damage:

.1 longitudinal extent:

.1.1 ships of 75,000 tonnes deadweight and above:
$0.6L$ measured from the forward perpendicular;

.1.2 ships of less than 75,000 tonnes deadweight:
$0.4L$ measured from the forward perpendicular;

.2 transverse extent: $\frac{B}{3}$ anywhere in the bottom;

.3 vertical extent: breach of the outer hull.

Regulation 29
Slop tanks

1 Subject to the provisions of paragraph 4 of regulation 3 of this Annex, oil tankers of 150 gross tonnage and above shall be provided with slop tank arrangements in accordance with the requirements of paragraphs 2.1 to 2.3 of this regulation. In oil tankers delivered on or before 31 December 1979, as defined in regulation 1.28.1, any cargo tank may be designated as a slop tank.

2.1 Adequate means shall be provided for cleaning the cargo tanks and transferring the dirty ballast residue and tank washings from the cargo tanks into a slop tank approved by the Administration.

2.2 In this system arrangements shall be provided to transfer the oily waste into a slop tank or combination of slop tanks in such a way that any effluent discharged into the sea will be such as to comply with the provisions of regulation 34 of this Annex.

* Refer to part B, chapter 4, of the International Code on Intact Stability, 2008 (2008 IS Code), as amended; the Guidelines for the Approval of Stability Instruments (MSC.1/Circ.1229), annex, section 4, as amended; and the technical standards defined in part 1 of the Guidelines for verification of damage stability requirements for tankers (MSC.1/Circ.1461).

2.3 The arrangements of the slop tank or combination of slop tanks shall have a capacity necessary to retain the slop generated by tank washings, oil residues and dirty ballast residues. The total capacity of the slop tank or tanks shall not be less than 3% of the oil-carrying capacity of the ship, except that the Administration may accept:

.1 2% for such oil tankers where the tank washing arrangements are such that once the slop tank or tanks are charged with washing water, this water is sufficient for tank washing and, where applicable, for providing the driving fluid for eductors, without the introduction of additional water into the system;

.2 2% where segregated ballast tanks or dedicated clean ballast tanks are provided in accordance with regulation 18 of this Annex, or where a cargo tank cleaning system using crude oil washing is fitted in accordance with regulation 33 of this Annex. This capacity may be further reduced to 1.5% for such oil tankers where the tank washing arrangements are such that once the slop tank or tanks are charged with washing water, this water is sufficient for tank washing and, where applicable, for providing the driving fluid for eductors, without the introduction of additional water into the system; and

.3 1% for combination carriers where oil cargo is only carried in tanks with smooth walls. This capacity may be further reduced to 0.8% where the tank washing arrangements are such that once the slop tank or tanks are charged with washing water, this water is sufficient for tank washing and, where applicable, for providing the driving fluid for eductors, without the introduction of additional water into the system.

SEE INTERPRETATION 59

2.4 Slop tanks shall be so designed, particularly in respect of the position of inlets, outlets, baffles or weirs where fitted, so as to avoid excessive turbulence and entrainment of oil or emulsion with the water.

3 Oil tankers of 70,000 tonnes deadweight and above delivered after 31 December 1979, as defined in regulation 1.28.2, shall be provided with at least two slop tanks.

Regulation 30

Pumping, piping and discharge arrangement

1 In every oil tanker, a discharge manifold for connection to reception facilities for the discharge of dirty ballast water or oil-contaminated water shall be located on the open deck on both sides of the ship.

2 In every oil tanker of 150 gross tonnage and above, pipelines for the discharge to the sea of ballast water or oil-contaminated water from cargo tank areas which may be permitted under regulation 34 of this Annex shall be led to the open deck or to the ship's side above the waterline in the deepest ballast condition. Different piping arrangements to permit operation in the manner permitted in subparagraphs 6.1 to 6.5 of this regulation may be accepted.

SEE INTERPRETATION 60

3 In oil tankers of 150 gross tonnage and above delivered after 31 December 1979, as defined in regulation 1.28.2, means shall be provided for stopping the discharge into the sea of ballast water or oil-contaminated water from cargo tank areas, other than those discharges below the waterline permitted under paragraph 6 of this regulation, from a position on the upper deck or above located so that the manifold in use referred to in paragraph 1 of this regulation and the discharge to the sea from the pipelines referred to in paragraph 2 of this regulation may be visually observed. Means for stopping the discharge need not be provided at the observation position if a positive communication system such as a telephone or radio system is provided between the observation position and the discharge control position.

4 Every oil tanker delivered after 1 June 1982, as defined in regulation 1.28.4, required to be provided with segregated ballast tanks or fitted with a crude oil washing system, shall comply with the following requirements:

.1 it shall be equipped with oil piping so designed and installed that oil retention in the lines is minimized; and

.2 means shall be provided to drain all cargo pumps and all oil lines at the completion of cargo discharge, where necessary by connection to a stripping device. The line and pump draining shall be capable of being discharged both ashore and to a cargo tank or a slop tank. For discharge ashore a special small diameter line shall be provided and shall be connected outboard of the ship's manifold valves.

SEE INTERPRETATION 61

5 Every crude oil tanker delivered on or before 1 June 1982, as defined in regulation 1.28.3, required to be provided with segregated ballast tanks, or to be fitted with a crude oil washing system, shall comply with the provisions of paragraph 4.2 of this regulation.

6 On every oil tanker the discharge of ballast water or oil-contaminated water from cargo tank areas shall take place above the waterline, except as follows:

.1 Segregated ballast and clean ballast may be discharged below the waterline:

.1.1 in ports or at offshore terminals, or

.1.2 at sea by gravity, or

.1.3 at sea by pumps if the ballast water exchange is performed under the provisions of regulation D-1.1 of the International Convention for the Control and Management of Ships' Ballast Water and Sediments,

provided that the surface of the ballast water has been examined either visually or by other means immediately before the discharge to ensure that no contamination with oil has taken place.

.2 Oil tankers delivered on or before 31 December 1979, as defined in regulation 1.28.1, which, without modification, are not capable of discharging segregated ballast above the waterline may discharge segregated ballast below the waterline at sea, provided that the surface of the ballast water has been examined immediately before the discharge to ensure that no contamination with oil has taken place.

.3 Oil tankers delivered on or before 1 June 1982, as defined in regulation 1.28.3, operating with dedicated clean ballast tanks, which without modification are not capable of discharging ballast water from dedicated clean ballast tanks above the waterline, may discharge this ballast below the waterline provided that the discharge of the ballast water is supervised in accordance with regulation 18.8.3 of this Annex.

.4 On every oil tanker at sea, dirty ballast water or oil-contaminated water from tanks in the cargo area, other than slop tanks, may be discharged by gravity below the waterline, provided that sufficient time has elapsed in order to allow oil/water separation to have taken place and the ballast water has been examined immediately before the discharge with an oil/water interface detector referred to in regulation 32 of this Annex, in order to ensure that the height of the interface is such that the discharge does not involve any increased risk of harm to the marine environment.

.5 On oil tankers delivered on or before 31 December 1979, as defined in regulation 1.28.1, at sea dirty ballast water or oil-contaminated water from cargo tank areas may be discharged below the waterline, subsequent to or in lieu of the discharge by the method referred to in subparagraph 6.4 of this paragraph, provided that:

.5.1 a part of the flow of such water is led through permanent piping to a readily accessible location on the upper deck or above where it may be visually observed during the discharge operation; and

.5.2 such part flow arrangements comply with the requirements established by the Administration, which shall contain at least all the provisions of the Specifications for the Design, Installation and Operation of a Part Flow System for Control of Overboard Discharges adopted by the Organization.*

SEE INTERPRETATION 62

7 Every oil tanker of 150 gross tonnage and above delivered on or after 1 January 2010, as defined in regulation 1.28.8, which has installed a sea chest that is permanently connected to the cargo pipeline system, shall be equipped with both a sea chest valve and an inboard isolation valve. In addition to these valves, the sea chest shall be capable of isolation from the cargo piping system whilst the tanker is loading, transporting, or discharging cargo by use of a positive means that is to the satisfaction of the Administration. Such a positive means is a facility that is installed in the pipeline system in order to prevent, under all circumstances, the section of pipeline between the sea chest valve and the inboard valve being filled with cargo.

SEE INTERPRETATION 63

Part B – Equipment

Regulation 31

Oil discharge monitoring and control system

1 Subject to the provisions of paragraphs 4 and 5 of regulation 3 of this Annex, oil tankers of 150 gross tonnage and above shall be equipped with an oil discharge monitoring and control system approved by the Administration.

2 In considering the design of the oil content meter to be incorporated in the system, the Administration shall have regard to the specification recommended by the Organization.† The system shall be fitted with a recording device to provide a continuous record of the discharge in litres per nautical mile and total quantity discharged, or the oil content and rate of discharge. This record shall be identifiable as to time and date and shall be kept for at least three years. The oil discharge monitoring and control system shall come into operation when there is any discharge of effluent into the sea and shall be such as will ensure that any discharge of oily mixture is automatically stopped when the instantaneous rate of discharge of oil exceeds that permitted by regulation 34 of this Annex. Any failure of this monitoring and control system shall stop the discharge. In the event of failure of the oil discharge monitoring and control system a manually operated alternative method may be used, but the defective unit shall be made operable as soon as possible. Subject to allowance by the port State authority, a tanker with a defective oil discharge monitoring and control system may undertake one ballast voyage before proceeding to a repair port.

3 The oil discharge monitoring and control system shall be designed and installed in compliance with the guidelines and specifications for oil discharge monitoring and control systems for oil tankers developed by the Organization.‡ Administrations may accept such specific arrangements as detailed in the Guidelines and Specifications.

4 Instructions as to the operation of the system shall be in accordance with an operational manual approved by the Administration. They shall cover manual as well as automatic operations and shall be intended to ensure that at no time shall oil be discharged except in compliance with the conditions specified in regulation 34 of this Annex.

* See appendix 4 to Unified Interpretations.

† For oil content meters installed on oil tankers built prior to 2 October 1986, refer to the Recommendation on international performance and test specifications for oily-water separating equipment and oil content meters (resolution A.393(X)). For oil content meters as part of discharge monitoring and control systems installed on oil tankers built on or after 2 October 1986, refer to Guidelines and specifications for oil discharge monitoring and control systems for oil tankers (resolution A.586(14)). For oil content meters as part of discharge monitoring and control systems installed on oil tankers built on or after 1 January 2005, refer to Revised Guidelines and specifications for oil discharge monitoring and control systems for oil tankers (resolution MEPC.108(49), as amended by resolution MEPC.240(65)).

‡ Refer to Guidelines and specifications for oil discharge monitoring and control systems for oil tankers (resolution A.496(XII)), or Revised Guidelines and specifications for oil discharge monitoring and control systems for oil tankers (resolution A.586(14)), or Revised Guidelines and specifications for oil discharge monitoring and control systems for oil tankers (resolution MEPC.108(49), as amended by resolution MEPC.240(65)), as applicable.

Regulation 32

*Oil/water interface detector**

Subject to the provisions of paragraphs 4 and 5 of regulation 3 of this Annex, oil tankers of 150 gross tonnage and above shall be provided with effective oil/water interface detectors approved by the Administration for a rapid and accurate determination of the oil/water interface in slop tanks and shall be available for use in other tanks where the separation of oil and water is effected and from which it is intended to discharge effluent direct to the sea.

Regulation 33

Crude oil washing requirements

SEE INTERPRETATION 34

1 Every crude oil tanker of 20,000 tonnes deadweight and above delivered after 1 June 1982, as defined in regulation 1.28.4, shall be fitted with a cargo tank cleaning system using crude oil washing. The Administration shall ensure that the system fully complies with the requirements of this regulation within one year after the tanker was first engaged in the trade of carrying crude oil or by the end of the third voyage carrying crude oil suitable for crude oil washing, whichever occurs later.

2 Crude oil washing installation and associated equipment and arrangements shall comply with the requirements established by the Administration. Such requirements shall contain at least all the provisions of the Specifications for the Design, Operation and Control of Crude Oil Washing Systems adopted by the Organization.† When a ship is not required, in accordance with paragraph 1 of this regulation, to be, but is equipped with crude oil washing equipment, it shall comply with the safety aspects of the above-mentioned Specifications.

3 Every crude oil washing system required to be provided in accordance with regulation 18.7 of this Annex shall comply with the requirements of this regulation.

Part C – Control of operational discharges of oil

Regulation 34

Control of discharge of oil

A Discharges outside special areas except in Arctic waters

1 Subject to the provisions of regulation 4 of this Annex and paragraph 2 of this regulation, any discharge into the sea of oil or oily mixtures from the cargo area of an oil tanker shall be prohibited except when all the following conditions are satisfied:

.1 the tanker is not within a special area;

.2 the tanker is more than 50 nautical miles from the nearest land;

.3 the tanker is proceeding en route;

.4 the instantaneous rate of discharge of oil content does not exceed 30 litres per nautical mile;

.5 the total quantity of oil discharged into the sea does not exceed for tankers delivered on or before 31 December 1979, as defined in regulation 1.28.1, $\frac{1}{15,000}$ of the total quantity of the particular cargo of which the residue formed a part, and for tankers delivered after 31 December 1979, as defined in regulation 1.28.2, $\frac{1}{30,000}$ of the total quantity of the particular cargo of which the residue formed a part; and

SEE INTERPRETATION 64

.6 the tanker has in operation an oil discharge monitoring and control system and a slop tank arrangement as required by regulations 29 and 31 of this Annex.

* Refer to Specifications for oil/water interface detectors (resolution MEPC.5(XIII)).

† Refer to Revised specifications for the design, operation and control of crude oil washing systems (resolution A.446(XI), amended by resolutions A.497(XII) and A.897(21)).

2 The provisions of paragraph 1 of this regulation shall not apply to the discharge of clean or segregated ballast.

B Discharges in special areas

3 Subject to the provisions of paragraph 4 of this regulation, any discharge into the sea of oil or oily mixture from the cargo area of an oil tanker shall be prohibited while in a special area.*

4 The provisions of paragraph 3 of this regulation shall not apply to the discharge of clean or segregated ballast.

5 Nothing in this regulation shall prohibit a ship on a voyage only part of which is in a special area from discharging outside the special area in accordance with paragraph 1 of this regulation.

C Requirements for oil tankers of less than 150 gross tonnage

6 The requirements of regulations 29, 31 and 32 of this Annex shall not apply to oil tankers of less than 150 gross tonnage, for which the control of discharge of oil under this regulation shall be effected by the retention of oil on board with subsequent discharge of all contaminated washings to reception facilities. The total quantity of oil and water used for washing and returned to a storage tank shall be discharged to reception facilities unless adequate arrangements are made to ensure that any effluent which is allowed to be discharged into the sea is effectively monitored to ensure that the provisions of this regulation are complied with.

D General requirements

7 Whenever visible traces of oil are observed on or below the surface of the water in the immediate vicinity of a ship or its wake, the Governments of Parties to the present Convention should, to the extent they are reasonably able to do so, promptly investigate the facts bearing on the issue of whether there has been a violation of the provisions of this regulation. The investigation should include, in particular, the wind and sea conditions, the track and speed of the ship, other possible sources of the visible traces in the vicinity, and any relevant oil discharge records.

8 No discharge into the sea shall contain chemicals or other substances in quantities or concentrations which are hazardous to the marine environment or chemicals or other substances introduced for the purpose of circumventing the conditions of discharge specified in this regulation.

9 The oil residues which cannot be discharged into the sea in compliance with paragraphs 1 and 3 of this regulation shall be retained on board for subsequent discharge to reception facilities.

Regulation 35
Crude oil washing operations

SEE INTERPRETATION 34

1 Every oil tanker operating with crude oil washing systems shall be provided with an Operations and Equipment Manual† detailing the system and equipment and specifying operational procedures. Such a Manual shall be to the satisfaction of the Administration and shall contain all the information set out in the specifications referred to in paragraph 2 of regulation 33 of this Annex. If an alteration affecting the crude oil washing system is made, the Operations and Equipment Manual shall be revised accordingly.

* Refer to regulation 38.8.

† Refer to the Standard format of the Crude Oil Washing Operation and Equipment Manual (resolution MEPC.3(XII), as amended by resolution MEPC.81(43)).

2 With respect to the ballasting of cargo tanks, sufficient cargo tanks shall be crude oil washed prior to each ballast voyage in order that, taking into account the tanker's trading pattern and expected weather conditions, ballast water is put only into cargo tanks which have been crude oil washed.

3 Unless an oil tanker carries crude oil which is not suitable for crude oil washing, the oil tanker shall operate the crude oil washing system in accordance with the Operations and Equipment Manual.

Regulation 36

Oil Record Book Part II – Cargo/ballast operations

1 Every oil tanker of 150 gross tonnage and above shall be provided with an Oil Record Book Part II (Cargo/Ballast Operations). The Oil Record Book Part II, whether as a part of the ship's official logbook or otherwise, shall be in the form specified in appendix III to this Annex.

2 The Oil Record Book Part II shall be completed on each occasion, on a tank-to-tank basis if appropriate, whenever any of the following cargo/ballast operations take place in the ship:

- **.1** loading of oil cargo;
- **.2** internal transfer of oil cargo during voyage;
- **.3** unloading of oil cargo;
- **.4** ballasting of cargo tanks and dedicated clean ballast tanks;
- **.5** cleaning of cargo tanks including crude oil washing;
- **.6** discharge of ballast except from segregated ballast tanks;
- **.7** discharge of water from slop tanks;
- **.8** closing of all applicable valves or similar devices after slop tank discharge operations;
- **.9** closing of valves necessary for isolation of dedicated clean ballast tanks from cargo and stripping lines after slop tank discharge operations; and
- **.10** disposal of residues.

3 For oil tankers referred to in regulation 34.6 of this Annex, the total quantity of oil and water used for washing and returned to a storage tank shall be recorded in the Oil Record Book Part II.

4 In the event of such discharge of oil or oily mixture as is referred to in regulation 4 of this Annex or in the event of accidental or other exceptional discharge of oil not excepted by that regulation, a statement shall be made in the Oil Record Book Part II of the circumstances of, and the reasons for, the discharge.

5 Each operation described in paragraph 2 of this regulation shall be fully recorded without delay in the Oil Record Book Part II so that all entries in the book appropriate to that operation are completed. Each completed operation shall be signed by the officer or officers in charge of the operations concerned and each completed page shall be signed by the master of ship. The entries in the Oil Record Book Part II shall be at least in English, French or Spanish. Where entries in an official language of the State whose flag the ship is entitled to fly are also used, this shall prevail in case of dispute or discrepancy.

6 Any failure of the oil discharge monitoring and control system shall be noted in the Oil Record Book Part II.

7 The Oil Record Book shall be kept in such a place as to be readily available for inspection at all reasonable times and, except in the case of unmanned ships under tow, shall be kept on board the ship. It shall be preserved for a period of three years after the last entry has been made.

8 The competent authority of the Government of a Party to the Convention may inspect the Oil Record Book Part II on board any ship to which this Annex applies while the ship is in its port or offshore terminals and may make a copy of any entry in that book and may require the master of the ship to certify that the copy is a true copy of such entry. Any copy so made which has been certified by the master of the ship as a true copy of an entry in the ship's Oil Record Book Part II shall be made admissible in any judicial proceedings as evidence of the facts stated in the entry. The inspection of an Oil Record Book Part II and the taking of a certified copy by the competent authority under this paragraph shall be performed as expeditiously as possible without causing the ship to be unduly delayed.

9 For oil tankers of less than 150 gross tonnage operating in accordance with regulation 34.6 of this Annex, an appropriate Oil Record Book should be developed by the Administration.

Chapter 5 – Prevention of pollution arising from an oil pollution incident

Regulation 37
Shipboard oil pollution emergency plan

1 Every oil tanker of 150 gross tonnage and above and every ship other than an oil tanker of 400 gross tonnage and above shall carry on board a shipboard oil pollution emergency plan approved by the Administration.

SEE INTERPRETATION 65

2 Such a plan shall be prepared based on guidelines* developed by the Organization and written in the working language of the master and officers. The plan shall consist at least of:

- **.1** the procedure to be followed by the master or other persons having charge of the ship to report an oil pollution incident, as required in article 8 and Protocol I of the present Convention, based on the guidelines developed by the Organization;†
- **.2** the list of authorities or persons to be contacted in the event of an oil pollution incident;
- **.3** a detailed description of the action to be taken immediately by persons on board to reduce or control the discharge of oil following the incident; and
- **.4** the procedures and point of contact on the ship for coordinating shipboard action with national and local authorities in combating the pollution.

3 In the case of ships to which regulation 17 of Annex II of the present Convention also applies, such a plan may be combined with the shipboard marine pollution emergency plan for noxious liquid substances required under regulation 17 of Annex II of the present Convention. In this case, the title of such a plan shall be "Shipboard marine pollution emergency plan".

4 All oil tankers of 5,000 tonnes deadweight or more shall have prompt access to computerized shore-based damage stability and residual structural strength calculation programs.

* Refer to Guidelines for the development of shipboard oil pollution emergency plans (resolution MEPC.54(32) as amended by resolution MEPC.86(44)).

† Refer to General principles for ship reporting systems and ship reporting requirements, including guidelines for reporting incidents involving dangerous goods, harmful substances and/or marine pollutants (resolution A.851(20), as amended by resolution MEPC.138(53)).

Chapter 6 – Reception facilities

Regulation 38
Reception facilities

SEE INTERPRETATION 66

A *Reception facilities outside special areas*

1 The Government of each Party to the present Convention undertakes to ensure the provision at oil loading terminals, repair ports, and in other ports in which ships have oily residues to discharge, of facilities for the reception of such residues and oily mixtures as remain from oil tankers and other ships adequate* to meet the needs of the ships using them without causing undue delay to ships.

2 Reception facilities in accordance with paragraph 1 of this regulation shall be provided in:

.1 all ports and terminals in which crude oil is loaded into oil tankers where such tankers have immediately prior to arrival completed a ballast voyage of not more than 72 h or not more than 1,200 nautical miles;

.2 all ports and terminals in which oil other than crude oil in bulk is loaded at an average quantity of more than 1,000 tonnes per day;

.3 all ports having ship repair yards or tank cleaning facilities;

.4 all ports and terminals which handle ships provided with the oil residue (sludge) tank(s) required by regulation 12 of this Annex;

.5 all ports in respect of oily bilge waters and other residues that cannot be discharged in accordance with regulations 15 and 34 of this Annex and paragraph 1.1.1 of part II-A of the Polar Code; and

.6 all loading ports for bulk cargoes in respect of oil residues from combination carriers which cannot be discharged in accordance with regulation 34 of this Annex.

3 The capacity for the reception facilities shall be as follows:

.1 Crude oil loading terminals shall have sufficient reception facilities to receive oil and oily mixtures which cannot be discharged in accordance with the provisions of regulation 34.1 of this Annex from all oil tankers on voyages as described in paragraph 2.1 of this regulation.

.2 Loading ports and terminals referred to in paragraph 2.2 of this regulation shall have sufficient reception facilities to receive oil and oily mixtures which cannot be discharged in accordance with the provisions of regulation 34.1 of this Annex from oil tankers which load oil other than crude oil in bulk.

.3 All ports having ship repair yards or tank cleaning facilities shall have sufficient reception facilities to receive all residues and oily mixtures which remain on board for disposal from ships prior to entering such yards or facilities.

.4 All facilities provided in ports and terminals under paragraph 2.4 of this regulation shall be sufficient to receive all residues retained according to regulation 12 of this Annex from all ships that may reasonably be expected to call at such ports and terminals.

* Refer to Guidelines for ensuring the adequacy of port waste reception facilities (resolution MEPC.83(44)).

.5 All facilities provided in ports and terminals under this regulation shall be sufficient to receive oily bilge waters and other residues which cannot be discharged in accordance with regulation 15 of this Annex and paragraph 1.1.1 of part II-A of the Polar Code.

.6 The facilities provided in loading ports for bulk cargoes shall take into account the special problems of combination carriers as appropriate.

4 Small Island Developing States may satisfy the requirements in paragraphs 1 to 3 of this regulation through regional arrangements when, because of those States' unique circumstances, such arrangements are the only practical means to satisfy these requirements. Parties participating in a regional arrangement shall develop a Regional Reception Facilities Plan, taking into account the guidelines developed by the Organization.*

The Government of each Party participating in the arrangement shall consult with the Organization, for circulation to the Parties of the present Convention:

.1 how the Regional Reception Facilities Plan takes into account the Guidelines;

.2 particulars of the identified Regional Ships Waste Reception Centres; and

.3 particulars of those ports with only limited facilities.

B Reception facilities within special areas

5 The Government of each Party to the present Convention the coastline of which borders on any given special area shall ensure that all oil loading terminals and repair ports within the special area are provided with facilities adequate for the reception and treatment of all the dirty ballast and tank washing water from oil tankers. In addition, all ports within the special area shall be provided with adequate† reception facilities for other residues and oily mixtures from all ships. Such facilities shall have adequate capacity to meet the needs of the ships using them without causing undue delay.

6 Small Island Developing States may satisfy the requirements in paragraph 5 of this regulation through regional arrangements when, because of those States' unique circumstances, such arrangements are the only practical means to satisfy these requirements. Parties participating in a regional arrangement shall develop a Regional Reception Facilities Plan, taking into account the guidelines developed by the Organization.

The Government of each Party participating in the arrangement shall consult with the Organization for circulation to the Parties of the present Convention:

.1 how the Regional Reception Facilities Plan takes into account the Guidelines;

.2 particulars of the identified Regional Ships Waste Reception Centres; and

.3 particulars of those ports with only limited facilities.

7 The Government of each Party to the present Convention having under its jurisdiction entrances to seawater courses with low depth contour which might require a reduction of draught by the discharge of ballast shall ensure the provision of the facilities referred to in paragraph 5 of this regulation but with the proviso that ships required to discharge slops or dirty ballast could be subject to some delay.

8 With regard to the Red Sea area, Gulfs area,‡ Gulf of Aden area and Oman area of the Arabian Sea:

.1 Each Party concerned shall notify the Organization of the measures taken pursuant to provisions of paragraphs 5 and 7 of this regulation. Upon receipt of sufficient notifications, the Organization shall establish a date from which the discharge requirements of regulations 15 and 34 of this Annex in respect of the area in question shall take effect. The Organization shall notify all Parties of the date so established no less than twelve months in advance of that date.

* Refer to 2012 Guidelines for the development of a regional reception facilities plan (resolution MEPC.221(63)).

† Refer to Guidelines for ensuring the adequacy of port waste reception facilities (resolution MEPC.83(44))

‡ In accordance with resolution MEPC.168(56), the discharge requirements for the Gulfs area special area set out in regulations 15 and 34 of this Annex took effect on 1 August 2008.

.2 During the period between the entry into force of the present Convention and the date so established, ships while navigating in the special area shall comply with the requirements of regulations 15 and 34 of this Annex as regards discharges outside special areas.

.3 After such date, oil tankers loading in ports in these special areas where such facilities are not yet available shall also fully comply with the requirements of regulations 15 and 34 of this Annex as regards discharges within special areas. However, oil tankers entering these special areas for the purpose of loading shall make every effort to enter the area with only clean ballast on board.

.4 After the date on which the requirements for the special area in question take effect, each Party shall notify the Organization for transmission to the Parties concerned of all cases where the facilities are alleged to be inadequate.

.5 At least the reception facilities as prescribed in paragraphs 1, 2 and 3 of this regulation shall be provided one year after the date of entry into force of the present Convention.

9 Notwithstanding paragraphs 5, 7 and 8 of this regulation, the following rules apply to the Antarctic area:

.1 The Government of each Party to the present Convention at whose ports ships depart en route to or arrive from the Antarctic area undertakes to ensure that as soon as practicable adequate facilities are provided for the reception of all oil residue (sludge), dirty ballast, tank washing water, and other oily residues and mixtures from all ships, without causing undue delay, and according to the needs of the ships using them.

.2 The Government of each Party to the present Convention shall ensure that all ships entitled to fly its flag, before entering the Antarctic area, are fitted with a tank or tanks of sufficient capacity on board for the retention of all oil residue (sludge), dirty ballast, tank washing water and other oily residues and mixtures while operating in the area and have concluded arrangements to discharge such oily residues at a reception facility after leaving the area.

C *General requirements*

10 Each Party shall notify the Organization for transmission to the Parties concerned of all cases where the facilities provided under this regulation are alleged to be inadequate.

Chapter 7 – Special requirements for fixed or floating platforms

Regulation 39

Special requirements for fixed or floating platforms

SEE INTERPRETATION 67

1 This regulation applies to fixed or floating platforms including drilling rigs, floating production, storage and offloading facilities (FPSOs) used for the offshore production and storage of oil, and floating storage units (FSUs) used for the offshore storage of produced oil.

2 Fixed or floating platforms when engaged in the exploration, exploitation and associated offshore processing of sea-bed mineral resources and other platforms shall comply with the requirements of this Annex applicable to ships of 400 gross tonnage and above other than oil tankers, except that:

.1 they shall be equipped as far as practicable with the installations required in regulations 12 and 14 of this Annex;

.2 they shall keep a record of all operations involving oil or oily mixture discharges, in a form approved by the Administration; and

.3 subject to the provisions of regulation 4 of this Annex, the discharge into the sea of oil or oily mixture shall be prohibited except when the oil content of the discharge without dilution does not exceed 15 ppm.

3 In verifying compliance with this Annex in relation to platforms configured as FPSOs or FSUs, in addition to the requirements of paragraph 2, Administrations should take account of the Guidelines developed by the Organization.*

* Refer to Guidelines for the application of the revised MARPOL Annex I requirements to FPSOs and FSUs (resolution MEPC.139(53), as amended by resolution MEPC.142(54)).

Chapter 8 – Prevention of pollution during transfer of oil cargo between oil tankers at sea

Regulation 40

Scope of application

1 The regulations contained in this chapter apply to oil tankers of 150 gross tonnage and above engaged in the transfer of oil cargo between oil tankers at sea (STS operations) and their STS operations conducted on or after 1 April 2012. However, STS operations conducted before that date but after the approval of the Administration of STS operations Plan required under regulation 41.1 shall be in accordance with the STS operations Plan as far as possible.

2 The regulations contained in this chapter shall not apply to oil transfer operations associated with fixed or floating platforms including drilling rigs; floating production, storage and offloading facilities (FPSOs) used for the offshore production and storage of oil; and floating storage units (FSUs) used for the offshore storage of produced oil.*

3 The regulations contained in this chapter shall not apply to bunkering operations.

4 The regulations contained in this chapter shall not apply to STS operations necessary for the purpose of securing the safety of a ship or saving life at sea, or for combating specific pollution incidents in order to minimize the damage from pollution.

5 The regulations contained in this chapter shall not apply to STS operations where either of the ships involved is a warship, naval auxiliary or other ship owned or operated by a State and used, for the time being, only on government non-commercial service. However, each State shall ensure, by the adoption of appropriate measures not impairing operations or operational capabilities of such ships that the STS operations are conducted in a manner consistent, so far as is reasonable and practicable, with this chapter.

Regulation 41

General rules on safety and environmental protection

1 Any oil tanker involved in STS operations shall carry on board a Plan prescribing how to conduct STS operations (STS operations Plan) not later than the date of the first annual, intermediate or renewal survey of the ship to be carried out on or after 1 January 2011. Each oil tanker's STS operations Plan shall be approved by the Administration. The STS operations Plan shall be written in the working language of the ship.

2 The STS operations Plan shall be developed taking into account the information contained in the best practice guidelines for STS operations identified by the Organization.† The STS operations Plan may be incorporated into an existing Safety Management System required by chapter IX of the International Convention for the Safety of Life at Sea, 1974, as amended, if that requirement is applicable to the oil tanker in question.

3 Any oil tanker subject to this chapter and engaged in STS operations shall comply with its STS operations Plan.

4 The person in overall advisory control of STS operations shall be qualified to perform all relevant duties, taking into account the qualifications contained in the best practice guidelines for STS operations identified by the Organization.

---

* Revised Annex I of MARPOL, chapter 7 (resolution MEPC.117(52)) and UNCLOS article 56 are applicable and address these operations.

† Refer to *Manual on Oil Pollution, Section I, Prevention, 2011 Edition,* and the OCIMF *Ship to Ship Transfer Guide for Petroleum, Chemicals and Liquefied Gases* (2013).

5 Records* of STS operations shall be retained on board for three years and be readily available for inspection by a Party to the present Convention.

Regulation 42

Notification

1 Each oil tanker subject to this chapter that plans STS operations within the territorial sea, or the exclusive economic zone of a Party to the present Convention shall notify that Party not less than 48 h in advance of the scheduled STS operations. Where, in an exceptional case, all of the information specified in paragraph 2 is not available not less than 48 h in advance, the oil tanker discharging the oil cargo shall notify the Party to the present Convention, not less than 48 h in advance that an STS operation will occur and the information specified in paragraph 2 shall be provided to the Party at the earliest opportunity.

2 The notification specified in paragraph 1 of this regulation† shall include at least the following:

.1 name, flag, call sign, IMO Number and estimated time of arrival of the oil tankers involved in the STS operations;

.2 date, time and geographical location at the commencement of the planned STS operations;

.3 whether STS operations are to be conducted at anchor or underway;

.4 oil type and quantity;

.5 planned duration of the STS operations;

.6 identification of STS operations service provider or person in overall advisory control and contact information; and

.7 confirmation that the oil tanker has on board an STS operations plan meeting the requirements of regulation 41.

3 If the estimated time of arrival of an oil tanker at the location or area for the STS operations changes by more than six hours, the master, owner or agent of that oil tanker shall provide a revised estimated time of arrival to the Party to the present Convention specified in paragraph 1 of this regulation.

* Revised Annex I of MARPOL chapters 3 and 4 (resolution MEPC.117(52)); requirements for recording bunkering and oil cargo transfer operations in the Oil Record Book, and any records required by the STS operations Plan.

† The national operational contact point as listed in document MSC-MEPC.6/Circ.15 of 31 December 2016 or its subsequent amendments.

Chapter 9 – Special requirements for the use or carriage of oils in the Antarctic area

Regulation 43
Special requirements for the use or carriage of oils in the Antarctic area

1 With the exception of vessels engaged in securing the safety of ships or in a search and rescue operation, the carriage in bulk as cargo, use as ballast, or carriage and use as fuel of the following:

.1 crude oils having a density at 15°C higher than 900 kg/m^3;

.2 oils, other than crude oils, having a density at 15°C higher than 900 kg/m^3 or a kinematic viscosity at 50°C higher than 180 mm^2/s; or

.3 bitumen, tar and their emulsions,

shall be prohibited in the Antarctic area, as defined in Annex I, regulation 1.11.7.

2 When prior operations have included the carriage or use of oils listed in paragraphs 1.1 to 1.3 of this regulation, the cleaning or flushing of tanks or pipelines is not required.

Chapter 10 – Verification of compliance with the provisions of this Convention

Regulation 44
Application

Parties shall use the provisions of the Code for Implementation in the execution of their obligations and responsibilities contained in this Annex.

Regulation 45
Verification of compliance

1 Every Party shall be subject to periodic audits by the Organization in accordance with the audit standard to verify compliance with and implementation of this Annex.

2 The Secretary-General of the Organization shall have responsibility for administering the Audit Scheme, based on the guidelines developed by the Organization.*

3 Every Party shall have responsibility for facilitating the conduct of the audit and implementation of a programme of actions to address the findings, based on the guidelines developed by the Organization.

4 Audit of all Parties shall be:

- **.1** based on an overall schedule developed by the Secretary-General of the Organization, taking into account the guidelines developed by the Organization;* and
- **.2** conducted at periodic intervals, taking into account the guidelines developed by the Organization.

* Refer to the Framework and Procedures for the IMO Member State Audit Scheme (resolution A.1067(28)).

Chapter 11 – International Code for Ships Operating in Polar Waters

Regulation 46

Definitions

For the purpose of this Annex,

1 *Polar Code* means the International Code for Ships Operating in Polar Waters, consisting of an introduction, parts I-A and II-A and parts I-B and II-B, adopted by resolutions MSC.385(94) and MEPC.264(68), as may be amended, provided that:

.1 amendments to the environment-related provisions of the introduction and chapter 1 of part II-A of the Polar Code are adopted, brought into force and take effect in accordance with the provisions of article 16 of the present Convention concerning the amendment procedures applicable to an appendix to an annex; and

.2 amendments to part II-B of the Polar Code are adopted by the Marine Environment Protection Committee in accordance with its Rules of Procedure.

2 *Arctic waters* means those waters which are located north of a line from the latitude 58°00′.0 N and longitude 042°00′.0 W to latitude 64°37′.0 N, longitude 035°27′.0 W and thence by a rhumb line to latitude 67°03′.9 N, longitude 026°33′.4 W and thence by a rhumb line to the latitude 70°49′.56 N and longitude 008°59′.61 W (Sørkapp, Jan Mayen) and by the southern shore of Jan Mayen to 73°31′.6 N and 019°01′.0 E by the Island of Bjørnøya, and thence by a great circle line to the latitude 68°38′.29 N and longitude 043°23′.08 E (Cap Kanin Nos) and hence by the northern shore of the Asian Continent eastward to the Bering Strait and thence from the Bering Strait westward to latitude 60° N as far as Il'pyrskiy and following the 60th North parallel eastward as far as and including Etolin Strait and thence by the northern shore of the North American continent as far south as latitude 60° N and thence eastward along parallel of latitude 60° N, to longitude 056°37′.1 W and thence to the latitude 58°00′.0 N, longitude 042°00′.0 W.

3 *Polar waters* means Arctic waters and/or the Antarctic area.

Regulation 47

Application and requirements

1 This chapter applies to all ships operating in polar waters.

2 Unless expressly provided otherwise, any ship covered by paragraph 1 of this regulation shall comply with the environment-related provisions of the introduction and with chapter 1 of part II-A of the Polar Code, in addition to any other applicable requirements of this Annex.

3 In applying chapter 1 of part II-A of the Polar Code, consideration should be given to the additional guidance in part II-B of the Polar Code.

Appendices to Annex I

Appendix I
List of oils*

Asphalt solutions
Blending stocks
Roofers flux
Straight run residue

Oils
Clarified
Crude oil
Mixtures containing crude oil
Diesel oil
Fuel oil no. 4
Fuel oil no. 5
Fuel oil no. 6
Residual fuel oil
Road oil
Transformer oil
Aromatic oil (excluding vegetable oil)
Lubricating oils and blending stocks
Mineral oil
Motor oil
Penetrating oil
Spindle oil
Turbine oil

Distillates
Straight run
Flashed feed stocks

Gas oil
Cracked

Gasoline blending stocks
Alkylates – fuel
Reformates
Polymer – fuel

Gasolines
Casinghead (natural)
Automotive
Aviation
Straight run
Fuel oil no. 1 (kerosene)
Fuel oil no. 1-D
Fuel oil no. 2
Fuel oil no. 2-D

Jet fuels
JP-1 (kerosene)
JP-3
JP-4
JP-5 (kerosene, heavy)
Turbo fuel
Kerosene
Mineral spirit

Naphtha
Solvent
Petroleum
Heartcut distillate oil

* This list of oils shall not necessarily be considered as comprehensive.

Appendix II

Form of IOPP Certificate and Supplements

INTERNATIONAL OIL POLLUTION PREVENTION CERTIFICATE*

(*Note:* This Certificate shall be supplemented by a Record of Construction and Equipment)

Issued under the provisions of the International Convention for the Prevention of Pollution from Ships, 1973, as modified by the Protocol of 1978 relating thereto, as amended, (hereinafter referred to as "the Convention") under the authority of the Government of:

. . . .

(full designation of the country)

by. . . . .

(full designation of the competent person or organization authorized under the provisions of the Convention)

Particulars of ship†

Name of ship. . . . .

Distinctive number or letters. . . . .

Port of registry . . . .

Gross tonnage. . . . .

Deadweight of ship (tonnes)‡ . . . .

IMO Number§ . . . .

Type of ship:¶

Oil tanker

Ship other than an oil tanker with cargo tanks coming under regulation 2.2 of Annex I of the Convention

Ship other than any of the above

THIS IS TO CERTIFY:

1. That the ship has been surveyed in accordance with regulation 6 of Annex I of the Convention; and
2. That the survey shows that the structure, equipment, systems, fittings, arrangement and material of the ship and the condition thereof are in all respects satisfactory and that the ship complies with the applicable requirements of Annex I of the Convention.

This certificate is valid until (dd/mm/yyyy)** . . . .
subject to surveys in accordance with regulation 6 of Annex I of the Convention.

Completion date of the survey on which this certificate is based (dd/mm/yyyy) . . . .

Issued at . . . .

(place of issue of certificate)

Date (dd/mm/yyyy) . . . .

(date of issue) *(signature of duly authorized official issuing the certificate)*

(seal or stamp of the authority, as appropriate)

* The IOPP Certificate shall be at least in English, French or Spanish. If an official language of the issuing country is also used, this shall prevail in case of a dispute or discrepancy.

† Alternatively, the particulars of the ship may be placed horizontally in boxes.

‡ For oil tankers.

§ Refer to the IMO Ship Identification Number Scheme (resolution A.1078(28)).

¶ Delete as appropriate.

** Insert the date of expiry as specified by the Administration in accordance with regulation 10.1 of Annex I of the Convention. The day and the month of this date correspond to the anniversary date as defined in regulation 1.27 of Annex I of the Convention, unless amended in accordance with regulation 10.8 of Annex I of the Convention.

ENDORSEMENT FOR ANNUAL AND INTERMEDIATE SURVEYS

THIS IS TO CERTIFY that at a survey required by regulation 6 of Annex I of the Convention the ship was found to comply with the relevant provisions of the Convention:

Annual survey

Signed. . . . .
(signature of duly authorized official)

Place . . . .

Date (dd/mm/yyyy) . . . .

(seal or stamp of the authority, as appropriate)

Annual/Intermediate* survey

Signed. . . . .
(signature of duly authorized official)

Place . . . .

Date (dd/mm/yyyy) . . . .

(seal or stamp of the authority, as appropriate)

Annual/Intermediate* survey

Signed. . . . .
(signature of duly authorized official)

Place . . . .

Date (dd/mm/yyyy) . . . .

(seal or stamp of the authority, as appropriate)

Annual survey

Signed. . . . .
(signature of duly authorized official)

Place . . . .

Date (dd/mm/yyyy) . . . .

(seal or stamp of the authority, as appropriate)

ANNUAL/INTERMEDIATE SURVEY IN ACCORDANCE WITH REGULATION 10.8.3

THIS IS TO CERTIFY that, at an annual/intermediate* survey in accordance with regulation 10.8.3 of Annex I of the Convention, the ship was found to comply with the relevant provisions of the Convention:

. . . .

Signed. . . . .
(signature of duly authorized official)

Place . . . .

Date (dd/mm/yyyy) . . . .

(seal or stamp of the authority, as appropriate)

ENDORSEMENT TO EXTEND THE CERTIFICATE IF VALID FOR LESS THAN 5 YEARS WHERE REGULATION 10.3 APPLIES

The ship complies with the relevant provisions of the Convention, and this Certificate shall, in accordance with regulation 10.3 of Annex I of the Convention, be accepted as valid until (dd/mm/yyyy) . . . .

Signed. . . . .
(signature of duly authorized official)

Place . . . .

Date (dd/mm/yyyy) . . . .

(seal or stamp of the authority, as appropriate)

* Delete as appropriate.

ENDORSEMENT WHERE THE RENEWAL SURVEY HAS BEEN COMPLETED AND REGULATION 10.4 APPLIES

The ship complies with the relevant provisions of the Convention, and this Certificate shall, in accordance with regulation 10.4 of Annex I of the Convention, be accepted as valid until (dd/mm/yyyy) . . . .

Signed. . . . .
(signature of duly authorized official)

Place . . . .

Date (dd/mm/yyyy) . . . .

(seal or stamp of the authority, as appropriate)

ENDORSEMENT TO EXTEND THE VALIDITY OF THE CERTIFICATE UNTIL REACHING THE PORT OF SURVEY OR FOR A PERIOD OF GRACE WHERE REGULATION 10.5 OR 10.6 APPLIES

This Certificate shall, in accordance with regulation 10.5 or 10.6* of Annex I of the Convention, be accepted as valid until (dd/mm/yyyy) . . . .

Signed. . . . .
(signature of duly authorized official)

Place . . . .

Date (dd/mm/yyyy) . . . .

(seal or stamp of the authority, as appropriate)

ENDORSEMENT FOR ADVANCEMENT OF ANNIVERSARY DATE WHERE REGULATION 10.8 APPLIES

In accordance with regulation 10.8 of Annex I of the Convention, the new anniversary date is (dd/mm/yyyy) . . . .

Signed. . . . .
(signature of duly authorized official)

Place . . . .

Date (dd/mm/yyyy) . . . .

(seal or stamp of the authority, as appropriate)

In accordance with regulation 10.8 of Annex I of the Convention, the new anniversary date is (dd/mm/yyyy) . . . .

Signed. . . . .
(signature of duly authorized official)

Place . . . .

Date (dd/mm/yyyy) . . . .

(seal or stamp of the authority, as appropriate)

* Delete as appropriate.

Appendix

FORM A

Supplement to the International Oil Pollution Prevention Certificate (IOPP Certificate)

RECORD OF CONSTRUCTION AND EQUIPMENT FOR SHIPS OTHER THAN OIL TANKERS

in respect of the provisions of Annex I of the International Convention for the Prevention of Pollution from Ships, 1973, as modified by the Protocol of 1978 relating thereto (hereinafter referred to as "the Convention").

Notes

1 This Form is to be used for the third type of ships as categorized in the IOPP Certificate, i.e. "ship other than any of the above". For oil tankers and ships other than oil tankers with cargo tanks coming under regulation 2.2 of Annex I of the Convention, Form B shall be used.

2 This Record shall be permanently attached to the IOPP Certificate. The IOPP Certificate shall be available on board the ship at all times.

3 The language of the original Record shall be at least in English, French or Spanish. If an official language of the issuing country is also used, this shall prevail in case of a dispute or discrepancy.

4 Entries in boxes shall be made by inserting either a cross (x) for the answers "yes" and "applicable" or a dash (–) for the answers "no" and "not applicable" as appropriate.

5 Regulations mentioned in this Record refer to regulations of Annex I of the Convention and resolutions refer to those adopted by the International Maritime Organization.

1 Particulars of ship

1.1 Name of ship . . . .

1.2 Distinctive number or letters . . . .

1.3 Port of registry . . . .

1.4 Gross tonnage . . . .

1.5 Date of build:

1.5.1 Date of building contract . . . .

1.5.2 Date on which keel was laid or ship was at a similar stage of construction . . . .

1.5.3 Date of delivery . . . .

1.6 Major conversion (if applicable):

1.6.1 Date of conversion contract . . . .

1.6.2 Date on which conversion was commenced . . . .

1.6.3 Date of completion of conversion . . . .

1.7 The ship has been accepted by the Administration as a "ship delivered on or before 31 December 1979" under regulation 1.28.1 due to unforeseen delay in delivery . . . . ☐

2 Equipment for the control of oil discharge from machinery space bilges and oil fuel tanks (regulations 16 and 14)

2.1 Carriage of ballast water in oil fuel tanks:

2.1.1 The ship may under normal conditions carry ballast water in oil fuel tanks . . . . ☐

2.2 Type of oil filtering equipment fitted:

2.2.1 Oil filtering (15 ppm) equipment (regulation 14.6) . . . . ☐

2.2.2 Oil filtering (15 ppm) equipment with alarm and automatic stopping device (regulation 14.7) . . . . . ☐

2.3 Approval standards:*

2.3.1 The separating/filtering equipment:

.1 has been approved in accordance with resolution A.393(X) . . . . ☐

.2 has been approved in accordance with resolution MEPC.60(33) . . . . ☐

.3 has been approved in accordance with resolution MEPC.107(49). . . . . ☐

.4 has been approved in accordance with resolution A.233(VII) . . . . ☐

.5 has been approved in accordance with national standards not based upon resolution A.393(X) or A.233(VII) . . . . ☐

.6 has not been approved. . . . . ☐

2.3.2 The process unit has been approved in accordance with resolution A.444(XI). . . . . ☐

2.3.3 The oil content meter:

.1 has been approved in accordance with resolution A.393(X) . . . . ☐

.2 has been approved in accordance with resolution MEPC.60(33) . . . . ☐

.3 has been approved in accordance with resolution MEPC.107(49). . . . . ☐

2.4 Maximum throughput of the system is . . . . m^3/h.

2.5 Waiver of regulation 14:

2.5.1 The requirements of regulation 14.1 or 14.2 are waived in respect of the ship in accordance with regulation 14.5.

2.5.1.1 The ship is engaged exclusively on voyages within special area(s) . . . . ☐

2.5.1.2 The ship is certified under the International Code of Safety for High-Speed Craft and engaged on a scheduled service with a turn-around time not exceeding 24 h . . . . ☐

2.5.2 The ship is fitted with holding tank(s) for the total retention on board of all oily bilge water as follows:

Tank identification	Tank location		Volume (m^3)
	Frames (from)–(to)	Lateral position	
		Total volume: . . . .m^3	

2A.1 The ship is required to be constructed according to regulation 12A and complies with the requirements of:

paragraphs 6 and either 7 or 8 (double hull construction). . . . . ☐

paragraph 11 (accidental oil fuel outflow performance) . . . . ☐

2A.2 The ship is not required to comply with the requirements of regulation 12A. . . . . ☐

3 Means for retention and disposal of oil residues (sludge) (regulation 12) **and oily bilge water holding tank(s)**†

3.1 The ship is provided with oil residue (sludge) tanks for retention of oil residues (sludge) on board as follows:

Tank identification	Tank location		Volume (m^3)
	Frames (from)–(to)	Lateral position	
		Total volume: . . . .m^3	

* Refer to the Recommendation on international performance and test specifications of oily-water separating equipment and oil content meters (resolution A.393(X)), Guidelines and specifications for pollution prevention equipment for machinery space bilges (resolution MEPC.60(33)), 2011 Guidelines and specifications for add-on equipment for upgrading resolution MEPC.60(33)-compliant oil filtering equipment (resolution MEPC.205(62)), and Revised guidelines and specifications for pollution prevention equipment for machinery spaces of ships (resolution MEPC.107(49)).

† Oily bilge water holding tank(s) are not required by the Convention; if such tank(s) are provided they shall be listed in table 3.3.

3.2 Means for the disposal of oil residues (sludge) retained in oil residue (sludge) tanks:

3.2.1 Incinerator for oil residues (sludge) . . . . ☐

3.2.2 Auxiliary boiler suitable for burning oil residues (sludge) . . . . ☐

3.2.3 Other acceptable means, state which . . . . ☐

3.3 The ship is provided with holding tank(s) for the retention on board of oily bilge water as follows:

Tank identification	Tank location		Volume (m^3)
	Frames (from)–(to)	Lateral position	
		Total volume: . . . . m^3	

4 **Standard discharge connection** (regulation 13)

4.1 The ship is provided with a pipeline for the discharge of residues from machinery bilges and sludges to reception facilities, fitted with a standard discharge connection in accordance with regulation 13 . . . . ☐

5 **Shipboard oil/marine pollution emergency plan** (regulation 37)

5.1 The ship is provided with a shipboard oil pollution emergency plan in compliance with regulation 37 . . . . ☐

5.2 The ship is provided with a shipboard marine pollution emergency plan in compliance with regulation 37.3 . . . . ☐

6 **Exemption**

6.1 Exemptions have been granted by the Administration from the requirements of chapter 3 of Annex I of the Convention in accordance with regulation 3.1 on those items listed under paragraph(s) . . . . of this Record. . . . . ☐

7 **Equivalents** (regulation 5)

7.1 Equivalents have been approved by the Administration for certain requirements of Annex I on those items listed under paragraph(s) . . . . of this Record . . . . ☐

8 **Compliance with part II-A – chapter 1 of the Polar Code**

8.1 The ship is in compliance with additional requirements in the environment-related provisions of the Introduction and section 1.2 of chapter 1 of part II-A of the Polar Code. . . . . ☐

THIS IS TO CERTIFY that this Record is correct in all respects.

Issued at . . . .
(place of issue of the Record)

Date (dd/mm/yyyy) . . . .
(date of issue)   *(signature of duly authorized official issuing the Record)*

(seal or stamp of the issuing authority, as appropriate)

FORM B

Supplement to the International Oil Pollution Prevention Certificate (IOPP Certificate)

RECORD OF CONSTRUCTION AND EQUIPMENT FOR OIL TANKERS

in respect of the provisions of Annex I of the International Convention for the Prevention of Pollution from Ships, 1973, as modified by the Protocol of 1978 relating thereto (hereinafter referred to as "the Convention").

Notes

1 This form is to be used for the first two types of ships as categorized in the IOPP Certificate, i.e. "oil tankers" and "ships other than oil tankers with cargo tanks coming under regulation 2.2 of Annex I of the Convention". For the third type of ships as categorized in the IOPP Certificate, Form A shall be used.

2 This Record shall be permanently attached to the IOPP Certificate. The IOPP Certificate shall be available on board the ship at all times.

3 The language of the original Record shall be at least in English, French or Spanish. If an official language of the issuing country is also used, this shall prevail in case of a dispute or discrepancy.

4 Entries in boxes shall be made by inserting either a cross (×) for the answers "yes" and "applicable" or a dash (–) for the answers "no" and "not applicable" as appropriate.

5 Unless otherwise stated, regulations mentioned in this Record refer to regulations of Annex I of the Convention and resolutions refer to those adopted by the International Maritime Organization.

1 Particulars of ship

1.1 Name of ship . . . .

1.2 Distinctive number or letters . . . .

1.3 Port of registry. . . . .

1.4 Gross tonnage. . . . .

1.5 Carrying capacity of ship . . . . (m^3)

1.6 Deadweight of ship . . . . (tonnes) (regulation 1.23)

1.7 Length of ship . . . .(m) (regulation 1.19)

1.8 Date of build:

1.8.1 Date of building contract . . . .

1.8.2 Date on which keel was laid or ship was at a similar stage of construction . . . .

1.8.3 Date of delivery. . . . .

1.9 Major conversion (if applicable):

1.9.1 Date of conversion contract . . . .

1.9.2 Date on which conversion was commenced. . . . .

1.9.3 Date of completion of conversion . . . .

1.10 Unforeseen delay in delivery:

1.10.1 The ship has been accepted by the Administration as a "ship delivered on or before 31 December 1979" under regulation 1.28.1 due to unforeseen delay in delivery . . . . ☐

1.10.2 The ship has been accepted by the Administration as an "oil tanker delivered on or before 1 June 1982" under regulation 1.28.3 due to unforeseen delay in delivery. ☐

1.10.3 The ship is not required to comply with the provisions of regulation 26 due to unforeseen delay in delivery . . . . ☐

1.11 Type of ship:

1.11.1 Crude oil tanker ☐

1.11.2 Product carrier ☐

1.11.3 Product carrier not carrying fuel oil or heavy diesel oil as referred to in regulation 20.2, or lubricating oil ☐

1.11.4 Crude oil/product carrier ☐

1.11.5 Combination carrier ☐

1.11.6 Ship, other than an oil tanker, with cargo tanks coming under regulation 2.2 of Annex I of the Convention ☐

1.11.7 Oil tanker dedicated to the carriage of products referred to in regulation 2.4 ☐

1.11.8 The ship, being designated as a "crude oil tanker" operating with COW, is also designated as a "product carrier" operating with CBT, for which a separate IOPP Certificate has also been issued. ☐

1.11.9 The ship, being designated as a "product carrier" operating with CBT, is also designated as a "crude oil tanker" operating with COW, for which a separate IOPP Certificate has also been issued. ☐

2 Equipment for the control of oil discharge from machinery space bilges and oil fuel tanks (regulations 16 and 14)

2.1 Carriage of ballast water in oil fuel tanks:

2.1.1 The ship may under normal conditions carry ballast water in oil fuel tanks ☐

2.2 Type of oil filtering equipment fitted:

2.2.1 Oil filtering (15 ppm) equipment (regulation 14.6). ☐

2.2.2 Oil filtering (15 ppm) equipment with alarm and automatic stopping device (regulation 14.7) ☐

2.3 Approval standards:*

2.3.1 The separating/filtering equipment:

.1 has been approved in accordance with resolution A.393(X) ☐

.2 has been approved in accordance with resolution MEPC.60(33) ☐

.3 has been approved in accordance with resolution MEPC.107(49)

.4 has been approved in accordance with resolution A.233(VII) ☐

.5 has been approved in accordance with national standards not based upon resolution A.393(X) or A.233(VII) ☐

.6 has not been approved. ☐

2.3.2 The process unit has been approved in accordance with resolution A.444(XI) ☐

2.3.3 The oil content meter:

.1 has been approved in accordance with resolution A.393(X); ☐

.2 has been approved in accordance with resolution MEPC.60(33); ☐

.3 has been approved in accordance with resolution MEPC.107(49). ☐

2.4 Maximum throughput of the system is .... m^3/h.

2.5 Waiver of regulation 14:

2.5.1 The requirements of regulation 14.1 or 14.2 are waived in respect of the ship in accordance with regulation 14.5.

The ship is engaged exclusively on voyages within special area(s): ☐

* Refer to the Recommendation on international performance and test specifications of oily-water separating equipment and oil content meters (resolution A.393(X)), Guidelines and specifications for pollution prevention equipment for machinery space bilges (resolution MEPC.60(33)), 2011 Guidelines and specifications for add-on equipment for upgrading resolution MEPC.60(33)-compliant oil filtering equipment (resolution MEPC.205(62)), and Revised guidelines and specifications for pollution prevention equipment for machinery spaces of ships (resolution MEPC.107(49), as amended by resolution MEPC.285(70)).

2.5.2 The ship is fitted with holding tank(s) for the total retention on board of all oily bilge water as follows: ☐

Tank identification	Tank location		Volume (m^3)
	Frames (from)–(to)	Lateral position	
		Total volume: .... m^3	

2.5.3 In lieu of the holding tank(s) the ship is provided with arrangements to transfer bilge water to the slop tank. . . . . ☐

2A.1 The ship is required to be constructed according to regulation 12A and complies with the requirements of:

paragraphs 6 and either 7 or 8 (double hull construction). . . . . ☐

paragraph 11 (accidental oil fuel outflow performance) . . . . ☐

2A.2 The ship is not required to comply with the requirements of regulation 12A. ☐

3 **Means for retention and disposal of oil residues (sludge)** (regulation 12) **and oily bilge water holding tank(s)***

3.1 The ship is provided with oil residue (sludge) tanks for retention of oil residues (sludge) on board as follows:

Tank identification	Tank location		Volume (m^3)
	Frames (from)–(to)	Lateral position	
		Total volume: .... m^3	

3.2 Means for the disposal of oil residues (sludge) retained in oil residue (sludge) tanks:

3.2.1 Incinerator for oil residues (sludge) . . . . ☐

3.2.2 Auxiliary boiler suitable for burning oil residues (sludge) . . . . ☐

3.2.3 Other acceptable means, state which . . . . ☐

3.3 The ship is provided with holding tank(s) for the retention on board of oily bilge water as follows:

Tank identification	Tank location		Volume (m^3)
	Frames (from)–(to)	Lateral position	
		Total volume: .... m^3	

4 **Standard discharge connection** (regulation 13)

4.1 The ship is provided with a pipeline for the discharge of residues from machinery bilges and sludges to reception facilities, fitted with a standard discharge connection in compliance with regulation 13 . . . . ☐

5 **Construction** (regulations 18, 19, 20, 23, 26, 27 and 28)

5.1 In accordance with the requirements of regulation 18, the ship is:

5.1.1 required to be provided with SBT, PL and COW . . . . ☐

5.1.2 required to be provided with SBT and PL . . . . ☐

5.1.3 required to be provided with SBT . . . . ☐

5.1.4 required to be provided with SBT or COW . . . . ☐

5.1.5 required to be provided with SBT or CBT . . . . ☐

5.1.6 not required to comply with the requirements of regulation 18. . . . . ☐

* Oily bilge water holding tank(s) are not required by the Convention; if such tank(s) are provided they shall be listed in table 3.3.

5.2 Segregated ballast tanks (SBT):

5.2.1 The ship is provided with SBT in compliance with regulation 18 . . . . ☐

5.2.2 The ship is provided with SBT, in compliance with regulation 18, which are arranged in protective locations (PL) in compliance with regulations 18.12 to 18.15 . . . . ☐

5.2.3 SBT are distributed as follows:

Tank	Volume (m^3)	Tank	Volume (m^3)
		Total volume: . . . . m^3	

5.3 Dedicated clean ballast tanks (CBT):

5.3.1 The ship is provided with CBT in compliance with regulation 18.8, and may operate as a product carrier . . . . ☐

5.3.2 CBT are distributed as follows:

Tank	Volume (m^3)	Tank	Volume (m^3)
		Total volume: . . . . m^3	

5.3.3 The ship has been supplied with a valid Dedicated Clean Ballast Tank Operation Manual, which is dated. . . . . ☐

5.3.4 The ship has common piping and pumping arrangements for ballasting the CBT and handling cargo oil . . . . ☐

5.3.5 The ship has separate independent piping and pumping arrangements for ballasting the CBT. . . ☐

5.4 Crude oil washing (COW):

5.4.1 The ship is equipped with a COW system in compliance with regulation 33 . . . . ☐

5.4.2 The ship is equipped with a COW system in compliance with regulation 33 except that the effectiveness of the system has not been confirmed in accordance with regulation 33.1 and paragraph 4.2.10 of the Revised COW Specifications (resolution A.446(XI) as amended by resolutions A.497(XII) and A.897(21)) . . . . ☐

5.4.3 The ship has been supplied with a valid Crude Oil Washing Operations and Equipment Manual, which is dated. . . . . ☐

5.4.4 The ship is not required to be but is equipped with COW in compliance with the safety aspects of the Revised COW Specifications (resolution A.446(XI) as amended by resolutions A.497(XII) and A.897(21)) . . . . ☐

5.5 Exemption from regulation 18:

5.5.1 The ship is solely engaged in trade between. . . . . in accordance with regulation 2.5 and is therefore exempted from the requirements of regulation 18 . . . . ☐

5.5.2 The ship is operating with special ballast arrangements in accordance with regulation 18.10 and is therefore exempted from the requirements of regulation 18 . . . . ☐

5.6 Limitation of size and arrangements of cargo tanks (regulation 26):

5.6.1 The ship is required to be constructed according to, and complies with, the requirements of regulation 26 . . . . ☐

5.6.2 The ship is required to be constructed according to, and complies with, the requirements of regulation 26.4 (see regulation 2.2). . . . . ☐

5.7 Subdivision and stability (regulation 28):

5.7.1 The ship is required to be constructed according to, and complies with, the requirements of regulation 28 . . . . . . ☐

5.7.2 Information and data required under regulation 28.5 have been supplied to the ship in an approved form . . . . . . ☐

5.7.3 The ship is required to be constructed according to, and complies with, the requirements of regulation 27 . . . . . . ☐

5.7.4 Information and data required under regulation 27 for combination carriers have been supplied to the ship in a written procedure approved by the Administration . . . . . . ☐

5.7.5 The ship is provided with an Approved Stability Instrument in accordance with regulation 28(6) . . . . . . ☐

5.7.6 The requirements of regulation 28(6) are waived in respect of the ship in accordance with regulation 3.6. Stability is verified by one or more of the following means:

- .1 loading only to approved conditions defined in the stability information provided to the master in accordance with regulation 28(5) . . . . . . ☐
- .2 verification is made remotely by a means approved by the Administration . . . . . . ☐
- .3 loading within an approved range of loading conditions defined in the stability information provided to the master in accordance with regulation 28(5) . . . . . . ☐
- .4 loading in accordance with approved limiting KG/GM curves covering all applicable intact and damage stability requirements defined in the stability information provided to the master in accordance with regulation 28(5) . . . . . . ☐

5.8 Double-hull construction:

5.8.1 The ship is required to be constructed according to regulation 19 and complies with the requirements of:

- .1 paragraph 3 (double-hull construction) . . . . . . ☐
- .2 paragraph 4 (mid-height deck tankers with double side construction) . . . . . . ☐
- .3 paragraph 5 (alternative method approved by the Marine Environment Protection Committee) . . . . . . ☐

5.8.2 The ship is required to be constructed according to and complies with the requirements of regulation 19.6 . . . . . . ☐

5.8.3 The ship is not required to comply with the requirements of regulation 19 . . . . . . ☐

5.8.4 The ship is subject to regulation 20 and:

- .1 is required to comply with paragraphs 2 to 5, 7 and 8 of regulation 19 and regulation 28 in respect of paragraph 28.7 not later than . . . . . . ☐
- .2 is allowed to continue operation in accordance with regulation 20.5 until . . . . . . ☐
- .3 is allowed to continue operation in accordance with regulation 20.7 until . . . . . . ☐

5.8.5 The ship is not subject to regulation 20 and:

- .1 the ship is less than 5,000 tonnes deadweight . . . . . . ☐
- .2 the ship complies with regulation 20.1.2 . . . . . . ☐
- .3 the ship complies with regulation 20.1.3. . . . . . . ☐

5.8.6 The ship is subject to regulation 21 and:

- .1 is required to comply with regulation 21.4 not later than . . . . . . ☐
- .2 is allowed to continue operation in accordance with regulation 21.5 until . . . . . . ☐
- .3 is allowed to continue operation in accordance with regulation 21.6.1 until . . . . . . ☐
- .4 is allowed to continue operation in accordance with regulation 21.6.2 until . . . . . . ☐
- .5 is exempted from the provisions of regulation 21 in accordance with regulation 21.7.2 . . . . ☐

5.8.7 The ship is not subject to regulation 21 and:

- .1 the ship is less than 600 tonnes deadweight . . . . . . ☐
- .2 the ship complies with regulation 19 (tonnes deadweight $\geq$ 5,000) . . . . . . ☐
- .3 the ship complies with regulation 21.1.2. . . . . . . ☐
- .4 the ship complies with regulation 21.4.2 (600 $\leq$ tonnes deadweight $<$ 5,000) . . . . . . ☐
- .5 the ship does not carry "heavy grade oil" as defined in regulation 21.2 of MARPOL Annex I . . . . . . ☐

5.8.8 The ship is subject to regulation 22 and:

.1 complies with the requirements of regulation 22.2 . . . . ☐

.2 complies with the requirements of regulation 22.3 . . . . ☐

.3 complies with the requirements of regulation 22.5 . . . . ☐

5.8.9 The ship is not subject to regulation 22 . . . . ☐

5.9 Accidental oil outflow performance:

5.9.1 The ship complies with the requirements of regulation 23 . . . . ☐

6 Retention of oil on board (regulations 29, 31 and 32)

6.1 Oil discharge monitoring and control system:

6.1.1 The ship comes under category . . . . oil tanker as defined in resolution A.496(XII) or A.586(14)* *(delete as appropriate)* . . . . ☐

6.1.2 The oil discharge monitoring and control system has been approved in accordance with resolution MEPC.108(49). . . . . ☐

6.1.3 The system comprises:

.1 control unit . . . . ☐

.2 computing unit . . . . ☐

.3 calculating unit . . . . ☐

6.1.4 The system is:

.1 fitted with a starting interlock . . . . ☐

.2 fitted with automatic stopping device . . . . ☐

6.1.5 The oil content meter is approved under the terms of resolution A.393(X) or A.586(14) or MEPC.108(49)† *(delete as appropriate)* suitable for:

.1 crude oil . . . . ☐

.2 black products . . . . ☐

.3 white products . . . . ☐

6.1.6 The ship has been supplied with an operations manual for the oil discharge monitoring and control system . . . . ☐

6.2 Slop tanks:

6.2.1 The ship is provided with . . . . dedicated slop tank(s) with the total capacity of . . . . m^3, which is . . . . % of the oil carrying capacity, in accordance with:

.1 regulation 29.2.3 . . . . ☐

.2 regulation 29.2.3.1 . . . . ☐

.3 regulation 29.2.3.2 . . . . ☐

.4 regulation 29.2.3.3 . . . . ☐

6.2.2 Cargo tanks have been designated as slop tanks . . . . ☐

6.3 Oil/water interface detectors:

6.3.1 The ship is provided with oil/water interface detectors approved under the terms of resolution MEPC.5(XIII)‡ . . . . ☐

6.4 Exemptions from regulations 29, 31 and 32:

6.4.1 The ship is exempted from the requirements of regulations 29, 31 and 32 in accordance with regulation 2.4 . . . . ☐

6.4.2 The ship is exempted from the requirements of regulations 29, 31 and 32 in accordance with regulation 2.2 . . . . ☐

---

* Oil tankers the keels of which are laid, or which are at a similar stage of construction, on or after 2 October 1986 should be fitted with a system approved under resolution A.586(14).

† For oil content meters installed on tankers built prior to 2 October 1986, refer to the Recommendation on international performance and test specifications for oily-water separating equipment and oil content meters (resolution A.393(X)). For oil content meters as part of discharge monitoring and control systems installed on tankers built on or after 2 October 1986, refer to Guidelines and specifications for oil discharge monitoring and control systems for oil tankers (resolution A.586(14)). For oil content meters as part of discharge monitoring and control systems installed on tankers built on or after 1 January 2005, refer to Revised guidelines and specifications for oil discharge monitoring and control systems for oil tankers (resolution MEPC.108(49), as amended by resolution MEPC.240(65)).

‡ Refer to the Specification for oil/water interface detectors (resolution MEPC.5(XIII)).

6.5 Waiver of regulations 31 and 32:

6.5.1 The requirements of regulations 31 and 32 are waived in respect of the ship in accordance with regulation 3.5. The ship is engaged exclusively on:

.1 specific trade under regulation 2.5 . . . . ☐

.2 voyages within special area(s) . . . . ☐

.3 voyages within 50 nautical miles of the nearest land outside special area(s) of 72 h or less in duration restricted to . . . . ☐

7 Pumping, piping and discharge arrangements (regulation 30)

7.1 The overboard discharge outlets for segregated ballast are located:

7.1.1 Above the waterline . . . . ☐

7.1.2 Below the waterline . . . . ☐

7.2 The overboard discharge outlets, other than the discharge manifold, for clean ballast are located:*

7.2.1 Above the waterline . . . . ☐

7.2.2 Below the waterline . . . . ☐

7.3 The overboard discharge outlets, other than the discharge manifold, for dirty ballast water or oil-contaminated water from cargo tank areas are located:*

7.3.1 Above the waterline . . . . ☐

7.3.2 Below the waterline in conjunction with the part flow arrangements in compliance with regulation 30.6.5 . . . . ☐

7.3.3 Below the waterline . . . . ☐

7.4 Discharge of oil from cargo pumps and oil lines (regulations 30.4 and 30.5):

7.4.1 Means to drain all cargo pumps and oil lines at the completion of cargo discharge:

.1 drainings capable of being discharged to a cargo tank or slop tank . . . . ☐

.2 for discharge ashore a special small-diameter line is provided . . . . ☐

8 Shipboard oil/marine pollution emergency plan (regulation 37)

8.1 The ship is provided with a shipboard oil pollution emergency plan in compliance with regulation 37 . . . . ☐

8.2 The ship is provided with a shipboard marine pollution emergency plan in compliance with regulation 37.3 . . . . ☐

8A Ship-to-ship oil transfer operations at sea (regulation 41)

8A.1 The oil tanker is provided with an STS operations Plan in compliance with regulation 41 . . . . ☐

9 Exemption

9.1 Exemptions have been granted by the Administration from the requirements of chapter 3 of Annex I of the Convention in accordance with regulation 3.1 on those items listed under paragraph(s) . . . . of this Record . . . . ☐

10 Equivalents (regulation 5)

10.1 Equivalents have been approved by the Administration for certain requirements of Annex I on those items listed under paragraph(s) . . . . of this Record . . . . ☐

11 Compliance with part II-A – chapter 1 of the Polar Code

11.1 The ship is in compliance with additional requirements in the environment-related provisions of the Introduction and section 1.2 of chapter 1 of part II-A of the Polar Code . . . . ☐

* Only those outlets which can be monitored are to be indicated.

THIS IS TO CERTIFY that this Record is correct in all respects.

Issued at . . . .
(place of issue of the Record)

Date (dd/mm/yyyy) . . . .
(date of issue) *(signature of duly authorized official issuing the Record)*

(seal or stamp of the issuing authority, as appropriate)

Appendix III
Form of Oil Record Book

OIL RECORD BOOK*

PART I – Machinery space operations

(All ships)

Name of ship. . . . .

Distinctive number or letters. . . . .

Gross tonnage. . . . .

Period from: . . . . to. . . . .

Note: Oil Record Book Part I shall be provided to every oil tanker of 150 gross tonnage and above and every ship of 400 gross tonnage and above, other than oil tankers, to record relevant machinery space operations. For oil tankers, Oil Record Book Part II shall also be provided to record relevant cargo/ballast operations.

Introduction

The following pages of this section show a comprehensive list of items of machinery space operations which are, when appropriate, to be recorded in the Oil Record Book in accordance with regulation 17 of Annex I of the International Convention for the Prevention of Pollution from Ships, 1973, as modified by the Protocol of 1978 relating thereto (MARPOL 73/78). The items have been grouped into operational sections, each of which is denoted by a letter Code.

When making entries in the Oil Record Book Part I, the date, operational code and item number shall be inserted in the appropriate columns and the required particulars shall be recorded chronologically in the blank spaces.

Each completed operation shall be signed for and dated by the officer or officers in charge. The master of the ship shall sign each completed page.

The Oil Record Book Part I contains many references to oil quantity. The limited accuracy of tank measurement devices, temperature variations and clingage will affect the accuracy of these readings. The entries in the Oil Record Book Part I should be considered accordingly.

In the event of accidental or other exceptional discharge of oil, statement shall be made in the Oil Record Book Part I of the circumstances of, and the reasons for, the discharge.

Any failure of the oil filtering equipment shall be noted in the Oil Record Book Part I.

The entries in the Oil Record Book Part I, for ships holding an IOPP Certificate, shall be at least in English, French or Spanish. Where entries in official language of the State whose flag the ship is entitled to fly are also used, this shall prevail in case of a dispute or discrepancy.

The Oil Record Book Part I shall be kept in such a place as to be readily available for inspection at all reasonable times and, except in the case of unmanned ships under tow, shall be kept on board the ship. It shall be preserved for a period of three years after the last entry has been made.

The competent authority of the Government of a Party to the Convention may inspect the Oil Record Book Part I on board any ship to which this Annex applies while the ship is in its port or offshore terminals and may make a copy of any entry in that book and may require the master of the ship to certify that the copy is a true copy of such entry. Any copy so made which has been certified by the master of the ship as a true copy of an entry in the Oil Record Book Part I shall be made admissible in any judicial proceedings as evidence of the facts stated in the entry. The inspection of an Oil Record Book Part I and the taking of a certified copy by the competent authority under this paragraph shall be performed as expeditiously as possible without causing the ship to be unduly delayed.

---

* Refer to Guidance for the recording of operations in the Oil Record Book Part I – Machinery space operations (all ships) (MEPC.1/Circ.736/Rev.2).

LIST OF ITEMS TO BE RECORDED

(A) Ballasting or cleaning of oil fuel tanks

1 Identity of tank(s) ballasted.

2 Whether cleaned since they last contained oil and, if not, type of oil previously carried.

3 Cleaning process:

.1 position of ship and time at the start and completion of cleaning;

.2 identify tank(s) in which one or another method has been employed (rinsing through, steaming, cleaning with chemicals; type and quantity of chemicals used, in m^3);

.3 identity of tank(s) into which cleaning water was transferred and the quantity in m^3.

4 Ballasting:

.1 position of ship and time at start and end of ballasting;

.2 quantity of ballast if tanks are not cleaned, in m^3.

(B) Discharge of dirty ballast or cleaning water from oil fuel tanks referred to under Section (A)

5 Identity of tank(s).

6 Position of ship at start of discharge.

7 Position of ship on completion of discharge.

8 Ship's speed(s) during discharge.

9 Method of discharge:

.1 through 15 ppm equipment;

.2 to reception facilities.

10 Quantity discharged, in m^3.

(C) Collection, transfer and disposal of oil residues (sludge)

11 Collection of oil residues (sludge).

Quantities of oil residues (sludge) retained on board. The quantity should be recorded weekly:* (this means that the quantity must be recorded once a week even if the voyage lasts more than one week):

.1 identity of tank(s)

.2 capacity of tank(s) .... m^3

.3 total quantity of retention .... m^3

.4 quantity of residue collected by manual operation .... m^3

(Operator initiated manual collections where oil residue (sludge) is transferred into the oil residue (sludge) holding tank(s).)

12 Methods of transfer or disposal of oil residues (sludge).

State quantity of oil residues transferred or disposed of, the tank(s) emptied and the quantity of contents retained in m^3:

.1 to reception facilities (identify port);†

.2 to another (other) tank(s) (indicate tank(s) and the total content of tank(s));

.3 incinerated (indicate total time of operation);

.4 other method (state which).

(D) Non-automatic starting of discharge overboard, transfer or disposal otherwise of bilge water which has accumulated in machinery spaces

13 Quantity discharged, transferred or disposed of, in m^3.‡

14 Time of discharge, transfer or disposal (start and stop).

---

* Only those tanks listed in item 3.1 of Forms A and B of the Supplement to the IOPP Certificate used for oil residues (sludge).

† The ship's master should obtain from the operator of the reception facilities, which includes barges and tank trucks, a receipt or certificate detailing the quantity of tank washings, dirty ballast, residues or oily mixtures transferred, together with the time and date of the transfer. This receipt or certificate, if attached to the Oil Record Book Part I, may aid the master of the ship in proving that the ship was not involved in an alleged pollution incident. The receipt or certificate should be kept together with the Oil Record Book Part I.

‡ In case of discharge or disposal of bilge water from holding tank(s), state identity and capacity of holding tank(s) and quantity retained in holding tank.

15 Method of discharge, transfer, or disposal:

.1 through 15 ppm equipment (state position at start and end);

.2 to reception facilities (identify port);*

.3 to slop tank or holding tank or other tank(s) (indicate tank(s); state quantity retained in tank(s), in m^3).

(E) Automatic starting of discharge overboard, transfer or disposal otherwise of bilge water which has accumulated in machinery spaces

16 Time and position of ship at which the system has been put into automatic mode of operation for discharge overboard, through 15 ppm equipment.

17 Time when the system has been put into automatic mode of operation for transfer of bilge water to holding tank (identify tank).

18 Time when the system has been put into manual operation.

(F) Condition of the oil filtering equipment

19 Time of system failure.†

20 Time when system has been made operational.

21 Reasons for failure.

(G) Accidental or other exceptional discharges of oil

22 Time of occurrence.

23 Place or position of ship at time of occurrence.

24 Approximate quantity and type of oil.

25 Circumstances of discharge or escape, the reasons therefor and general remarks.

(H) Bunkering of fuel or bulk lubricating oil

26 Bunkering:

.1 Place of bunkering.

.2 Time of bunkering.

.3 Type and quantity of fuel oil and identity of tank(s) (state quantity added, in tonnes and total content of tank(s)).

.4 Type and quantity of lubricating oil and identity of tank(s) (state quantity added, in tonnes and total content of tank(s)).

* The ship's master should obtain from the operator of the reception facilities, which includes barges and tank trucks, a receipt or certificate detailing the quantity of tank washings, dirty ballast, residues or oily mixtures transferred, together with the time and date of the transfer. This receipt or certificate, if attached to the Oil Record Book Part I, may aid the master of the ship in proving that the ship was not involved in an alleged pollution incident. The receipt or certificate should be kept together with the Oil Record Book Part I.

† The condition of the oil filtering equipment covers also the alarm and automatic stopping devices, if applicable.

(I) Additional operational procedures and general remarks

Name of ship . . . .

Distinctive number or letters . . . .

MACHINERY SPACE OPERATIONS

Date	Code (letter)	Item (number)	Record of operations/signature of officer in charge

Signature of master . . . .

OIL RECORD BOOK

PART II – Cargo/ballast operations

(Oil tankers)

Name of ship. . . . .

Distinctive number or letters. . . . .

Gross tonnage. . . . .

Period from: . . . . to. . . . .

Note: Every oil tanker of 150 gross tonnage and above shall be provided with Oil Record Book Part II to record relevant cargo/ballast operations. Such a tanker shall also be provided with Oil Record Book Part I to record relevant machinery space operations.

Name of ship. . . . .

Distinctive number or letters. . . . .

PLAN VIEW OF CARGO AND SLOP TANKS
(to be completed on board)

Identification of tanks	Capacity
Depth of slop tank(s):	

(Give the capacity of each tank and the depth of slop tank(s))

Introduction

The following pages of this section show a comprehensive list of items of cargo and ballast operations which are, when appropriate, to be recorded in the Oil Record Book Part II in accordance with regulation 36 of Annex I of the International Convention for the Prevention of Pollution from Ships, 1973, as modified by the Protocol of 1978 relating thereto (MARPOL 73/78). The items have been grouped into operational sections, each of which is denoted by a code letter.

When making entries in the Oil Record Book Part II, the date, operational code and item number shall be inserted in the appropriate columns and the required particulars shall be recorded chronologically in the blank spaces.

Each completed operation shall be signed for and dated by the officer or officers in charge. Each completed page shall be countersigned by the master of the ship.

In respect of the oil tankers engaged in specific trades in accordance with regulation 2.5 of Annex I of MARPOL 73/78, appropriate entry in the Oil Record Book Part II shall be endorsed by the competent port State authority.*

The Oil Record Book Part II contains many references to oil quantity. The limited accuracy of tank measurement devices, temperature variations and clingage will affect the accuracy of these readings. The entries in the Oil Record Book Part II should be considered accordingly.

In the event of accidental or other exceptional dicharge of oil, a statement shall be made in the Oil Record Book Part II of the circumstances of, and the reasons for, the discharge.

Any failure of the oil discharge monitoring and control system shall be noted in the Oil Record Book Part II.

The entries in the Oil Record Book Part II, for ships holding an IOPP Certificate, shall be in at least English, French or Spanish. Where entries in an official language of the State whose flag the ship is entitled to fly are also used, this shall prevail in case of a dispute or discrepancy.

The Oil Record Book Part II shall be kept in such a place as to be readily available for inspection at all reasonable times and, except in the case of unmanned ships under tow, shall be kept on board the ship. It shall be preserved for a period of three years after the last entry has been made.

The competent authority of the Government of a Party to the Convention may inspect the Oil Record Book Part II on board the ship to which this Annex applies while the ship is in its port or offshore terminals and may make a copy of any entry in that book and may require the master of the ship to certify that the copy is a true copy of such entry. Any copy so made which has been certified by the master of the ship as a true copy of an entry in the Oil Record Book Part II shall be made admissible in any judicial proceedings as evidence of the facts stated in the entry. The inspection of an Oil Record Book Part II and taking of a certified copy by the competent authority under this paragraph shall be performed as expeditiously as possible without causing the ship to be unduly delayed.

LIST OF ITEMS TO BE RECORDED

(A) Loading of oil cargo

1 Place of loading.

2 Type of oil loaded and identity of tank(s).

3 Total quantity of oil loaded (state quantity added, in cubic metres, at 15°C and the total content of tank(s), in cubic metres).

(B) Internal transfer of oil cargo during voyage

4 Identity of tank(s):

.1 from:

.2 to: (state quantity transferred and total quantity of tank(s), in cubic metres).

5 Was (were) the tank(s) in 4.1 emptied? (If not, state quantity retained, in cubic metres.)

(C) Unloading of oil cargo

6 Place of unloading.

7 Identity of tank(s) unloaded.

8 Was (were) the tank(s) emptied? (If not, state quantity retained, in cubic metres.)

* This sentence should only be inserted for the Oil Record Book of a tanker engaged in a specific trade.

(D) Crude oil washing (COW tankers only)
(To be completed for each tank being crude oil washed)

9 Port where crude oil washing was carried out or ship's position if carried out between two discharge ports.

10 Identity of tank(s) washed.*

11 Number of machines in use.

12 Time of start of washing.

13 Washing pattern employed.†

14 Washing line pressure.

15 Time washing was completed or stopped.

16 State method of establishing that tank(s) was (were) dry.

17 Remarks.‡

(E) Ballasting of cargo tanks

18 Position of ship at start and end of ballasting.

19 Ballasting process:

.1 identity of tank(s) ballasted;

.2 time of start and end;

.3 quantity of ballast received. Indicate total quantity of ballast for each tank involved in the operation, in cubic metres.

(F) Ballasting of dedicated clean ballast tanks (CBT tankers only)

20 Identity of tank(s) ballasted.

21 Position of ship when water intended for flushing, or port ballast was taken to dedicated clean ballast tank(s).

22 Position of ship when pump(s) and lines were flushed to slop tank.

23 Quantity of the oily water which, after line flushing, is transferred to the slop tank(s) or cargo tank(s) in which slop is preliminarily stored (identify tank(s)). State total quantity, in cubic metres.

24 Position of ship when additional ballast water was taken to dedicated clean ballast tank(s).

25 Time and position of ship when valves separating the dedicated clean ballast tanks from cargo and stripping lines were closed.

26 Quantity of clean ballast taken on board, in cubic metres.

(G) Cleaning of cargo tanks

27 Identity of tank(s) cleaned.

28 Port or ship's position.

29 Duration of cleaning

30 Method of cleaning.§

31 Tank washings transferred to:

.1 reception facilities (state port and quantity, in cubic metres);¶

.2 slop tank(s) or cargo tank(s) designated as slop tank(s) (identify tank(s); state quantity transferred and total quantity, in cubic metres).

---

* When an individual tank has more machines than can be operated simultaneously, as described in the Operations and Equipment Manual, then the section being crude oil washed should be identified, e.g. No. 2 centre, forward section.

† In accordance with the Operations and Equipment Manual, enter whether single-stage or multi-stage method of washing is employed. If multi-stage method is used, give the vertical arc covered by the machines and the number of times that arc is covered for that particular stage of the programme.

‡ If the programmes given in the Operations and Equipment Manual are not followed, then the reasons must be given under Remarks.

§ Hand-hosing, machine washing and/or chemical cleaning. Where chemically cleaned, the chemical concerned and amount used should be stated.

¶ Ships' masters should obtain from the operator of the reception facilities, which includes barges and tank trucks, a receipt or certificate detailing the quantity of tank washings, dirty ballast, residues or oily mixtures transferred, together with the time and date of the transfer. This receipt or certificate, if attached to the Oil Record Book Part II, may aid the master of the ship in proving that his ship was not involved in an alleged pollution incident. The receipt or certificate should be kept together with the Oil Record Book Part II.

(H) Discharge of dirty ballast

32 Identity of tank(s).

33 Time and position of ship at start of discharge into the sea.

34 Time and position of ship on completion of discharge into the sea.

35 Quantity discharged into the sea, in cubic metres.

36 Ship's speed(s) during discharge.

37 Was the discharge monitoring and control system in operation during the discharge?

38 Was a regular check kept on the effluent and the surface of the water in the locality of the discharge?

39 Quantity of oily water transferred to slop tank(s) (identify slop tank(s)). State total quantity, in cubic metres.

40 Discharged to shore reception facilities (identify port and quantity involved, in cubic metres).*

(I) Discharge of water from slop tanks into the sea

41 Identity of slop tanks.

42 Time of settling from last entry of residues, or

43 Time of settling from last discharge.

44 Time and position of ship at start of discharge.

45 Ullage of total contents at start of discharge.

46 Ullage of oil/water interface at start of discharge.

47 Bulk quantity discharged in cubic metres and rate of discharge in m^3/hour.

48 Final quantity discharged in cubic metres and rate of discharge in m^3/hour.

49 Time and position of ship on completion of discharge.

50 Was the discharge monitoring and control system in operation during the discharge?

51 Ullage of oil/water interface on completion of discharge, in metres.

52 Ship's speed(s) during discharge.

53 Was a regular check kept on the effluent and the surface of the water in the locality of the discharge?

54 Confirm that all applicable valves in the ship's piping system have been closed on completion of discharge from the slop tanks.

(J) Collection, transfer and disposal of residues and oily mixtures not otherwise dealt with

55 Identity of tanks.

56 Quantity transferred or disposed of from each tank. (State the quantity retained, in m^3.)

57 Method of transfer or disposal:

.1 disposal to reception facilities (identify port and quantity involved);*

.2 mixed with cargo (state quantity);

.3 transferred to or from (an)other tank(s) including transfer from machinery space oil residue (sludge) and oily bilge water tanks (identify tank(s); state quantity transferred and total quantity in tank(s), in m^3); and

.4 other method (state which); state quantity disposed of in m^3.

(K) Discharge of clean ballast contained in cargo tanks

58 Position of ship at start of discharge of clean ballast.

59 Identity of tank(s) discharged.

60 Was (were) the tank(s) empty on completion?

61 Position of ship on completion if different from 58.

62 Was a regular check kept on the effluent and the surface of the water in the locality of the discharge?

* Ships' masters should obtain from the operator of the reception facilities, which includes barges and tank trucks, a receipt or certificate detailing the quantity of tank washings, dirty ballast, residues or oily mixtures transferred, together with the time and date of the transfer. This receipt or certificate, if attached to the Oil Record Book Part II, may aid the master of the ship in proving that his ship was not involved in an alleged pollution incident. The receipt or certificate should be kept together with the Oil Record Book Part II.

(L) Discharge of ballast from dedicated clean ballast tanks (CBT tankers only)

63 Identity of tank(s) discharged.

64 Time and position of ship at start of discharge of clean ballast into the sea.

65 Time and position of ship on completion of discharge into the sea.

66 Quantity discharged, in cubic metres:

.1 into the sea; or

.2 to reception facility (identify port).*

67 Was there any indication of oil contamination of the ballast water before or during discharge into the sea?

68 Was the discharge monitored by an oil content meter?

69 Time and position of ship when valves separating dedicated clean ballast tanks from the cargo and stripping lines were closed on completion of deballasting.

(M) Condition of oil discharge monitoring and control system

70 Time of system failure.

71 Time when system has been made operational.

72 Reasons for failure.

(N) Accidental or other exceptional discharges of oil

73 Time of occurrence.

74 Port or ship's position at time of occurrence.

75 Approximate quantity, in cubic metres, and type of oil.

76 Circumstances of discharge or escape, the reasons therefor and general remarks.

(O) Additional operational procedures and general remarks

TANKERS ENGAGED IN SPECIFIC TRADES

(P) Loading of ballast water

77 Identity of tank(s) ballasted.

78 Position of ship when ballasted.

79 Total quantity of ballast loaded in cubic metres.

80 Remarks.

(Q) Re-allocation of ballast water within the ship

81 Reasons for re-allocation.

(R) Ballast water discharge to reception facility

82 Port(s) where ballast water was discharged.

83 Name or designation of reception facility.

84 Total quantity of ballast water discharged in cubic metres.

85 Date, signature and stamp of port authority official.

* Ships' masters should obtain from the operator of the reception facilities, which includes barges and tank trucks, a receipt or certificate detailing the quantity of tank washings, dirty ballast, residues or oily mixtures transferred, together with the time and date of the transfer. This receipt or certificate, if attached to the Oil Record Book Part II, may aid the master of the ship in proving that his ship was not involved in an alleged pollution incident. The receipt or certificate should be kept together with the Oil Record Book Part II.

Name of ship. . . . .

Distinctive number or letters. . . . .

CARGO/BALLAST OPERATIONS (OIL TANKERS)

Date	Code (letter)	Item (number)	Record of operations/signature of officer in charge

Signature of master . . . .

Unified Interpretations of Annex I

Notes: For the purposes of the Unified Interpretations, the following abbreviations are used:

MARPOL	The 1973 MARPOL Convention as modified by the 1978 and 1997 Protocols relating thereto
Regulation	Regulation in Annex I of MARPOL
IOPP Certificate	International Oil Pollution Prevention Certificate
SBT	Segregated ballast tanks
CBT	Dedicated clean ballast tanks
COW	Crude oil washing system
IGS	Inert gas systems
PL	Protective location of segregated ballast tanks
CAS	Condition Assessment Scheme

1 Definition of oil
Treatment for oily rags

Reg. 1.1

Oily rags, as defined in the Guidelines for the Implementation of Annex V of MARPOL, should be treated in accordance with Annex V and the procedures set out in the Guidelines.

2 Definition of an oil tanker

Reg. 1.5

FPSOs and FSUs are not *oil tankers* and are not to be used for the transport of oil except that, with the specific agreement by the flag and relevant coastal States on a voyage basis, produced oil may be transported to port in abnormal and rare circumstances.

3 Definition of major conversion

Reg. 1.9

3.1 The deadweight to be used for determining the application of provisions of Annex I is the deadweight assigned to an oil tanker at the time of the assignment of the load lines. If the load lines are reassigned for the purpose of altering the deadweight, without alteration of the structure of the ship, any substantial alteration of the deadweight consequential upon such reassignments should not be construed as a "major conversion" as defined in regulation 1.9. However, the IOPP Certificate should indicate only one deadweight of the ship and be renewed on every reassignment of load lines.

3.2 If a crude oil tanker of 40,000 tonnes deadweight and above delivered on or before 1 June 1982 as defined in regulation 1.28.3 satisfying the requirements of COW changes its trade for the carriage of product oil* conversion to CBT or SBT and reissuing of the IOPP Certificate will be necessary. Such conversion should not be considered as a "major conversion" as defined in regulation 1.9.

3.3 When an oil tanker is used solely for the storage of oil and is subsequently put into service in the transport of oil, such a change of function should not be construed as a "major conversion" as defined in regulation 1.9.

3.4 The conversion of an existing oil tanker to a combination carrier, or the shortening of a tanker by removing a transverse section of cargo tanks, should constitute a "major conversion" as defined in regulation 1.9.

3.5 The conversion of an existing oil tanker to a segregated ballast tanker by the addition of a transverse section of tanks should constitute a "major conversion" as defined in regulation 1.9 only when the cargo-carrying capacity of the tanker is increased.

3.6 When a ship built as a combination carrier operates exclusively in the bulk cargo trade, the ship may be treated as a ship other than an oil tanker and Form A of the Record of Construction and Equipment should be issued to the ship. The change of such a ship from the bulk trade to the oil trade should not be construed as a "major conversion" as defined in regulation 1.9.

* *Product oil* means any oil other than crude oil as defined in regulation 1.2.

4 Definition of "segregated ballast"

Reg. 1.18

4.1 The segregated ballast system should be a system which is "completely separated from the cargo oil and fuel systems" as required by regulation 1.18. Nevertheless, provision may be made for emergency discharge of the segregated ballast by means of a connection to a cargo pump through a portable spool piece. In this case non-return valves should be fitted on the segregated ballast connections to prevent the passage of oil to the segregated ballast tanks. The portable spool piece should be mounted in a conspicuous position in the pump-room and a permanent notice restricting its use should be prominently displayed adjacent to it.

4.2 Sliding type couplings should not be used for expansion purposes where lines for cargo oil or fuel oil pass through tanks for segregated ballast, and where lines for segregated ballast pass through cargo oil or fuel oil tanks. This interpretation is applicable to ships, the keel of which is laid, or which are at a similar stage of construction, on or after 1 July 1992.

5 Definition of lightweight

Reg. 1.24

The weight of mediums on board for the fixed fire-fighting systems (e.g. freshwater, CO_2, dry chemical powder, foam concentrate, etc.) should be included in the lightweight and lightship condition.

6 Unforeseen delay in delivery of ships

Reg. 1.28

6.1 For the purpose of defining the category of a ship under regulation 1.28, a ship for which the building contract (or keel laying) and delivery were scheduled before the dates specified in these regulations, but which has been subject to delay in delivery beyond the specific date due to unforeseen circumstances beyond the control of the builder and the owner, may be accepted by the Administration as a ship of the category related to the estimated date of delivery. The treatment of such ships should be considered by the Administration on a case-by-case basis, bearing in mind the particular circumstances.

6.2 It is important that ships delivered after the specified dates due to unforeseen delay and allowed to be treated as a ship of the category related to the estimated date of delivery by the Administration should also be accepted as such by port States. In order to ensure this, the following practice is recommended to Administrations when considering an application for such a ship:

.1 the Administration should thoroughly consider applications on a case-by-case basis, bearing in mind the particular circumstances. In doing so in the case of a ship built in a foreign country, the Administration may require a formal report from the authorities of the country in which the ship was built, stating that the delay was due to unforeseen circumstances beyond the control of the builder and the owner;

.2 when a ship is treated as a ship of the category related to the estimated date of delivery upon such an application, the IOPP Certificate for the ship should be endorsed to indicate that the ship is accepted by the Administration as such a ship; and

.3 the Administration should report to the Organization on the identity of the ship and the grounds on which the ship has been accepted as such a ship.

7 Definition of "a similar stage of construction"

Regs. 1.28, 1.30

A similar stage of construction means the stage at which:

.1 construction identifiable with a specific ship begins; and

.2 assembly of that ship has commenced comprising at least 50 tonnes or one per cent of the estimated mass of all structural material, whichever is less.

8 Definition of generation of ships

Regs. 1.28.2, 1.28.4, 1.28.6, 1.28.7, 1.28.8, 1.28.9

For the purpose of defining the ships in accordance with regulations 1.28.2, 1.28.4, 1.28.6, 1.28.7, 1.28.8 and 1.28.9, a ship which falls into any one of the categories listed in subparagraphs 1, 2, 3, 4.1, 4.2, or 4.3 of these paragraphs should be considered as a ship falling under the corresponding definition.

9 Annex I substances which through their physical properties inhibit effective product/water separation and monitoring

Reg. 2.4

9.1 The Government of the receiving Party should establish appropriate measures in order to ensure that provisions of 9.2 are complied with.

9.2 A tank which has been unloaded should, subject to the provisions of 9.3, be washed and all contaminated washings should be discharged to a reception facility before the ship leaves the port of unloading for another port.

9.3 At the request of the ship's master, the Government of the receiving Party may exempt the ship from the requirements referred to in 9.2, where it is satisfied that:

.1 the tank unloaded is to be reloaded with the same substance or another substance compatible with the previous one and that the tanker will not be washed or ballasted prior to loading; and

.2 the tank unloaded is neither washed nor ballasted at sea if the ship is to proceed to another port unless it has been confirmed in writing that a reception facility at that port is available and adequate for the purpose of receiving the residues and solvents necessary for the cleaning operations.

9.4 An exemption referred to in 9.3 should only be granted by the Government of the receiving Party to a ship engaged in voyages to ports or terminals under the jurisdiction of other Parties to the Convention. When such an exemption has been granted it should be certified in writing by the Government of the receiving Party.

9.5 In the case of ships retaining their residues on board and proceeding to ports or terminals under the jurisdiction of other Parties to the Convention, the Government of the receiving Party is advised to inform the next port of call of the particulars of the ship and cargo residues, for their information and appropriate action for the detection of violations and enforcement of the Convention.

10 Conditions for waiver

Regs. 3.4, 3.5, 14.5.3

The International Oil Pollution Prevention Certificate should contain sufficient information to permit the port State to determine if the ship complies with the waiver conditions regarding the phrase "restricted voyages as determined by the Administration". This may include a list of ports, the maximum duration of the voyage between ports having reception facilities, or similar conditions as established by the Administration.

11 Voyages of 72 h or less in duration

Regs. 3.4, 3.5.2.3.2

The time limitation "of 72 h or less in duration" in regulations 3.4 and 3.5.2.3.2 should be counted:

.1 from the time the tanker leaves the special area, when a voyage starts within a special area; or

.2 from the time the tanker leaves a port situated outside the special area to the time the tanker approaches a special area.

12 Definition of "all oily mixtures"

Regs. 3.4, 3.5.2.4

The phrase "all oily mixtures" in regulations 3.4 and 3.5.2.4 includes all ballast water and tank washing residues from cargo oil tanks.

13 Equivalents

Reg. 5

Acceptance by an Administration under regulation 5 of any fitting, material, appliance, or apparatus as an alternative to that required by Annex I includes type approval of pollution prevention equipment which is equivalent to that specified in resolution A.393(X).* An Administration that allows such type approval shall communicate particulars thereof, including the test results on which the approval of equivalency was based, to the Organization in accordance with regulation 5.2.

With regard to the term "appropriate action, if any" in regulation 5.2, any Party to the Convention that has an objection to an equivalency submitted by another Party should communicate this objection to the Organization and to the Party which allowed the equivalency within one year after the Organization circulates the equivalency to the Parties. The Party objecting to the equivalency should specify whether the objection pertains to ships entering its ports.

* For oily-water separating equipment for machinery space bilges of ships, refer to Guidelines and specifications for pollution prevention equipment for machinery space bilges (resolution MEPC.60(33)), 2011 Guidelines and specifications for add-on equipment for upgrading resolution MEPC.60(33)-compliant oil filtering equipment (resolution MEPC.205(62)), and Revised guidelines and specifications for pollution prevention equipment for machinery spaces of ships (resolution MEPC.107(49), as amended by resolution MEPC.285(70)). For oil discharge monitoring and control systems installed on oil tankers built before 2 October 1986, refer to Guidelines and specifications for oil discharge monitoring and control systems for oil tankers (resolution A.496(XII)), and for oil discharge monitoring and control systems installed on oil tankers built after 2 October 1986, refer to Revised guidelines and specifications for oil discharge monitoring and control systems (resolution A.586(14)). For oil discharge monitoring and control systems installed on oil tankers the keels of which are laid or are in a similar stage of construction on or after 1 January 2005, refer to Revised guidelines and specifications for oil discharge monitoring and control systems (resolution MEPC.108(49), as amended by resolution MEPC.240(65)).

14 Survey and inspection

Regs. 6.1.3, 6.1.4

Intermediate and annual survey for ships not required to hold an IOPP Certificate

The applicability of regulations 6.1.3 and 6.1.4 to ships which are not required to hold an International Oil Pollution Prevention Certificate should be determined by the Administration.

15 Designation of the type of oil tankers

Regs. 7, 19

15.1 Oil tankers must be designated on the Supplement Form B to the IOPP Certificate as either "crude oil tanker", "product carrier" or "crude oil/product carrier". Furthermore, the requirements contained in regulation 19 differ for different age categories of "crude oil tankers" and "product carriers", and compliance with these provisions is recorded on the IOPP Certificate. Oil trades in which different types of oil tankers are allowed to be engaged are as follows:

.1 *Crude oil/product carrier* is allowed to carry either crude oil or product oil, or both simultaneously;

.2 *Crude oil tanker* is allowed to carry crude oil but is prohibited from carrying product oil; and

.3 *Product carrier* is allowed to carry product oil but is prohibited from carrying crude oil.

15.2 In determining the designation of the type of oil tanker on the IOPP Certificate based on the compliance with the provisions for SBT, PL, CBT and COW, the following standards should apply.

15.3 *Oil tankers delivered after 1 June 1982 as defined in regulation 1.28.4 of less than 20,000 tonnes deadweight*

15.3.1 These oil tankers may be designated as "crude oil/product carriers".

15.4 *Oil tankers delivered after 1 June 1982 as defined in regulation 1.28.4 of 20,000 tonnes deadweight and above*

15.4.1 Oil tankers satisfying the requirements for SBT + PL + COW may be designated as "crude oil/product carrier".

15.4.2 Oil tankers satisfying the requirements for SBT + PL but not COW should be designated as "product carrier".

15.4.3 Oil tankers of 20,000 tonnes deadweight and above but less than 30,000 tonnes deadweight not carrying crude oil, fuel oil, heavy diesel oil or lubricating oil as cargo, not fitted with SBT + PL, should be designated as "product carrier".

15.5 *Oil tankers delivered on or before 1 June 1982 as defined in regulation 1.28.3 but delivered after 31 December 1979 as defined in regulation 1.28.2 of 70,000 tonnes deadweight and above*

15.5.1 The oil tankers satisfying the requirements for SBT may be designated as "crude oil/product carrier".

15.6 *Oil tankers delivered on or before 1 June 1982 as defined in regulation 1.28.3 of less than 40,000 tonnes deadweight*

15.6.1 These oil tankers may be designated as "crude oil/product carrier".

15.7 *Oil tankers delivered on or before 1 June 1982 as defined in regulation 1.28.3 of 40,000 tonnes deadweight and above*

15.7.1 Oil tankers satisfying the requirements for SBT should be designated as "crude oil/product carrier".

15.7.2 Oil tankers satisfying the requirements for COW only should be designated as "crude oil tanker".

15.7.3 Oil tankers satisfying the requirements for CBT should be designated as "product carrier".

16 New form of IOPP Certificate or its Supplement

Reg. 9

In the case where the form of the IOPP Certificate or its Supplement is amended, and this amendment does not cause a shortening of the validity of the ship's IOPP Certificate, the existing form of the certificate or supplement which is current when the amendment enters into force may remain valid until the expiry of that certificate, provided that, at the first survey after the date of entry into force of the amendment, necessary changes are indicated in the existing certificate or supplement by means of suitable corrections, e.g. striking over the invalid entry and typing the new entry.

17 Revalidation of an IOPP Certificate

Reg. 10

Where an annual or an intermediate survey required in regulation 6 of Annex I of MARPOL is not carried out within the period specified in that regulation, the IOPP Certificate ceases to be valid. When a survey corresponding to the requisite survey is carried out subsequently, the validity of the Certificate may be restored without altering the anniversary and expiry date of the original Certificate and the Certificate endorsed to this effect. The thoroughness and stringency of such survey will depend on the period for which the prescribed survey has elapsed and the conditions of the ship.

18 Capacity of oil residue (sludge) tanks

Reg. 12.3.1

18.1 To assist Administrations in determining the adequate capacity of oil residue (sludge) tanks, the following criteria may be used as guidance. These criteria should not be construed as determining the amount of oily residues which will be produced by the machinery installation in a given period of time. The capacity of oil residue (sludge) tanks may, however, be calculated upon any other reasonable assumptions. For a ship the keel of which is laid or which is at a similar stage of construction on or after 31 December 1990, the guidance given in items .4 and .5 below should be used in lieu of the guidance contained in items .1 and .2.

.1 For ships which do not carry ballast water in oil fuel tanks, the minimum oil residue (sludge) tank capacity (V_1) should be calculated by the following formula:

$$V_1 = K_1CD \qquad (m^3)$$

where:

K_1 = 0.01 for ships where heavy fuel oil is purified for main engine use, or 0.005 for ships using diesel oil or heavy fuel oil which does not require purification before use,

C = daily fuel oil consumption (metric tons); and

D = maximum period of voyage between ports where oil residue (sludge) can be discharged ashore (days). In the absence of precise data a figure of 30 days should be used.

.2 When such ships are fitted with homogenizers, oil residue (sludge) incinerators or other recognized means on board for the control of oil residue (sludge), the minimum oil residue (sludge) tank capacity (V_1) should, in lieu of the above, be:

V_1 = 1 m^3 for ships of 400 gross tonnage and above but less than 4,000 gross tonnage, or 2 m^3 for ships of 4,000 gross tonnage and above.

.3 For ships which carry ballast water in fuel oil tanks, the minimum oil residue (sludge) tank capacity (V_2) should be calculated by the following formula:

$$V_2 = V_1 + K_2B \qquad (m^3)$$

where:

V_1 = oil residue (sludge) tank capacity specified in .1 or .2 above in m^3;

K_2 = 0.01 for heavy fuel oil bunker tanks, or 0.005 for diesel oil bunker tanks; and

B = capacity of water ballast tanks which can also be used to carry oil fuel (tonnes).

.4 For ships which do not carry ballast water in fuel oil tanks, the minimum oil residue (sludge) tank capacity (V_1) should be calculated by the following formula:

$$V_1 = K_1CD \qquad (m^3)$$

where:

K_1 = 0.015 for ships where heavy fuel oil is purified for main engine use or 0.005 for ships using diesel oil or heavy fuel oil which does not require purification before use;

C = daily fuel oil consumption (m^3); and

D = maximum period of voyage between ports where oil residue (sludge) can be discharged ashore (days). In the absence of precise data a figure of 30 days should be used.

.5 For ships where the building contract is placed, or in the absence of a building contract, the keel of which is laid before 1 July 2010, and which are fitted with homogenizers, oil residue (sludge) incinerators or other recognized means on board for the control of oil residue (sludge), the minimum oil residue (sludge) tank capacity should be:

.5.1 50% of the value calculated according to item .4 above; or

.5.2 1 m^3 for ships of 400 gross tonnage and above but less than 4,000 gross tonnage or 2 m^3 for ships of 4,000 gross tonnage and above; whichever is the greater.

18.2 Administrations should establish that in a ship the keel of which is laid or which is at a similar stage of construction on or after 31 December 1990, adequate tank capacity, which may include the oil residue (sludge) tank(s) referred to under 18.1 above, is available also for leakage, drain and waste oils from the machinery installations. In existing installations this should be taken into consideration as far as reasonable and practicable.

19 Designated pump for disposal

Reg. 12.3.2 A designated pump should be interpreted as any pump used for the disposal of oil residue (sludge) through the standard discharge connection referred to in regulation 13, or any pump used to transfer oil residue (sludge) to any other approved means of disposal such as an incinerator, auxiliary boiler suitable for burning oil residues (sludge) or other acceptable means which are prescribed in paragraph 3.2 of the Supplement to IOPP Certificate Form A or B.

20 No discharge connection

Reg. 12.3.3 A screw-down non-return valve, arranged in lines connecting to common piping leading to the standard discharge connection required by regulation 13, provides an acceptable means to prevent oil residue (sludge) from being transferred or discharged to the bilge system, oily bilge water holding tank(s), tank top or oily water separators.

21 Overboard connection of oil residue (sludge) tanks

Reg. 12.3.4 Ships having piping to and from oil residue (sludge) tanks to overboard discharge outlets, other than the standard discharge connection referred to in regulation 13 installed prior to 4 April 1993 may comply with regulation 12.3.4 by the installation of blanks in this piping.

22 Cleaning of oil residue (sludge) tanks and discharge of residues

Reg. 12.3.5 To assist Administrations in determining the adequacy of the design and construction of oil residue (sludge) tanks to facilitate their cleaning and the discharge of residues to reception facilities, the following guidance is provided, having effect on ships the keel of which is laid or which is at a similar stage of construction on or after 31 December 1990:

.1 sufficient man-holes should be provided such that, taking into consideration the internal structure of the oil residue (sludge) tanks, all parts of the tank can be reached to facilitate cleaning;

.2 oil residue (sludge) tanks in ships operating with heavy oil, that needs to be purified for use, should be fitted with adequate heating arrangements or other suitable means to facilitate the pump ability and discharge of the tank content;

.3 the oil residue (sludge) tank should be provided with a designated pump for the discharge of the tank content to reception facilities. The pump should be of a suitable type, capacity and discharge head, having regard to the characteristics of the liquid being pumped and the size and position of tank(s) and the overall discharge time.

.4 where any oil residue (sludge) tank (i.e. oil residue (sludge) service tank)* that directly supplies oil residue (sludge) to the means of the disposal of oil residues (sludge) prescribed in paragraph 3.2 of the Supplement to IOPP Certificate Form A or B is equipped with suitable means for drainage, the requirements in subparagraph .3 above may not be applied to the oil residue (sludge) tank.

23 Oil fuel tank protection

Regs. 12A.6, 12A.7, 12A.8 23.1 Valves for oil fuel tanks located in accordance with the provisions of paragraphs 6, 7 and 8 of MARPOL Annex I, regulation 12A, may be treated in a manner similar to the treatment of suction wells as per MARPOL regulation 12A.10 and, therefore, arranged at a distance from the ship's bottom of not less than $h/2$.

23.2 Valves for tanks which are permitted to be located at a distance from the ship's bottom or side at a distance less than h or w, respectively, in accordance with the accidental oil fuel outflow performance standard of MARPOL Annex I, regulation 12A.11, may be arranged at the distance less than h or w, respectively.

* *Oil residue (sludge) service tank* means a tank for preparation of oil residue (sludge) for incineration as defined in paragraph 5.3.3 of the appendix to the annex to the 2008 Revised guidelines for systems for handling oily wastes in machinery spaces of ships incorporating guidance notes for an integrated bilge water treatment system (IBTS) (MEPC.1/Circ.642, as amended by MEPC.1/Circ.676 and MEPC.1/Circ.760).

23.3 Fuel tank air escape pipes and overflow pipes are not considered as part of "*lines of fuel oil piping*" and, therefore, may be located at a distance from the ship's side of less than *w*.

23.4 In addition to being as small as practicable, the size of the suction wells mentioned in MARPOL Annex I, regulation 12A.10, should be appropriate to the size of the suction pipe and area covered.

24 Measuring distance "*h*"

Regs. 12A.6, 12A.7, 12A.8, 12A.11.8

24.1 The distance "*h*" should be measured from the moulded line of the bottom shell plating at right angle to it (regulation12A, Figure 1).

.1 For vessels designed with a skeg, the skeg should not be considered as offering protection for the FO tanks. For the area within skeg's width the distance "*h*" should be measured perpendicular to a line parallel to the baseline at the intersection of the skeg and the moulded line of the bottom shell plating as indicated in Figure A.

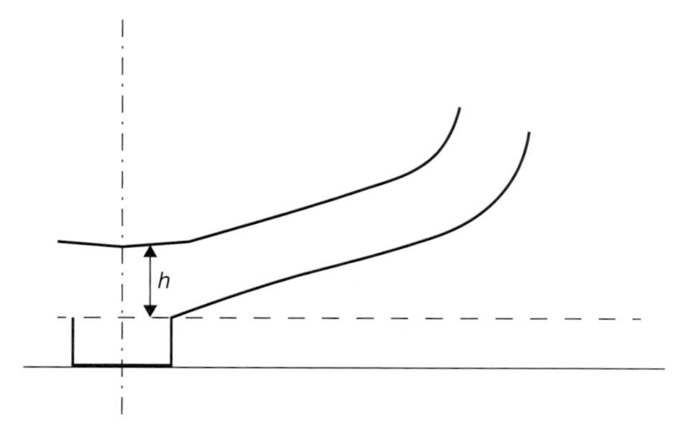

Figure A

.2 For vessels designed with a permanent trim, the baseline should not be used as a reference point. The distance "*h*" should be measured perpendicular to the moulded line of the bottom shell plating at the relevant frames where fuel tanks are to be protected.

24.2 For vessels designed with deadrising bottom, the distance "1.5*h*" should be measured from the moulded line of the bottom shell plating but at right angle to the baseline, as indicated in Figure B.

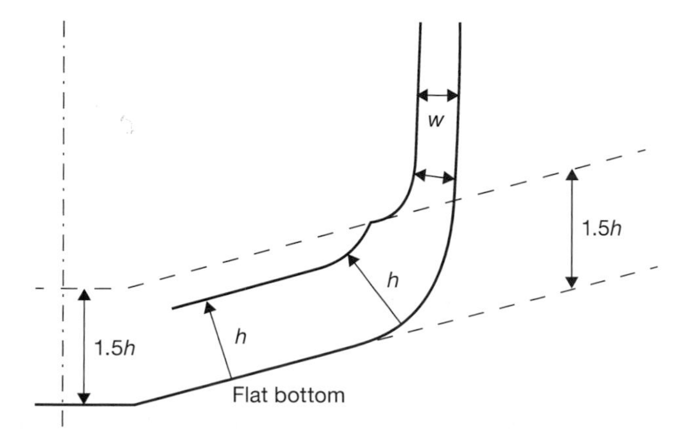

Figure B

24.3 Paragraphs 1 and 2, above also apply to the reference to the distance "*h*" in regulation 12A.11.8.

25 Application of regulation 12A to MODUs

Regs. 12A.7, 12A.8

In applying regulation 12A of MARPOL Annex I to column-stabilized units (MODUs) as defined in the MODU Code, for the purpose of placing the oil fuel tanks, the location limitations of paragraphs 7 and 8 of the regulation apply to those areas subject to damage as follows:

.1 only those columns, underwater hulls and braces on the periphery of the unit shall be assumed to be damaged and the damage shall be assumed in the exposed portions of the columns, underwater hulls and braces;

.2 columns and braces shall be assumed to be damaged at any level between 5.0 m above and 3.0 m below the range of draughts in the MODUs operating manual for normal and severe weather operations; and

.3 underwater hull and footings shall be assumed to be damaged when operating in a transit condition in the same manner as indicated in .1 and .2, having regard to their shape.

26 Automatic stopping device required by regulation 15.3.2

Regs. 14, 15

Regulation 15.3.2 includes a reference to regulation 14.7 which requires both a 15 ppm bilge alarm and a stopping device which will ensure that the discharge is automatically stopped when the oil content of the effluent exceeds 15 ppm. Since, however, this is not a requirement of regulation 14 for ships of less than 10,000 gross tonnage, such ships need not be required to be equipped with such alarm and stopping device if no effluent from machinery space bilge is to be discharged within special areas. Conversely, the discharge of effluent within special areas from ships without 15ppm bilge alarm and an automatic stopping device is a contravention of the Convention even if the oil content of the effluent is below 15 ppm.

27 Control of discharge of ballast water from oil fuel tanks

Reg. 14.1

27.1 The second sentence of regulation 14.1 should be interpreted as follows:

Any ship of 400 gross tonnage and above but less than 10,000 gross tonnage:

.1 which does not carry water ballast in oil fuel tanks should be fitted with 15 ppm oil filtering equipment for the control of discharge of machinery space bilges;

.2 which carries water ballast in oil fuel tanks should be fitted with the equipment required by regulation 14.2 for the control of machinery space bilges and dirty ballast water from oil fuel tanks. Ships on which it is not reasonable to fit this equipment should retain on board dirty ballast water from oil fuel tanks and discharge it to reception facilities.

27.2 The above equipment should be of adequate capacity to deal with the quantities of effluent to be discharged.

28 Oil filtering equipment

Regs. 14.1, 14.2

Oil filtering equipment referred to in regulations 14.1 and 14.2 is a 15 ppm bilge separator and may include any combination of a separator, filter or coalescer and also a single unit designed to produce an effluent with oil content not exceeding 15 ppm.

29 Waivers for restricted voyages

Reg. 14.5.3.4

The International Oil Pollution Prevention Certificate should contain sufficient information to permit the port State to determine if the ship complies with the waiver conditions regarding the phrase "restricted voyages as determined by the Administration". This may include a list of ports, the maximum duration of the voyage between ports having reception facilities, or similar conditions as established by the Administration.

30 Controls of discharge of oil

Reg. 15

Transfer of non-oil-cargo related oily residues to slop tanks of oil tankers

30.1 If non-oil-cargo related oily residues are transferred to slop tanks of oil tankers, the discharge of such residues should be in compliance with regulation 34.

30.2 The above interpretation should not be construed as relaxing any existing prohibition of piping arrangements connecting the engine room and slop tanks which may permit cargo to enter the machinery spaces. Any arrangements provided for machinery space bilge discharges into slop tanks should incorporate adequate means to prevent any backflow of liquid cargo and gases into the machinery spaces. Any such arrangements do not constitute a relaxing of the requirements of regulation 14 with respect to oil filtering equipment.

31 Definition of "en route"

Reg. 15.2.1

En route means that the ship is underway at sea on a course or courses, including deviation from the shortest direct route, which, as far as practicable for navigation purposes, will cause any discharge to be spread over as great an area of the sea as is reasonable and practicable.

32 Oil fuel

Reg. 16.2

Large quantities of oil fuel

32.1 The phrase "large quantities of oil fuel" in regulation 16.2 refers to ships which are required to stay at sea for extended periods because of the particular nature of their operation and trade. Under the circumstances considered, these ships would be required to fill their empty oil fuel tanks with water ballast in order to maintain sufficient stability and safe navigation conditions.

32.2 Such ships may include inter alia certain large fishing vessels or ocean-going tugs. Certain other types of ships which for reasons of safety, such as stability, may be required to carry ballast in oil fuel tanks may also be included in this category.

33 Application of regulation 16.4

Reg. 16.4

When the separation of oil fuel tanks and water ballast tanks is unreasonable or impracticable for ships covered by regulation 16.4, ballast water may be carried in oil fuel tanks, provided that such ballast water is discharged into the sea in compliance with regulations 15.2, 15.3, 15.5 and 15.6 or into reception facilities in compliance with regulation 15.9.

34 Oil tankers used for the storage of dirty ballast

Regs. 18, 19, 20, 33, 35

When an oil tanker is used as a floating facility to receive dirty ballast discharged from oil tankers, such a tanker is not required to comply with the provisions of regulations 18, 19, 20, 33 and 35.

35 SBT, CBT, COW and PL requirements

Reg. 18.3.2

Capacity of SBT

For the purpose of application of regulation 18.3.2, the following operations of oil tankers are regarded as falling within the category of exceptional cases:

.1 when combination carriers are required to operate beneath loading or unloading gantries;

.2 when tankers are required to pass under a low bridge;

.3 when local port or canal regulations require specific draughts for safe navigation;

.4 when loading and unloading arrangements require the tanker to be at a draught deeper than that achieved when all segregated ballast tanks are full;

.5 close-up inspection or/and steel thickness measurement using rafts where permitted by the rules; and

.6 tank hydrostatic pressure tests.

36 Segregated ballast conditions for oil tankers less than 150 m in length

Reg. 18.5

36.1 In determining the minimum draught and trim of oil tankers less than 150 m in length to be qualified as SBT oil tankers, the Administration should follow the guidance set out in appendix 1.

36.2 The formulae set out in appendix 1 replace those set out in regulation 18.2, and these oil tankers should also comply with the conditions laid down in regulations 18.3 and 18.4 in order to be qualified as SBT oil tankers.

37 Oil tankers as defined in regulation 1.28.3 of 40,000 tonnes deadweight and above with CBT and COW

Regs. 18.7, 18.8

37.1 Oil tankers as defined in regulation 1.28.3 of 40,000 tonnes deadweight and above which are fitted with CBT and COW and designated as "crude oil/product carriers" in the Supplement to the IOPP Certificate operate as follows:

.1 They should always operate with CBT and neither crude oil nor product oil should be carried in dedicated clean ballast tanks; and

.2 When carrying a complete or partial cargo of crude oil they should, in the crude carrying tanks, also operate with COW for sludge control.

37.2 Approved procedures by the Administration for changeover between COW and CBT modes on tankers with common or separate independent piping and pump arrangements for cargo and (CBT) ballast handling should be continuously acceptable as long as carriage of crude oil in CBT mode is not given as permissible.

38 Capacity of CBT

Reg. 18.8

For the purposes of determining the capacity of CBT, the following tanks may be included:

.1 segregated ballast tanks; and

.2 cofferdams and fore and after peak tanks, provided that they are exclusively used for the carriage of ballast water and are connected with permanent piping to ballast water pumps.

39 CBT oil content meter

Reg. 18.8.3

The discharge of ballast from the dedicated clean ballast tanks should be continuously monitored (but not necessarily recorded) by the oil content meter required by regulation 18.8.3 so that the oil content, if any, in the ballast water can be observed from time to time. This oil content meter is not required to come into operation automatically.

40 Protective location of SBT

Regs. 18.12 to 18.15

40.1 The measurement of the minimum width of wing tanks and of the minimum vertical depth of double bottom tanks should be taken and values of protective areas (PA_c and PA_s) should be calculated in accordance with the "Interim recommendation for a unified interpretation of regulations 18.12–18.15 – Protective location of segregated ballast spaces" set out in appendix 2.

40.2 Ships being built in accordance with this interpretation should be regarded as meeting the requirements of regulations 18.12 to 18.15 and would not need to be altered if different requirements were to result from a later interpretation.

40.3 If, in the opinion of the Administration, any oil tanker the keel of which was laid or which was at a similar stage of construction before 1 July 1980 complies with the requirements of regulation 18.12–18.15 without taking into account the above Interim Recommendation, the Administration may accept such tanker as complying with regulations 18.12–18.15.

41 Oil tankers with independent tanks

Reg. 19

Oil tankers with independent tanks are considered as double-hull oil tankers, provided that they are designed and constructed to be such that the minimum distances between the cargo tank boundaries and ship bottom and side-shell plating comply with the provisions of regulation 19.

42 Width of wing tanks and height of double bottom tanks at turn of the bilge area

Reg. 19.3.3

The requirements of regulation 19.3.3 at turn of the bilge areas are applicable throughout the entire tank length.

43 Aggregate capacity of ballast tanks

Reg. 19.4

43.1 Any ballast carried in localized inboard extensions, indentations or recesses of the double hull, such as bulkhead stools, should be excess ballast above the minimum requirement for segregated ballast capacity according to regulation 18.

43.2 In calculating the aggregate capacity under regulation 19.3.4, the following should be taken into account:

.1 the capacity of engine-room ballast tanks should be excluded from the aggregate capacity of ballast tanks;

.2 the capacity of ballast tank located inboard of double hull should be excluded from the aggregate capacity of ballast tanks (see figure 1).

.3 spaces such as void spaces located in the double hull within the cargo tank length should be included in the aggregate capacity of ballast tanks (see figure 2).

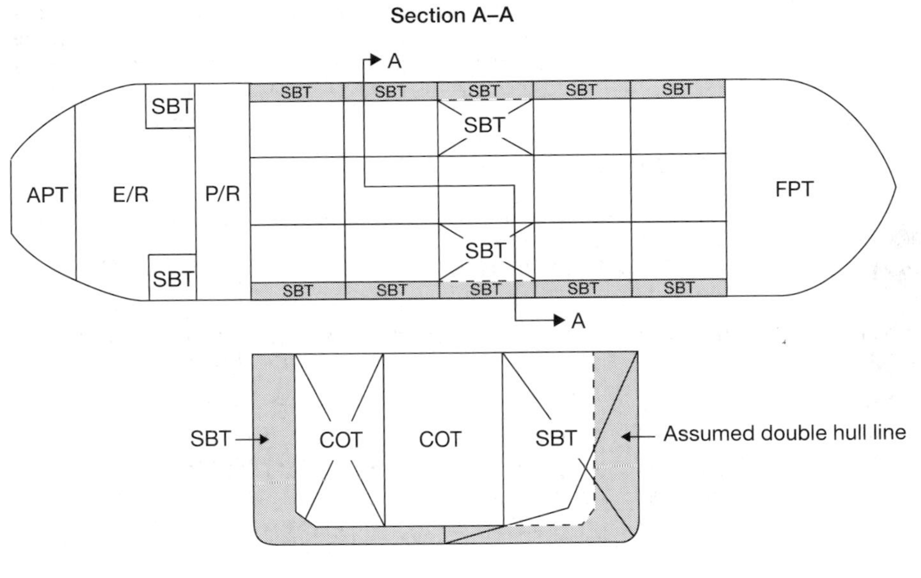

SBT: Segregated ballast tank
COT: Cargo oil tank
FPT: Fore peak tank
APT: After peak tank
E/R: Engine room
P/R: Pump room

Figure 1

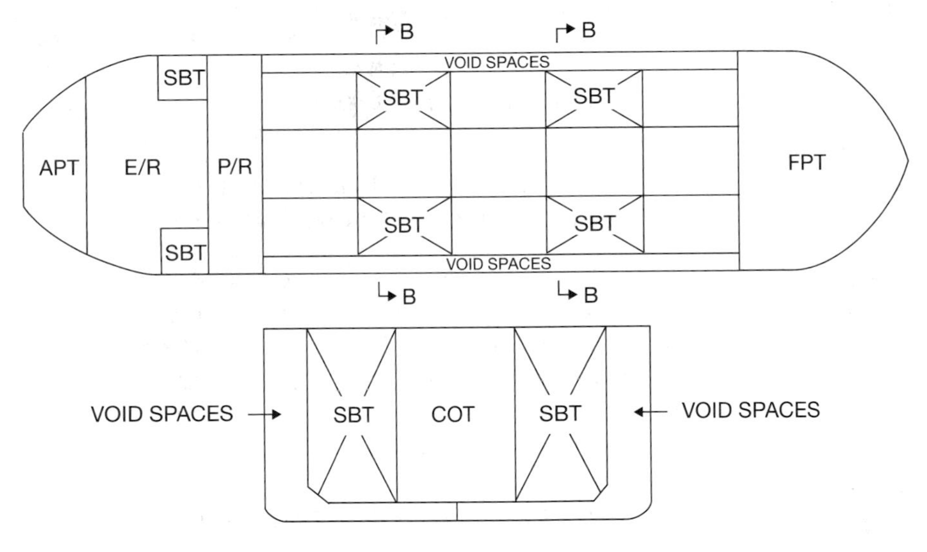

SBT: Segregated ballast tank
COT: Cargo oil tank
FPT: Fore peak tank
APT: After peak tank
E/R: Engine room
P/R: Pump room

Figure 2

44 Definition of double side wing tanks

Reg. 19.6.2 Wing tanks required for the protection of the entire cargo tank length by regulation 19.6.2, for the purpose of compliance with regulation 21.4.2, can be used as cargo tanks for the carriage of oil other than heavy grade oils when the ship is provided with cargo tanks so arranged that the capacity of each cargo tank does not exceed 700 m^3.

45 Definition of Category 2 oil tanker

Reg. 20.3.2 Any Category 2 oil tanker must be provided with segregated ballast tanks protectively located (SBT/PL).

46 Major conversion in respect of regulation 20.4

Reg. 20.4

For the purpose of determining the application date for the requirements of regulation 20.4 of MARPOL Annex I, where an oil tanker has undergone a major conversion, as defined in regulation 1 of MARPOL Annex I, that has resulted in the replacement of the fore-body, including the entire cargo carrying section, the major conversion completion date of the oil tanker shall be deemed to be the date of delivery of the ship referred to in regulation 20.4 of MARPOL Annex I, provided that:

.1 the oil tanker conversion was completed before 6 July 1996;

.2 the conversion included the replacement of the entire cargo section and fore-body and the tanker complies with all the relevant provisions of MARPOL Annex I applicable at the date of completion of the major conversion; and

.3 the original delivery date of the oil tanker will apply when considering the 15 years of age threshold relating to the first CAS survey to be completed in accordance with regulation 20.6 of MARPOL Annex I.

47 Wing tanks and double bottom spaces of tankers as defined in regulation 1.28.5 used for water ballast

Reg. 20.6

If the wing tanks and double bottom tanks referred to in regulation 20.6 are used for water ballast, the ballast arrangement should at least be in compliance with the Revised specifications for oil tankers with dedicated CBT (resolution A.495(XII)).

48 Requirements for the Condition Assessment Scheme (CAS)

Reg. 21.6.1

The first CAS survey shall be carried out concurrent with the first intermediate or renewal survey:

– after 5 April 2005, or

– after the date when the ship reaches 15 years of age,

whichever occurs later.

49 Pump-room bottom protection

Reg. 22.5

49.1 The term "pump-room" means a cargo pump-room. Ballast piping is permitted to be located within the pump-room double bottom provided any damage to that piping does not render the ship's pumps located in the "pump-room" ineffective.

49.2 The double bottom protecting the "pump-room" can be a void tank, a ballast tank or, unless prohibited by other regulations, a fuel oil tank.

49.3 Bilge wells may be accepted within the double bottom provided that such wells are as small as practicable and the distance between the well bottom and the ship's baseline measured at right angles to the ship's baseline is not less than $0.5h$.

49.4 Where a portion of the pump-room is located below the minimum height required in regulation 22.2, then only that portion of the pump-room is required to be a double bottom.

50 Accidental oil outflow performance
Overpressure in kPa

Reg. 23.7.3.2

If an inert gas system is fitted, the normal overpressure, in kPa, is to be taken as 5 kPa.

51 Tank size limitation and damage stability

Reg. 24.1.2

Bottom damage assumptions

When applying the figures for bottom damage within the forward part of the ship as specified in regulation 24.1.2 for the purpose of calculating both oil outflow and damage stability, $0.3L$ from the forward perpendicular should be the aftermost point of the extent of damage.

52 Hypothetical oil outflow for combination carriers

Reg. 25

For the purpose of calculation of the hypothetical oil outflow for combination carriers:

.1 the volume of a cargo tank should include the volume of the hatchway up to the top of the hatchway coamings, regardless of the construction of the hatch, but may not include the volume of any hatch cover; and

.2 for the measurement of the volume to moulded lines, no deduction should be made for the volume of internal structures.

53 Calculation of hypothetical oil outflow

Reg. 25.1.2

In a case where the width b_i is not constant along the length of a particular wing tank, the smallest b_i value in the tank should be used for the purposes of assessing the hypothetical outflows of oil O_c and O_s.

54 Hypothetical outflow of oil
Location of valves

Reg. 25.3.3

54.1 Valves or other closing arrangements located in accordance with the provisions of MARPOL Annex I, regulation 25.3.3, may be treated in a manner similar to the treatment of suction wells as per MARPOL regulation 12A.10 and, therefore, arranged at a distance from the ship's bottom of not less than $h/2$.

54.2 In addition to being not excessive in area, the size of the suction wells mentioned in MARPOL Annex I, regulation 25.3.3, should be appropriate to the size of the suction pipe and area covered.

55 Intact stability

Reg. 27

55.1 For proving compliance with regulation 27, either subparagraph .1 or .2, below, should be applied:

.1 The ship should be loaded with all cargo tanks filled to a level corresponding to the maximum combined total of vertical moment of volume plus free surface inertia moment at 0° heel, for each individual tank. Cargo density should correspond to the available cargo deadweight at the displacement at which transverse KM reaches a minimum value, assuming full departure consumables and 1% of the total water ballast capacity. The maximum free surface moment should be assumed in all ballast conditions. For the purpose of calculating GM_o, liquid free surface corrections should be based on the appropriate upright free surface inertia moment. The righting lever curve may be corrected on the basis of liquid transfer moments.

.2 An extensive analysis covering all possible combinations of cargo and ballast tank loading should be carried out. For such extensive analysis conditions, it is considered that:

.2.1 weight, centre of gravity coordinates and free surface moment for all tanks should be according to the actual content considered in the calculations; and

.2.2 the extensive calculations should be carried out in accordance with the following:

.2.2.1 the draughts should be varied between light ballast and scantling draught;

.2.2.2 consumables including, but not restricted to, fuel oil, diesel oil and fresh water corresponding to 97%, 50% and 10% content should be considered;

.2.2.3 for each draught and variation of consumables, the available deadweight should comprise ballast water and cargo, such that combinations between maximum ballast and minimum cargo and vice versa, are covered. In all cases the number of ballast and cargo tanks loaded is to be chosen to reflect the worst combination of VCG and free surface effects. Operational limits on the number of tanks considered to be simultaneously slack and exclusion of specific tanks should not be permitted. All ballast tanks should have at least 1% content;

.2.2.4 cargo densities between the lowest and highest intended to be carried should be considered; and

.2.2.5 sufficient steps between all limits should be examined to ensure that the worst conditions are identified. A minimum of 20 steps for the range of cargo and ballast content, between 1% and 99% of total capacity, should be examined. More closely spaced steps near critical parts of the range may be necessary.

At every stage, the criteria described in regulations 27.1.1 and 27.1.2 of MARPOL Annex I are to be met.

55.2 In applying θ_f, openings which "cannot be closed weathertight" include ventilators (complying with regulation 19(4) of the International Convention on Load Lines, 1966) that for operational reasons have to remain open to supply air to the engine room or emergency generator room (if the same is considered buoyant in the stability calculation or protecting openings leading below) for the effective operation of the ship.

56 Operating draught

Reg. 28.1

With regard to the term "any operating draught reflecting actual partial or full load conditions", the information required should enable the damage stability to be assessed under conditions the same as or similar to those under which the ship is expected to operate.

57 Suction wells

Reg. 28.2

For the purpose of determining the extent of assumed damage under regulation 28.2, suction wells may be neglected, provided such wells are not excessive in area and extend below the tank for a minimum distance and in no case more than half the height of the double bottom.

58 Subdivision and damage stability

Reg. 28.3

Other openings capable of being closed weathertight do not include ventilators (complying with regulation 19(4) of the International Convention on Load Lines, 1966) that for operational reasons have to remain open to supply air to the engine room or emergency generator room (if the same is considered buoyant in the stability calculation or protecting openings leading below) for the effective operation of the ship.

59 Tanks with smooth walls

Reg. 29.2.3.3

The term "tanks with smooth walls" should be taken to include the main cargo tanks of oil/bulk/ore carriers which may be constructed with vertical framing of a small depth. Vertically corrugated bulkheads are considered smooth walls.

60 Pumping and piping arrangements

Reg. 30.2

Piping arrangements for discharge above the waterline

60.1 Under regulation 30.2, lines for discharge to the sea above the waterline must be led either:

.1 to a ship's discharge outlet located above the waterline in the deepest ballast condition; or

.2 to a midship discharge manifold or, where fitted, a stern or bow loading/discharge facility above the upper deck.

60.2 The ship's side discharge outlet referred to in 60.1.1 should be so located that its lower edge will not be submerged when the ship carries the maximum quantity of ballast during its ballast voyages, having regard to the type and trade of the ship. The discharge outlet located above the waterline in the following ballast condition will be accepted as complying with this requirement:

.1 on oil tankers not provided with SBT or CBT, the ballast condition when the ship carries both normal departure ballast and normal clean ballast simultaneously; and

.2 on oil tankers provided with SBT or CBT, the ballast condition when the ship carries ballast water in segregated or dedicated clean ballast tanks, together with additional ballast in cargo oil tanks in compliance with regulation 18.3.

60.3 The Administration may accept piping arrangements which are led to the ship's side discharge outlet located above the departure ballast waterline but not above the waterline in the deepest ballast condition, if such arrangements have been fitted before 1 January 1981.

60.4 Although regulation 30.2 does not preclude the use of the facility referred to in 60.1.2 for the discharge of ballast water, it is recognized that the use of this facility is not desirable, and it is strongly recommended that ships be provided with either the side discharge outlets referred to in 60.1.1 or the part flow arrangements referred to in regulation 30.6.5.

61 Small diameter line

Reg. 30.4.2

61.1 For the purpose of application of regulation 30.4.2, the cross-sectional area of the small diameter line should not exceed:

.1 10% of that of a main cargo discharge line for oil tankers delivered after 1 June 1982, as defined in regulation 1.28.4, or oil tankers delivered on or before 1 June 1982, as defined in regulation 1.28.3, not already fitted with a small diameter line; or

.2 25% of that of a main cargo discharge line for oil tankers delivered on or before 1 June 1982, as defined in regulation 1.28.3, already fitted with such a line. (See paragraph 4.4.5 of the revised COW Specifications contained in resolution A.446(XI) as amended by the Organization by resolutions A.497(XII) and A.897(21)).

61.2 *Connection of the small diameter line to the manifold valve*

The phrase "connected outboard of" with respect to the small diameter line for discharge ashore should be interpreted to mean a connection on the downstream side of the tanker's deck manifold valves, both port and starboard, when the cargo is being discharged. This arrangement would permit drainage back from the tanker's cargo lines to be pumped ashore with the tanker's manifold valves closed through the same connections as for main cargo lines (see the sketch shown in appendix 3).

62 Part flow system specifications

Reg. 30.6.5.2 The Specifications for the Design, Installation and Operation of a Part Flow System for Control of Overboard Discharges referred to in regulation 30.6.5.2 is set out in appendix 4.

63 Examples of positive means

Reg. 30.7 Examples of positive means may take the form of blanks, spectacle blanks, pipeline blinds, evacuation or vacuum systems, or air or water pressure systems. In the event that the evacuation or vacuum systems, or air or water pressure systems are used, then these systems are to be equipped with both a pressure gauge and alarm system to enable the continuous monitoring of the status of the pipeline section, and thereby the valve integrity, between the sea chest and inboard valves.

64 Total quantity of discharge

Reg. 34.1.5 The phrase "the total quantity of the particular cargo of which the residue formed a part" in regulation 34.1.5 relates to the total quantity of the particular cargo which was carried on the previous voyage and should not be construed as relating only to the total quantity of cargo which was contained in the cargo tanks into which water ballast was subsequently loaded.

65 Shipboard oil pollution emergency plan

Reg. 37.1 *Equivalent provision for application of requirement for oil pollution emergency plans*

Any fixed or floating drilling rig or other offshore installation when engaged in the exploration, exploitation or associated offshore processing of sea-bed mineral resources, which has an oil pollution emergency plan coordinated with, and approved in accordance with procedures established by, the coastal State, should be regarded as complying with regulation 37.

66 Adequate reception facilities for substances regulated by regulation 2.4

Reg. 38 Unloading ports receiving substances regulated by regulation 2.4 (which include inter alia high-density oils) should have adequate facilities dedicated for such products, allowing the entire tank-cleaning operation to be carried out in the port, and should have adequate reception facilities for the proper discharge and reception of cargo residues and solvent necessary for the cleaning operation in accordance with paragraph 9.2 of the Unified Interpretations.

67 Requirements for fixed or floating platforms

Reg. 39
Art. 2(3)(b)(ii)

Application of MARPOL

There are five categories of discharges that may be associated with the operation of fixed or floating platforms covered by this regulation when engaged in the exploration and exploitation of mineral resources, i.e.:

.1 machinery space drainage;

.2 offshore processing drainage;

.3 production water discharge;

.4 displacement water discharge; and

.5 contaminated seawater from operational purposes such as produced oil tank cleaning water, produced oil tank hydrostatic testing water, water from ballasting of produced oil tank to carry out inspection by rafting.

Only the discharge of machinery space drainage and contaminated ballast should be subject to MARPOL (see diagram shown in appendix 5).

Appendices to Unified Interpretations of Annex I

Appendix 1

Guidance to Administrations concerning draughts recommended for segregated ballast tankers below 150 m in length

Introduction

1 Three formulations are set forth as guidance to Administrations concerning minimum draught requirements for segregated ballast tankers below 150 m in length.

2 The formulations are based both on the theoretical research and surveys of actual practice on tankers of differing configuration reflecting varying degrees of concern with propeller emergence, vibration, slamming, speed loss, rolling, docking and other matters. In addition, certain information concerning assumed sea conditions is included.

3 Recognizing the nature of the underlying work, the widely varying arrangement of smaller tankers and each vessel's unique sensitivity to wind and sea conditions, no basis for recommending a single formulation is found.

Caution

4 It must be cautioned that the information presented should be used as general guidance for Administrations. With regard to the unique operating requirements of a particular vessel, the Administration should be satisfied that the tanker has sufficient ballast capacity for safe operation. In any case the stability should be examined independently.

5 *Formulation A*

.1 mean draught (m) $= 0.200 + 0.032L$

.2 maximum trim $= (0.024 - 6 \times 10^{-5}L)L$

6 These expressions were derived from a study of 26 tankers ranging in length from 50 to 150 m. The draughts, in some cases, were abstracted from ship's trim and stability books and represent departure ballast conditions. The ballast conditions represent sailing conditions in weather up to and including Beaufort 5.

7 *Formulation B*

.1 minimum draught at bow (m) $= 0.700 + 0.0170L$

.2 minimum draught at stern (m) $= 2.300 + 0.030L$

or

.3 minimum mean draught (m) $= 1.550 + 0.023L$

.4 maximum trim $= 1.600 + 0.013L$

8 These expressions resulted from investigations based on theoretical research, model and full scale tests. These formulae are based on a Sea 6 (International Sea Scale).

9 *Formulation C*

.1 minimum draught aft (m) $= 2.0000 + 0.0275L$

.2 minimum draught forward (m) $= 0.5000 + 0.0225L$

10 These expressions provide for certain increased draughts to aid in the prevention of propeller emergence and slamming in higher length ships.

Appendix 2

Interim recommendation for a unified interpretation of regulations 18.12 to 18.15 "Protective location of segregated ballast spaces"

1 Regulation 18.15 of Annex I of MARPOL relating to the measurement of the 2 m minimum width of wing tanks and the measurement of the minimum vertical depth of double bottom tanks of 2 m or $\frac{B}{15}$ in respect of tanks at the ends of the ship where no identifiable bilge area exists should be interpreted as given hereunder. No difficulty exists in the measurement of the tanks in the parallel middle body of the ship where the bilge area is clearly identified. The regulation does not explain how the measurements should be taken.

2 The minimum width of wing tanks should be measured at a height of $\frac{D}{5}$ above the base line providing a reasonable level above which the 2 m width of collision protection should apply, under the assumption that in all cases $\frac{D}{5}$ is above the upper turn of bilge amidships (see figure 1). The minimum height of double bottom tanks should be measured at a vertical plane measured $D/5$ inboard from the intersection of the shell with a horizontal line $D/5$ above the base line (see figure 2).

3 The PA_c value for a wing tank which does not have a minimum width of 2 m throughout its length would be zero; no credit should be given for that part of the tank in which the minimum width is in excess of 2 m. No credit should be given in the assessment of PA_s to any double bottom tank, part of which does not meet the minimum depth requirements anywhere within its length. If, however, the projected dimensions of the bottom of the cargo tank above the double bottom fall entirely within the area of the double bottom tank or space which meets the minimum height requirement and provided the side bulkheads bounding the cargo tank above are vertical or have a slope of not more than 45° from the vertical, credit may be given to the part of the double bottom tank defined by the projection of the cargo tank bottom. For similar cases where the wing tanks above the double bottom are segregated ballast tanks or void spaces, such credit may also be given. This would not, however, preclude in the above cases credit being given to a PA_s value in the first case and to a PA_c value in the second case where the respective vertical or horizontal protection complies with the minimum distances prescribed in regulation 18.15.

4 Projected dimensions should be used as shown in examples of figures 3 to 8. Figures 7 and 8 represent measurement of the height for the calculation of PA_c for double bottom tanks with sloping tank top. Figures 9 and 10 represent the cases where credit is given in calculation of PA_s to part or the whole of a double bottom tank.

Figure 1 – **Measurement of minimum width of wing ballast tank at ends of ship**

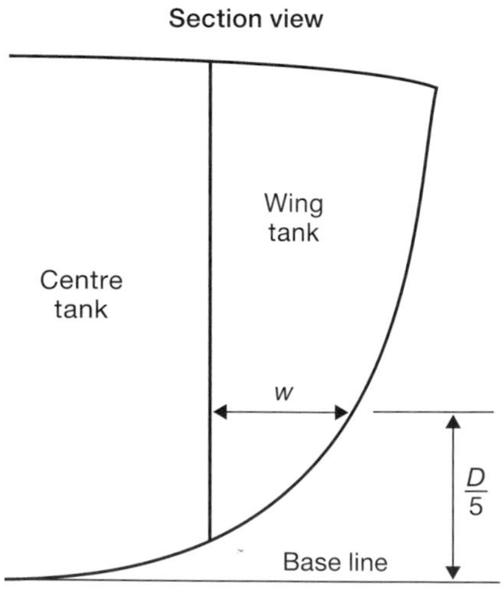

w must be at least 2 m along the entire length of the tank for the tank to be used in the calculation of PA_c

Figure 2 – **Measurement of minimum height of double bottom tank at ends of ship**

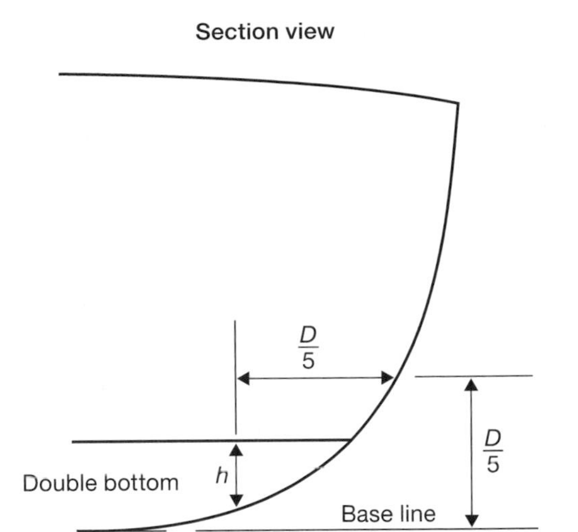

h must be at least 2 m or $\frac{B}{15}$, whichever is less, along the entire length of the tank for the tank to be used in the calculation of PA_s

Figure 3 – **Calculation of PA_c and PA_s for double bottom tank amidships**

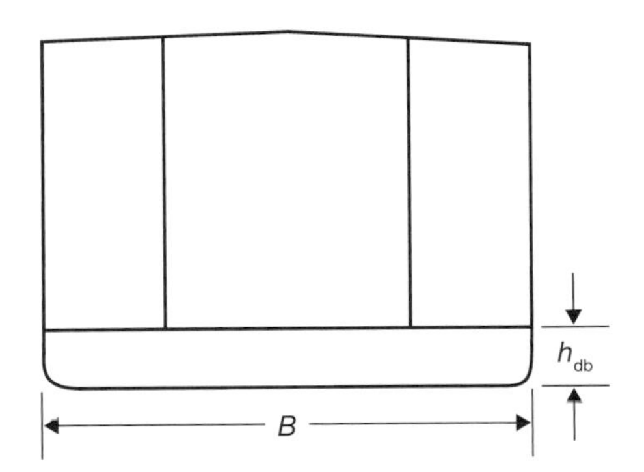

If h_{db} is at least 2 m or $\frac{B}{15}$, whichever is less, along entire tank length,

$PA_c = h_{db} \times$ double bottom tank length $\times$ 2

$PA_s = B \times$ double bottom tank length

If h_{db} is less than 2 m or $\frac{B}{15}$, whichever is less,

$PA_c = h_{db} \times$ double bottom tank length $\times$ 2

$PA_s = 0$

Figure 4 – **Calculation of PA_c and PA_s for double bottom tank at ends of ship**

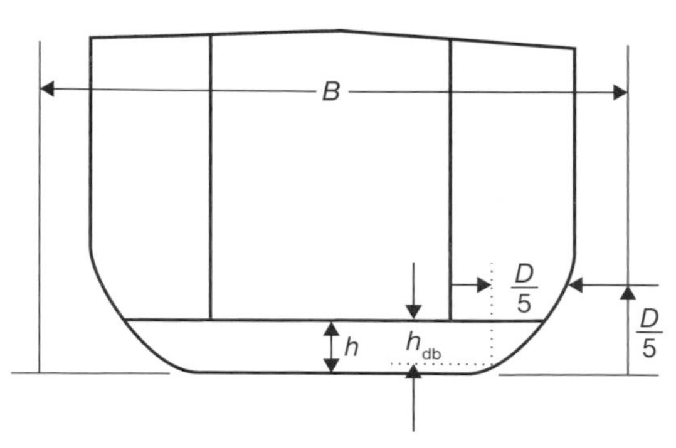

If h_{db} is at least 2 m or $\frac{B}{15}$, whichever is less, along entire tank length,

PA_c = h × double bottom tank length × 2

PA_s = B × double bottom tank length

If h_{db} is less than 2 m or $\frac{B}{15}$, whichever is less,

PA_c = h × double bottom tank length × 2

PA_s = 0

Figure 5 – **Calculation of PA_c and PA_s for wing tank amidships**

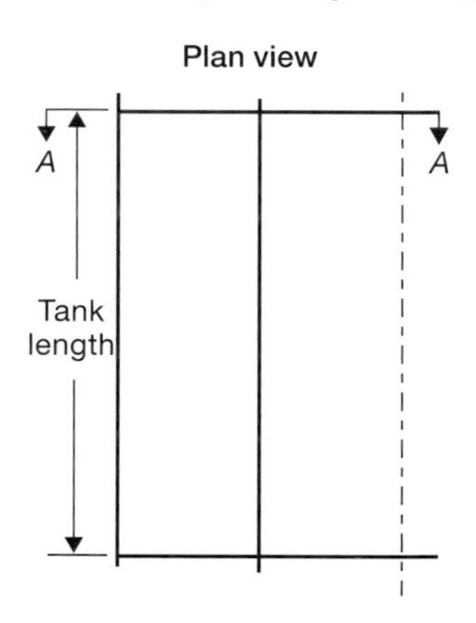

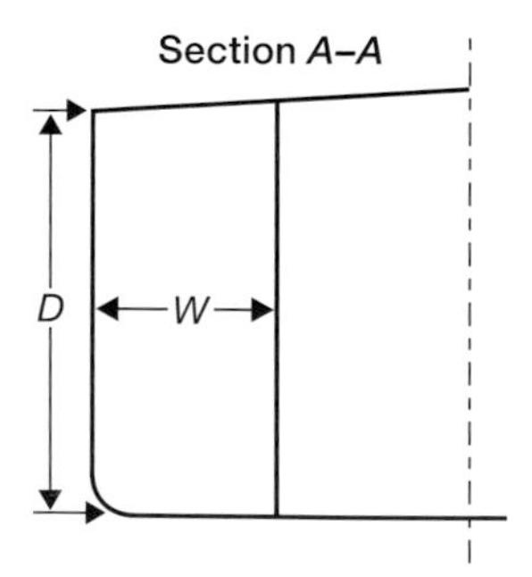

If W is 2 m or more,

PA_c = D × tank length × 2*

PA_s = W × tank length × 2*

If W is less than 2 m,

PA_c = 0

PA_s = W × tank length × 2*

* To include port and starboard.

Figure 6 – **Calculation of PA_c and PA_s for wing tank at end of ship**

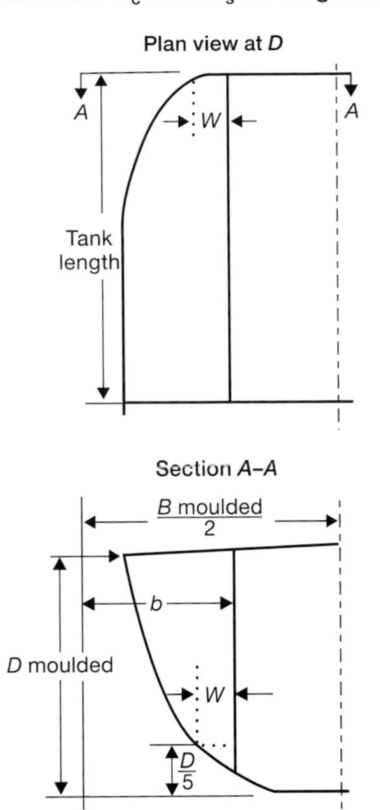

If W is 2 m or more,

$PA_c = D \times \text{tank length} \times 2$*

$PA_s = b \times \text{tank length} \times 2$*

If W is less than 2 m,

$PA_c = 0$

$PA_s = b \times \text{tank length} \times 2$*

Figure 7 – **Measurement of h for calculation of PA_c for double bottom tanks with sloping tank tops (1)**

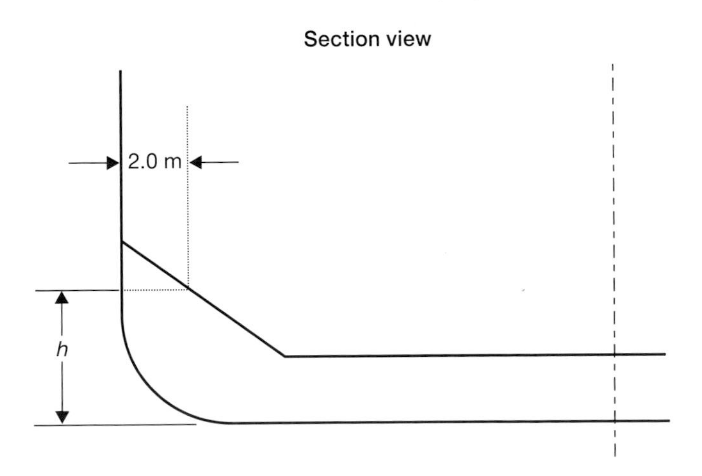

$PA_c = h \times \text{double bottom tank length} \times 2$*

* To include port and starboard.

Figure 8 – **Measurement of h for calculation of PA_c for double bottom tanks with sloping tank tops (2)**

Section view

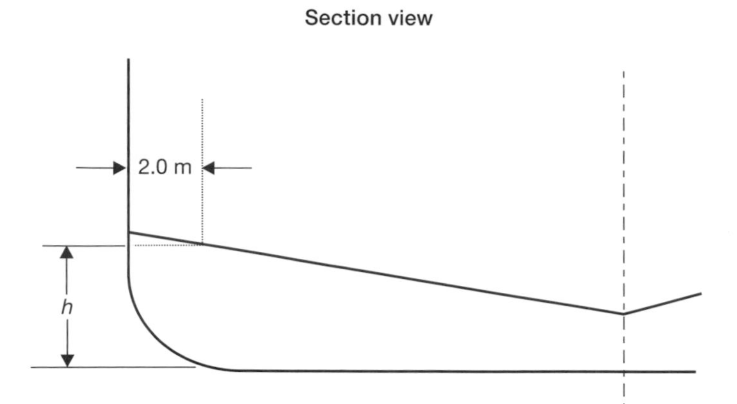

$PA_c \quad = \quad h \times \text{double bottom tank length} \times 2^*$

Figure 9 – **Calculation of PA_s for double bottom tank without clearly defined turn of bilge area – when wing tank is cargo tank**

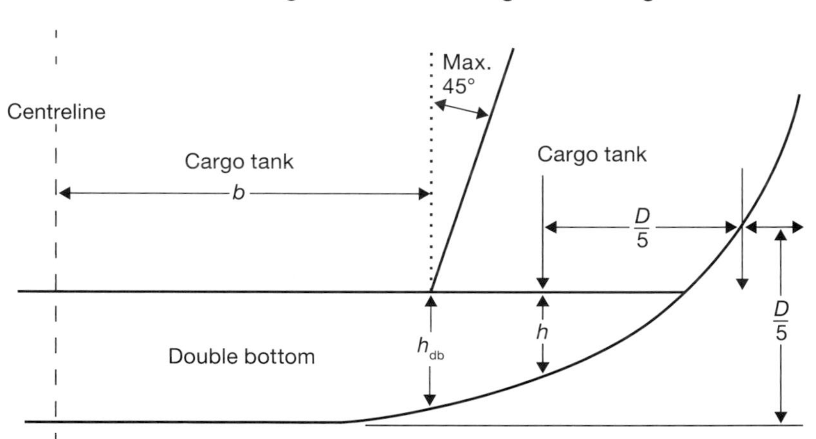

If h is less than 2 m or $\frac{B}{15}$, whichever is less, anywhere along the tank length, but h_{db} is at least 2 m or $\frac{B}{15}$, whichever is less, along the entire tank length within the width of $2b$, then:

$PA_s = 2b \times \text{cargo tank length}$

* To include port and starboard.

Figure 10 – **Calculation of PA_s for double bottom tank without clearly defined turn of bilge area – when wing tank is segregated ballast tank or void space**

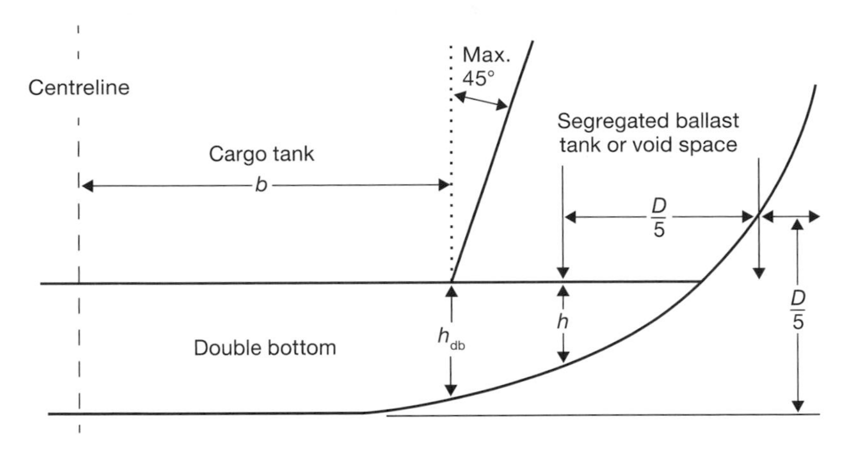

If h is less than 2 m or $\frac{B}{15}$, whichever is less, anywhere along the tank length,

but h_{db} is at least 2 m or $\frac{B}{15}$, whichever is less, along the entire tank length within the width of 2b, then:

$PA_s = B \times$ cargo tank length

Appendix 3

Connection of small diameter line to the manifold valve

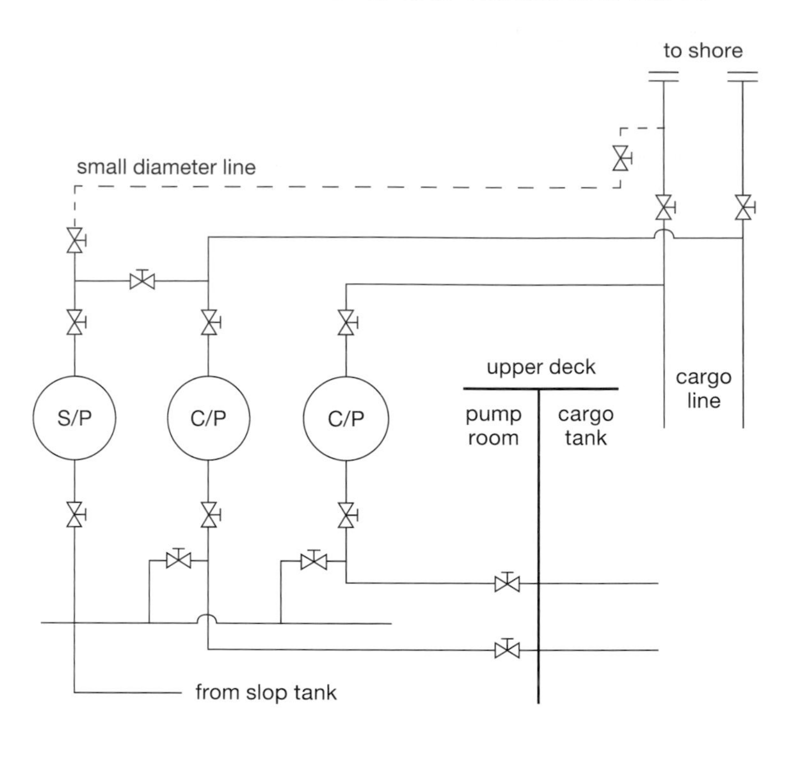

Appendix 4

Specifications for the design, installation and operation of a part flow system for control of overboard discharges

1 Purpose

1.1 The purpose of these specifications is to provide specific design criteria and installation and operational requirements for the part flow system referred to in regulation 30.6.5 of Annex I of the International Convention for the Prevention of Pollution from Ships, 1973, as modified by the Protocol of 1978 relating thereto (MARPOL).

2 Application

2.1 Oil tankers delivered on or before 31 December 1979, as defined in regulation 1.28.1, may, in accordance with regulation 30.6.5 of Annex I of MARPOL, discharge dirty ballast water and oil-contaminated water from cargo tank areas below the waterline, provided that a part of the flow is led through permanent piping to a readily accessible location on the upper deck or above where it may be visually observed during the discharge operation and provided that the arrangements comply with the requirements established by the Administration which shall at least contain all the provisions of these specifications.

2.2 The part flow concept is based on the principle that the observation of a representative part flow of the overboard effluent is equivalent to observing the entire effluent stream. These specifications provide the details of the design, installation and operation of a part flow system.

3 General provisions

3.1 The part flow system shall be so fitted that it can effectively provide a representative sample of the overboard effluent for visual display under all normal operating conditions.

3.2 The part flow system is in many respects similar to the sampling system for an oil discharge monitoring and control system but shall have pumping and piping arrangements separate from such a system, or combined equivalent arrangements acceptable to the Administration.

3.3 The display of the part flow shall be arranged in a sheltered and readily accessible location on the upper deck or above, approved by the Administration (e.g. the entrance to the pump-room). Regard should be given to effective communication between the location of the part flow display and the discharge control position.

3.4 Samples shall be taken from relevant sections of the overboard discharge piping and be passed to the display arrangement through a permanent piping system.

3.5 The part flow system shall include the following components:

- .1 sampling probes;
- .2 sample water piping system;
- .3 sample feed pump(s);
- .4 display arrangements;
- .5 sample discharge arrangements; and, subject to the diameter of the sample piping,
- .6 flushing arrangement.

3.6 The part flow system shall comply with the applicable safety requirements.

4 System arrangement

4.1 Sampling points

4.1.1 Sampling point location:

- .1 Sampling points shall be so located that relevant samples can be obtained of the effluent being discharged through outlets below the waterline which are used for operational discharges.
- .2 Sampling points shall as far as practicable be located in pipe sections where a turbulent flow is normally encountered.
- .3 Sampling points shall as far as practicable be arranged in accessible locations in vertical sections of the discharge piping.

4.1.2 Sampling probes:

- .1 Sampling probes shall be arranged to protrude into the pipe a distance of about one fourth of the pipe diameter.
- .2 Sampling probes shall be arranged for easy withdrawal for cleaning.

.3 The part flow system shall have a stop valve fitted adjacent to each probe, except that where the probe is mounted in a cargo line, two stop valves shall be fitted in series, in the sample line.

.4 Sampling probes should be of corrosion-resistant and oil-resistant material, of adequate strength, properly jointed and supported.

.5 Sampling probes shall have shape that is not prone to becoming clogged by particle contaminants and should not generate high hydrodynamic pressures at the sampling probe tip. Figure 1 is an example of one suitable shape of a sampling probe.

.6 Sampling probes shall have the same nominal bore as the sample piping.

4.2 **Sample piping**

.1 The sample piping shall be arranged as straight as possible between the sampling points and the display arrangement. Sharp bends and pockets where settled oil or sediment may accumulate should be avoided.

.2 The sample piping shall be so arranged that sample water is conveyed to the display arrangement within 20 s. The flow velocity in the piping should not be less than 2 m/s.

Figure 1 – **Sampling probe for a part flow display system**

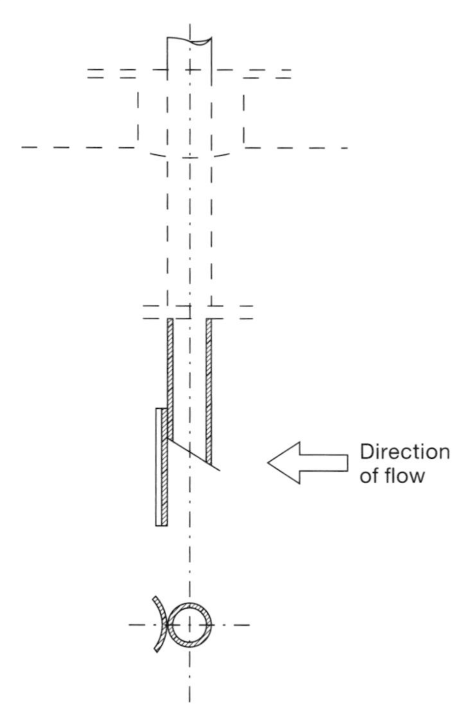

.3 The diameter of the piping shall not be less than 40 mm if no fixed flushing arrangement is provided and shall not be less than 25 mm if a pressurized flushing arrangement as detailed in paragraph 4.4 is installed.

.4 The sample piping should be of corrosion-resistant and oil-resistant material, of adequate strength, properly jointed and supported.

.5 Where several sampling points are installed, the piping shall be connected to a valve chest at the suction side of the sample feed pump.

4.3 **Sample feed pump**

.1 The sample feed pump capacity shall be suitable to allow the flow rate of the sample water to comply with 4.2.2.

4.4 **Flushing arrangement**

.1 If the diameter of sample piping is less than 40 mm, a fixed connection from a pressurized sea or fresh water piping system shall be installed for flushing of the sample piping system.

4.5 **Display arrangement**

.1 The display arrangement shall consist of a display chamber provided with a sight glass. The chamber should be of a size that will allow a free fall stream of the sample water to be clearly visible over a length of at least 200 mm. The Administration may approve equivalent arrangements.

.2 The display arrangement shall incorporate valves and piping in order to allow part of the sample flow to bypass the display chamber to obtain a laminar flow for display in the chamber.

.3 The display arrangement shall be designed to be easily opened and cleaned.

.4 The interior of the display chamber shall be white except for the background wall which shall be so coloured as to facilitate the observation of any change in the quality of the sample water.

.5 The lower part of the display chamber shall be shaped like a funnel for collection of the sample water.

.6 A test cock for taking a grab sample shall be provided in order that a sample of the water can be examined independent of that in the display chamber.

.7 The display arrangement shall be adequately lighted to facilitate visual observation of the sample water.

4.6 **Sample discharge arrangement**

.1 The sample water leaving the display chamber shall be routed to the sea or to a slop tank through fixed piping of adequate diameter.

5 Operation

5.1 When a discharge of dirty ballast water or other oil-contaminated water from the cargo tank area is taking place through an outlet below the waterline, the part flow system shall provide sample water from the relevant discharge outlet at all times.

5.2 The sample water should be observed particularly during those phases of the discharge operation when the greatest possibility of oil contamination occurs. The discharge shall be stopped whenever any traces of oil are visible in the flow and when the oil content meter reading indicates that the oil content exceeds permissible limits.

5.3 On those systems that are fitted with flushing arrangements, the sample piping should be flushed after contamination has been observed and, additionally, it is recommended that the sample piping be flushed after each period of usage.

5.4 The ship's cargo and ballast handling manuals and, where applicable, those manuals required for crude oil washing systems or dedicated clean ballast tanks operation shall clearly describe the use of the part flow system in conjunction with the ballast discharge and the slop tank decanting procedures.

Appendix 5
Discharges from fixed or floating platforms

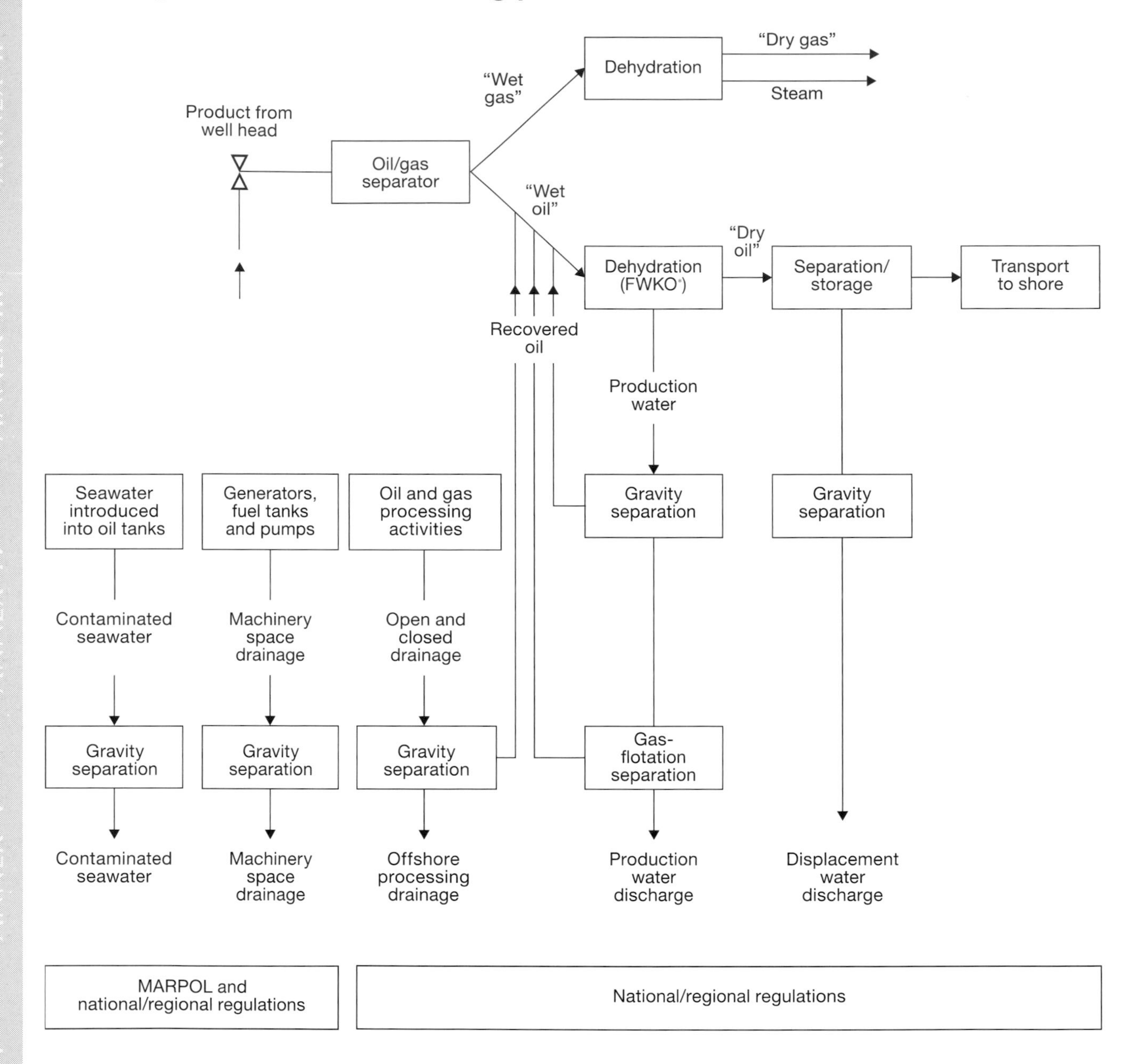

* FWKO means "free-water knock out".

MARPOL Annex II

Regulations for the control of pollution by noxious liquid substances in bulk

MARPOL Annex II

Regulations for the control of pollution by noxious liquid substances in bulk

Chapter 1 – General

Regulation 1
Definitions

For the purposes of this Annex:

1 *Anniversary date* means the day and the month of each year which will correspond to the date of expiry of the International Pollution Prevention Certificate for the Carriage of Noxious Liquid Substances in Bulk.

2 *Associated piping* means the pipeline from the suction point in a cargo tank to the shore connection used for unloading the cargo and includes all ship's piping, pumps and filters which are in open connection with the cargo unloading line.

3 *Ballast water*

Clean ballast means ballast water carried in a tank which, since it was last used to carry a cargo containing a substance in category X, Y or Z, has been thoroughly cleaned and the residues resulting therefrom have been discharged and the tank emptied in accordance with the appropriate requirements of this Annex.

Segregated ballast means ballast water introduced into a tank permanently allocated to the carriage of ballast or cargoes other than oil or noxious liquid substances as variously defined in the Annexes of the present Convention, and which is completely separated from the cargo and oil fuel system.

4 *Chemical Codes*

Bulk Chemical Code means the Code for the Construction and Equipment of Ships carrying Dangerous Chemicals in Bulk adopted by the Marine Environment Protection Committee of the Organization by resolution MEPC.20(22), as amended by the Organization, provided that such amendments are adopted and brought into force in accordance with the provisions of article 16 of the present Convention concerning amendment procedures applicable to an appendix to an Annex.

International Bulk Chemical Code means the International Code for the Construction and Equipment of Ships Carrying Dangerous Chemicals in Bulk adopted by the Marine Environment Protection Committee of the Organization by resolution MEPC.19(22), as amended by the Organization, provided that such amendments are adopted and brought into force in accordance with the provisions of article 16 of the present Convention concerning amendment procedures applicable to an appendix to an Annex.

5 *Depth of water* means the charted depth.

6 *En route* means that the ship is under way at sea on a course or courses, including deviation from the shortest direct route, which as far as practicable for navigational purposes, will cause any discharge to be spread over as great an area of the sea as is reasonable and practicable.

7 *Liquid substances* are those having a vapour pressure not exceeding 0.28 MPa absolute at a temperature of 37.8°C.

8 *Manual* means Procedures and Arrangements Manual in accordance with the model given in appendix IV of this Annex.

9 *Nearest land.* The term "from the nearest land" means from the baseline from which the territorial sea in question is established in accordance with international law, except that, for the purposes of the present Convention "from the nearest land" off the north-eastern coast of Australia shall mean from the line drawn from a point on the coast of Australia in:

latitude 11°00′ S, longitude 142°08′ E
to a point in latitude 10°35′ S, longitude 141°55′ E,
thence to a point latitude 10°00′ S, longitude 142°00′ E,
thence to a point latitude 09°10′ S, longitude 143°52′ E,
thence to a point latitude 09°00′ S, longitude 144°30′ E,
thence to a point latitude 10°41′ S, longitude 145°00′ E,
thence to a point latitude 13°00′ S, longitude 145°00′ E,
thence to a point latitude 15°00′ S, longitude 146°00′ E,
thence to a point latitude 17°30′ S, longitude 147°00′ E,
thence to a point latitude 21°00′ S, longitude 152°55′ E,
thence to a point latitude 24°30′ S, longitude 154°00′ E,
thence to a point on the coast of Australia
in latitude 24°42′ S, longitude 153°15′ E.

10 *Noxious liquid substance* means any substance indicated in the Pollution Category column of chapter 17 or 18 of the International Bulk Chemical Code or provisionally assessed under the provisions of regulation 6.3 as falling into category X, Y or Z.

11 *ppm* means mL/m^3.

12 *Residue* means any noxious liquid substance which remains for disposal.

13 *Residue/water mixture* means residue to which water has been added for any purpose (e.g. tank cleaning, ballasting, bilge slops).

14 *Ship construction*

14.1 *Ship constructed* means a ship the keel of which is laid or which is at a similar stage of construction. A ship converted to a chemical tanker, irrespective of the date of construction, shall be treated as a chemical tanker constructed on the date on which such conversion commenced. This conversion provision shall not apply to the modification of a ship which complies with all of the following conditions:

.1 the ship is constructed before 1 July 1986; and

.2 the ship is certified under the Bulk Chemical Code to carry only those products identified by the Code as substances with pollution hazards only.

14.2 *Similar stage of construction* means the stage at which:

.1 construction identifiable with a specific ship begins; and

.2 assembly of that ship has commenced comprising at least 50 tonnes or one per cent of the estimated mass of all structural material, whichever is less.

15 *Solidifying/non-solidifying*

15.1 *Solidifying substance* means a noxious liquid substance which:

.1 in the case of a substance with a melting point of less than 15°C, is at a temperature of less than 5°C above its melting point at the time of unloading; or

.2 in the case of a substance with a melting point of equal to or greater than 15°C, is at a temperature of less than 10°C above its melting point at the time of unloading.

15.2 *Non-solidifying substance* means a noxious liquid substance, which is not a solidifying substance.

16 *Tanker*

16.1 *Chemical tanker* means a ship constructed or adapted for the carriage in bulk of any liquid product listed in chapter 17 of the International Bulk Chemical Code.

16.2 *NLS tanker* means a ship constructed or adapted to carry a cargo of noxious liquid substances in bulk and includes an "oil tanker" as defined in Annex I of the present Convention when certified to carry a cargo or part cargo of noxious liquid substances in bulk.

17 *Viscosity*

17.1 *High-viscosity substance* means a noxious liquid substance in category X or Y with a viscosity equal to or greater than 50 mPa·s at the unloading temperature.

17.2 *Low-viscosity substance* means a noxious liquid substance which is not a high-viscosity substance.

18 *Audit* means a systematic, independent and documented process for obtaining audit evidence and evaluating it objectively to determine the extent to which audit criteria are fulfilled.

19 *Audit Scheme* means the IMO Member State Audit Scheme established by the Organization and taking into account the guidelines developed by the Organization.*

20 *Code for Implementation* means the IMO Instruments Implementation Code (III Code) adopted by the Organization by resolution A.1070(28).

21 *Audit Standard* means the Code for Implementation.

Regulation 2
Application

1 Unless expressly provided otherwise, the provisions of this Annex shall apply to all ships certified to carry noxious liquid substances in bulk.

2 Where a cargo subject to the provisions of Annex I of the present Convention is carried in a cargo space of an NLS tanker, the appropriate requirements of Annex I of the present Convention shall also apply.

Regulation 3
Exceptions

1 The discharge requirements of this Annex and chapter 2 of part II-A of the Polar Code shall not apply to the discharge into the sea of noxious liquid substances or mixtures containing such substances when such a discharge:

.1 is necessary for the purpose of securing the safety of a ship or saving life at sea; or

.2 results from damage to a ship or its equipment:

.2.1 provided that all reasonable precautions have been taken after the occurrence of the damage or discovery of the discharge for the purpose of preventing or minimizing the discharge; and

.2.2 except if the owner or the master acted either with intent to cause damage, or recklessly and with knowledge that damage would probably result; or

* Refer to the Framework and Procedures for the IMO Member State Audit Scheme (resolution A.1067(28)).

.3 is approved by the Administration, when being used for the purpose of combating specific pollution incidents in order to minimize the damage from pollution. Any such discharge shall be subject to the approval of any Government in whose jurisdiction it is contemplated the discharge will occur.

Regulation 4
Exemptions

1 With respect to amendments to carriage requirements due to the upgrading of the categorization of a substance, the following shall apply:

.1 where an amendment to this Annex and the International Bulk Chemical Code and Bulk Chemical Code involves changes to the structure or equipment and fittings due to the upgrading of the requirements for the carriage of certain substances, the Administration may modify or delay for a specified period the application of such an amendment to ships constructed before the date of entry into force of that amendment, if the immediate application of such an amendment is considered unreasonable or impracticable. Such relaxation shall be determined with respect to each substance;

.2 the Administration allowing a relaxation of the application of an amendment under this paragraph shall submit to the Organization a report giving details of the ship or ships concerned, the cargoes certified to carry, the trade in which each ship is engaged and the justification for the relaxation, for circulation to the Parties to the Convention for their information and appropriate action, if any, and reflect the exemption on the Certificate as referred to in regulation 7 or 9 of this Annex;

.3 Notwithstanding the above, an Administration may exempt ships from the carriage requirements under regulation 11 for ships certified to carry individually identified vegetable oils identified by the relevant footnote in chapter 17 of the IBC Code, provided the ship complies with the following conditions:

.3.1 subject to this regulation, the NLS tanker shall meet all requirements for ship type 3 as identified in the IBC Code except for cargo tank location;

.3.2 under this regulation, cargo tanks shall be located at the following distances inboard. The entire cargo tank length shall be protected by ballast tanks or spaces other than tanks that carry oil as follows:

.3.2.1 wing tanks or spaces shall be arranged such that cargo tanks are located inboard of the moulded line of the side shell plating nowhere less than 760 mm;

.3.2.2 double bottom tanks or spaces shall be arranged such that the distance between the bottom of the cargo tanks and the moulded line of the bottom shell plating measured at right angles to the bottom shell plating is not less than $B/15$ (m) or 2.0 m at the centreline, whichever is the lesser. The minimum distance shall be 1.0 m;

.3.3 the relevant certificate shall indicate the exemption granted.

2 Subject to the provisions of paragraph 3 of this regulation, the provisions of regulation 12.1 need not apply to a ship constructed before 1 July 1986 which is engaged in restricted voyages as determined by the Administration between:

.1 ports or terminals within a State Party to the present Convention; or

.2 ports or terminals of States Parties to the present Convention.

3 The provisions of paragraph 2 of this regulation shall only apply to a ship constructed before 1 July 1986 if:

.1 each time a tank containing category X, Y or Z substances or mixtures is to be washed or ballasted, the tank is washed in accordance with a prewash procedure approved by the Administration in compliance with appendix VI of this Annex, and the tank washings are discharged to a reception facility;

.2 subsequent washings or ballast water are discharged to a reception facility or at sea in accordance with other provisions of this Annex;

.3 the adequacy of the reception facilities at the ports or terminals referred to above, for the purpose of this paragraph, is approved by the Governments of the States Parties to the present Convention within which such ports or terminals are situated;

.4 in the case of ships engaged in voyages to ports or terminals under the jurisdiction of other States Parties to the present Convention, the Administration communicates to the Organization, for circulation to the Parties to the Convention, particulars of the exemption, for their information and appropriate action, if any; and

.5 the certificate required under this Annex is endorsed to the effect that the ship is solely engaged in such restricted voyages.

4 For a ship whose constructional and operational features are such that ballasting of cargo tanks is not required and cargo tank washing is only required for repair or dry-docking, the Administration may allow exemption from the provisions of regulation 12, provided that all of the following conditions are complied with:

.1 the design, construction and equipment of the ship are approved by the Administration, having regard to the service for which it is intended;

.2 any effluent from tank washings which may be carried out before a repair or dry-docking is discharged to a reception facility, the adequacy of which is ascertained by the Administration;

.3 the certificate required under this Annex indicates:

.3.1 that each cargo tank is certified for the carriage of a restricted number of substances which are comparable and can be carried alternately in the same tank without intermediate cleaning; and

.3.2 the particulars of the exemption;

.4 the ship carries a Manual approved by the Administration; and

.5 in the case of ships engaged in voyages to ports or terminals under the jurisdiction of other States Parties to the present Convention, the Administration communicates to the Organization, for circulation to the Parties to the Convention, particulars of the exemption, for their information and appropriate action, if any.

Regulation 5
Equivalents

1 The Administration may allow any fitting, material, appliance or apparatus to be fitted in a ship as an alternative to that required by this Annex if such fitting, material, appliance or apparatus is at least as effective as that required by this Annex. This authority of the Administration shall not extend to the substitution of operational methods to effect the control of discharge of noxious liquid substances as equivalent to those design and construction features which are prescribed by regulations in this Annex.

2 The Administration which allows a fitting, material, appliance or apparatus as alternative to that required by this Annex, under paragraph 1 of this regulation, shall communicate to the Organization, for circulation to the Parties to the Convention, particulars thereof, for their information and appropriate action, if any.

3 Notwithstanding the provisions of paragraphs 1 and 2 of this regulation, the construction and equipment of liquefied gas carriers certified to carry noxious liquid substances listed in the applicable Gas Carrier Code, shall be deemed to be equivalent to the construction and equipment requirements contained in regulations 11 and 12 of this Annex, provided that the gas carrier meets all following conditions:

.1 hold a Certificate of Fitness in accordance with the appropriate Gas Carrier Code for ships certified to carry liquefied gases in bulk;

.2 hold an International Pollution Prevention Certificate for the Carriage of Noxious Liquid Substances in Bulk, in which it is certified that the gas carrier may carry only those noxious liquid substances identified and listed in the appropriate Gas Carrier Code;

.3 be provided with segregated ballast arrangements;

.4 be provided with pumping and piping arrangements which, to the satisfaction of the Administration, ensure that the quantity of cargo residue remaining in the tank and its associated piping after unloading does not exceed the applicable quantity of residue as required by regulation 12.1, 12.2 or 12.3; and

.5 be provided with a Manual, approved by the Administration, ensuring that no operational mixing of cargo residues and water will occur and that no cargo residues will remain in the tank after applying the ventilation procedures prescribed in the Manual.

Chapter 2 – Categorization of noxious liquid substances

Regulation 6

Categorization and listing of noxious liquid substances and other substances

1 For the purpose of the regulations of this Annex, noxious liquid substances shall be divided into four categories as follows:

.1 Category X: Noxious liquid substances which, if discharged into the sea from tank cleaning or deballasting operations, are deemed to present a major hazard to either marine resources or human health and, therefore, justify the prohibition of the discharge into the marine environment;

.2 Category Y: Noxious liquid substances which, if discharged into the sea from tank cleaning or deballasting operations, are deemed to present a hazard to either marine resources or human health or cause harm to amenities or other legitimate uses of the sea and therefore justify a limitation on the quality and quantity of the discharge into the marine environment;

.3 Category Z: Noxious liquid substances which, if discharged into the sea from tank cleaning or deballasting operations, are deemed to present a minor hazard to either marine resources or human health and therefore justify less stringent restrictions on the quality and quantity of the discharge into the marine environment;

.4 Other substances: Substances indicated as OS (Other Substances) in the pollution category column of chapter 18 of the International Bulk Chemical Code which have been evaluated and found to fall outside category X, Y or Z as defined in regulation 6.1 of this Annex because they are, at present, considered to present no harm to marine resources, human health, amenities or other legitimate uses of the sea when discharged into the sea from tank cleaning or deballasting operations. The discharge of bilge or ballast water or other residues or mixtures containing only substances referred to as "Other Substances" shall not be subject to any requirements of the Annex.

2 Guidelines for use in the categorization of noxious liquid substances are given in appendix I to this Annex.

3 Where it is proposed to carry a liquid substance in bulk which has not been categorized under paragraph 1 of this regulation, the Governments of Parties to the Convention involved in the proposed operation shall establish and agree on a provisional assessment for the proposed operation on the basis of the guidelines referred to in paragraph 2 of this regulation. Until full agreement among the Governments involved has been reached, the substance shall not be carried. As soon as possible, but not later than 30 days after the agreement has been reached, the Government of the producing or shipping country, initiating the agreement concerned, shall notify the Organization and provide details of the substance and the provisional assessment for annual circulation to all Parties for their information. The Organization shall maintain a register of all such substances and their provisional assessment until such time as the substances are formally included in the IBC Code.

Chapter 3 – Surveys and certification

Regulation 7
Survey and certification of chemical tankers

Notwithstanding the provisions of regulations 8, 9, and 10 of this Annex, chemical tankers which have been surveyed and certified by States Parties to the present Convention in accordance with the provisions of the International Bulk Chemical Code or the Bulk Chemical Code, as applicable, shall be deemed to have complied with the provisions of the said regulations, and the certificate issued under that Code shall have the same force and receive the same recognition as the certificate issued under regulation 9 of this Annex.

Regulation 8
Surveys

1 Ships carrying noxious liquid substances in bulk shall be subject to the surveys specified below:

.1 An initial survey before the ship is put in service or before the Certificate required under regulation 9 of this Annex is issued for the first time, and which shall include a complete survey of its structure, equipment, systems, fittings, arrangements and material in so far as the ship is covered by this Annex. This survey shall be such as to ensure that the structure, equipment, systems, fittings, arrangements and material fully comply with the applicable requirements of this Annex.

.2 A renewal survey at intervals specified by the Administration, but not exceeding 5 years, except where regulation 10.2, 10.5, 10.6, or 10.7 of this Annex is applicable. The renewal survey shall be such as to ensure that the structure, equipment, systems, fittings, arrangements and material fully comply with applicable requirements of this Annex.

.3 An intermediate survey within 3 months before or after the second anniversary date or within 3 months before or after the third anniversary date of the Certificate which shall take the place of one of the annual surveys specified in paragraph 1.4 of this regulation. The intermediate survey shall be such as to ensure that the equipment and associated pump and piping systems fully comply with the applicable requirements of this Annex and are in good working order. Such intermediate surveys shall be endorsed on the Certificate issued under regulation 9 of this Annex.

.4 An annual survey within 3 months before or after each anniversary date of the Certificate including a general inspection of the structure, equipment, systems, fittings, arrangements and material referred to in paragraph 1.1 of this regulation to ensure that they have been maintained in accordance with paragraph 3 of this regulation and that they remain satisfactory for the service for which the ship is intended. Such annual surveys shall be endorsed on the Certificate issued under regulation 9 of this Annex.

.5 An additional survey either general or partial, according to the circumstances, shall be made after a repair resulting from investigations prescribed in paragraph 3 of this regulation, or whenever any important repairs or renewals are made. The survey shall be such as to ensure that the necessary repairs or renewals have been effectively made, that the material and workmanship of such repairs or renewals are in all respects satisfactory and that the ship complies in all respects with the requirements of this Annex.

2.1 Surveys of ships, as regards the enforcement of the provisions of this Annex, shall be carried out by officers of the Administration. The Administration may, however, entrust the surveys either to surveyors nominated for the purpose or to organizations recognized by it.

2.2 Such organizations, including classification societies, shall be authorized by the Administration in accordance with the provisions of the present Convention and with the Code for recognized organizations (RO Code), consisting of part 1 and part 2 (the provisions of which shall be treated as mandatory) and part 3 (the provisions of which shall be treated as recommendatory), as adopted by the Organization by resolution MEPC.237(65), as may be amended by the Organization, provided that:

- **.1** amendments to part 1 and part 2 of the RO Code are adopted, brought into force and take effect in accordance with the provisions of article 16 of the present Convention concerning the amendment procedures applicable to this annex;
- **.2** amendments to part 3 of the RO Code are adopted by the Marine Environment Protection Committee in accordance with its Rules of Procedure; and
- **.3** any amendments referred to in .1 and .2 adopted by the Maritime Safety Committee and the Marine Environment Protection Committee are identical and come into force or take effect at the same time, as appropriate.

2.3 An Administration nominating surveyors or recognizing organizations to conduct surveys as set forth in paragraph 2.1 of this regulation shall, as a minimum, empower any nominated surveyor or recognized organization to:

- **.1** require repairs to a ship; and
- **.2** carry out surveys if requested by the appropriate authorities of a port State.

2.4 The Administration shall notify the Organization of the specific responsibilities and conditions of the authority delegated to the nominated surveyors or recognized organizations, for circulation to Parties to the present Convention for the information of their officers.

2.5 When a nominated surveyor or recognized organization determines that the condition of the ship or its equipment does not correspond substantially with the particulars of the Certificate, or is such that the ship is not fit to proceed to sea without presenting an unreasonable threat of harm to the marine environment, such surveyor or organization shall immediately ensure that corrective action is taken and shall in due course notify the Administration. If such corrective action is not taken the Certificate should be withdrawn and the Administration shall be notified immediately, and if the ship is in a port of another Party, the appropriate authorities of the port State shall also be notified immediately. When an officer of the Administration, a nominated surveyor or a recognized organization has notified the appropriate authorities of the port State, the Government of the port State concerned shall give such officer, surveyor or organization any necessary assistance to carry out their obligations under this regulation. When applicable, the Government of the port State concerned shall take such steps as will ensure that the ship shall not sail until it can proceed to sea or leave the port for the purpose of proceeding to the nearest appropriate repair yard available without presenting an unreasonable threat of harm to the marine environment.

2.6 In every case, the Administration concerned shall fully guarantee the completeness and efficiency of the survey and shall undertake to ensure the necessary arrangements to satisfy this obligation.

3.1 The condition of the ship and its equipment shall be maintained to conform with the provisions of the present Convention to ensure that the ship in all respects will remain fit to proceed to sea without presenting an unreasonable threat of harm to the marine environment.

3.2 After any survey of the ship required under paragraph 1 of this regulation has been completed, no change shall be made in the structure, equipment, systems, fittings, arrangements or material covered by the survey, without the sanction of the Administration, except the direct replacement of such equipment and fittings.

3.3 Whenever an accident occurs to a ship or a defect is discovered which substantially affects the integrity of the ship or the efficiency or completeness of its equipment covered by this Annex, the master or owner of the ship shall report at the earliest opportunity to the Administration, the recognized organization or the nominated surveyor responsible for issuing the relevant Certificate, who shall cause investigations to be initiated to determine whether a survey as required by paragraph 1 of this regulation is necessary. If the ship is in a port of another Party, the master or owner shall also report immediately to the appropriate authorities of the port State and the nominated surveyor or recognized organization shall ascertain that such report has been made.

Regulation 9
Issue or endorsement of Certificate

1 An International Pollution Prevention Certificate for the Carriage of Noxious Liquid Substances in Bulk shall be issued, after an initial or renewal survey in accordance with the provisions of regulation 8 of this Annex, to any ship intended to carry noxious liquid substances in bulk and which is engaged in voyages to ports or terminals under the jurisdiction of other Parties to the Convention.

2 Such Certificate shall be issued or endorsed either by the Administration or by any person or organization duly authorized by it. In every case, the Administration assumes full responsibility for the Certificate.

3.1 The Government of a Party to the Convention may, at the request of the Administration, cause a ship to be surveyed and, if satisfied that the provisions of this Annex are complied with, shall issue or authorize the issue of an International Pollution Prevention Certificate for the Carriage of Noxious Liquid Substances in Bulk to the ship and, where appropriate, endorse or authorize the endorsement of that Certificate on the ship, in accordance with this Annex.

3.2 A copy of the Certificate and a copy of the survey report shall be transmitted as soon as possible to the requesting Administration.

3.3 A Certificate so issued shall contain a statement to the effect that it has been issued at the request of the Administration and it shall have the same force and receive the same recognition as the Certificate issued under paragraph 1 of this regulation.

3.4 No International Pollution Prevention Certificate for the Carriage of Noxious Liquid Substances in Bulk shall be issued to a ship which is entitled to fly the flag of a State which is not a party.

4 The International Pollution Prevention Certificate for the Carriage of Noxious Liquid Substances in Bulk shall be drawn up in the form corresponding to the model given in appendix III to this Annex and shall be at least in English, French or Spanish. Where entries in an official national language of the State whose flag the ship is entitled to fly are also used, this shall prevail in the case of a dispute or discrepancy.

Regulation 10
Duration and validity of Certificate

1 An International Pollution Prevention Certificate for the Carriage of Noxious Liquid Substances in Bulk shall be issued for a period specified by the Administration which shall not exceed 5 years.

2.1 Notwithstanding the requirements of paragraph 1 of this regulation, when the renewal survey is completed within 3 months before the expiry date of the existing Certificate, the new Certificate shall be valid from the date of completion of the renewal survey to a date not exceeding 5 years from the date of expiry of the existing Certificate.

2.2 When the renewal survey is completed after the expiry date of the existing Certificate, the new Certificate shall be valid from the date of completion of the renewal survey to a date not exceeding 5 years from the date of expiry of the existing Certificate.

2.3 When the renewal survey is completed more than 3 months before the expiry date of the existing Certificate, the new Certificate shall be valid from the date of completion of the renewal survey to a date not exceeding 5 years from the date of completion of the renewal survey.

3 If a Certificate is issued for a period of less than 5 years, the Administration may extend the validity of the Certificate beyond the expiry date to the maximum period specified in paragraph 1 of this regulation, provided that the surveys referred to in regulation 8.1.3 and 8.1.4 of this Annex applicable when a Certificate is issued for a period of 5 years are carried out as appropriate.

4 If a renewal survey has been completed and a new Certificate cannot be issued or placed on board the ship before the expiry date of the existing Certificate, the person or organization authorized by the Administration may endorse the existing Certificate and such a Certificate shall be accepted as valid for a further period which shall not exceed 5 months from the expiry date.

5 If a ship at the time when a Certificate expires is not in a port in which it is to be surveyed, the Administration may extend the period of validity of the Certificate but this extension shall be granted only for the purpose of allowing the ship to complete its voyage to the port in which it is to be surveyed, and then only in cases where it appears proper and reasonable to do so. No Certificates shall be extended for a period longer than 3 months, and a ship to which an extension is granted shall not, on its arrival in the port in which it is to be surveyed, be entitled by virtue of such extension to leave that port without having a new Certificate. When the renewal survey is completed, the new Certificate shall be valid to a date not exceeding 5 years from the date of expiry of the existing Certificate before the extension was granted.

6 A Certificate issued to a ship engaged on short voyages which has not been extended under the foregoing provisions of this regulation may be extended by the Administration for a period of grace of up to one month from the date of expiry stated on it. When the renewal survey is completed, the new Certificate shall be valid to a date not exceeding 5 years from the date of expiry of the existing Certificate before the extension was granted.

7 In special circumstances, as determined by the Administration, a new Certificate need not be dated from the date of expiry of the existing Certificate as required by paragraph 2.2, 5 or 6 of this regulation. In these special circumstances, the new Certificate shall be valid to a date not exceeding 5 years from the date of completion of the renewal survey.

8 If an annual or intermediate survey is completed before the period specified in regulation 8 of this Annex, then:

- **.1** the anniversary date shown on the Certificate shall be amended by endorsement to a date which shall not be more than 3 months later than the date on which the survey was completed;
- **.2** the subsequent annual or intermediate survey required by regulation 8 of this Annex shall be completed at the intervals prescribed by that regulation using the new anniversary date;
- **.3** the expiry date may remain unchanged provided one or more annual or intermediate surveys, as appropriate, are carried out so that the maximum intervals between the surveys prescribed by regulation 8 of this Annex are not exceeded.

9 A Certificate issued under regulation 9 of this Annex shall cease to be valid in any of the following cases:

- **.1** if the relevant surveys are not completed within the periods specified under regulation 8.1 of this Annex;
- **.2** if the Certificate is not endorsed in accordance with regulation 8.1.3 or 8.1.4 of this Annex;
- **.3** upon transfer of the ship to the flag of another State. A new Certificate shall only be issued when the Government issuing the new Certificate is fully satisfied that the ship is in compliance with the requirements of regulation 8.3.1 and 8.3.2 of this Annex. In the case of a transfer between Parties, if requested within 3 months after the transfer has taken place, the Government of the Party whose flag the ship was formerly entitled to fly shall, as soon as possible, transmit to the Administration copies of the Certificate carried by the ship before the transfer and, if available, copies of the relevant survey reports.

Chapter 4 – Design, construction, arrangement and equipment

Regulation 11

Design, construction, equipment and operations

1 The design, construction, equipment and operation of ships certified to carry noxious liquid substances in bulk identified in chapter 17 of the International Bulk Chemical Code, shall be in compliance with the following provisions to minimize the uncontrolled discharge into the sea of such substances:

.1 the International Bulk Chemical Code when the chemical tanker is constructed on or after 1 July 1986; or

.2 the Bulk Chemical Code as referred to in paragraph 1.7.2 of that Code for:

.2.1 ships for which the building contract is placed on or after 2 November 1973 but constructed before 1 July 1986, and which are engaged on voyages to ports or terminals under the jurisdiction of other States Parties to the Convention; and

.2.2 ships constructed on or after 1 July 1983 but before 1 July 1986, which are engaged solely on voyages between ports or terminals within the State the flag of which the ship is entitled to fly.

.3 The Bulk Chemical Code as referred to in paragraph 1.7.3 of that Code for:

.3.1 ships for which the building contract is placed before 2 November 1973 and which are engaged on voyages to ports or terminals under the jurisdiction of other States Parties to the Convention; and

.3.2 ships constructed before 1 July 1983 which are solely engaged on voyages between ports or terminals within the State the flag of which the ship is entitled to fly.

2 In respect of ships other than chemical tankers or liquefied gas carriers certified to carry noxious liquid substances in bulk identified in chapter 17 of the International Bulk Chemical Code, the Administration shall establish appropriate measures based on the Guidelines* developed by the Organization in order to ensure that the provisions shall be such as to minimize the uncontrolled discharge into the sea of such substances.

Regulation 12

Pumping, piping, unloading arrangements and slop tanks

1 Every ship constructed before 1 July 1986 shall be provided with a pumping and piping arrangement to ensure that each tank certified for the carriage of substances in category X or Y does not retain a quantity of residue in excess of 300 L in the tank and its associated piping and that each tank certified for the carriage of substances in category Z does not retain a quantity of residue in excess of 900 L in the tank and its associated piping. A performance test shall be carried out in accordance with appendix V of this Annex.

2 Every ship constructed on or after 1 July 1986 but before 1 January 2007 shall be provided with a pumping and piping arrangement to ensure that each tank certified for the carriage of substances in category X or Y does not retain a quantity of residue in excess of 100 L in the tank and its associated piping and that each tank certified for the carriage of substances in category Z does not retain a quantity of residue in excess of 300 L in the tank and its associated piping. A performance test shall be carried out in accordance with appendix V of this Annex.

* Refer to resolution A.673(16), as amended by resolution MEPC.158(55), and to resolution MEPC.148(54).

3 Every ship constructed on or after 1 January 2007 shall be provided with a pumping and piping arrangement to ensure that each tank certified for the carriage of substances in category X, Y or Z does not retain a quantity of residue in excess of 75 L in the tank and its associated piping. A performance test shall be carried out in accordance with appendix V of this Annex.

4 For a ship other than a chemical tanker constructed before 1 January 2007 which cannot meet the requirements for the pumping and piping arrangements for substances in category Z referred to in paragraphs 1 and 2 of this regulation no quantity requirement shall apply. Compliance is deemed to be reached if the tank is emptied to the most practicable extent.

5 Pumping performance tests referred to in paragraphs 1, 2 and 3 of this regulation shall be approved by the Administration. Pumping performance tests shall use water as the test medium.

6 Ships certified to carry substances of category X, Y or Z shall have an underwater discharge outlet (or outlets).

7 For ships constructed before 1 January 2007 and certified to carry substances in category Z an underwater discharge outlet as required under paragraph 6 of this regulation is not mandatory.

8 The underwater discharge outlet (or outlets) shall be located within the cargo area in the vicinity of the turn of the bilge and shall be so arranged as to avoid the re-intake of residue/water mixtures by the ship's seawater intakes.

9 The underwater discharge outlet arrangement shall be such that the residue/water mixture discharged into the sea will not pass through the ship's boundary layer. To this end, when the discharge is made normal to the ship's shell plating, the minimum diameter of the discharge outlet is governed by the following equation:

$$d = \frac{Q_d}{5L_d}$$

where

d = minimum diameter of the discharge outlet (m)

L_d = distance from the forward perpendicular to the discharge outlet (m)

Q_d = the maximum rate selected at which the ship may discharge a residue/water mixture through the outlet (m^3/h).

10 When the discharge is directed at an angle to the ship's shell plating, the above relationship shall be modified by substituting for Q_d the component of Q_d which is normal to the ship's shell plating.

11 *Slop tanks*

Although this Annex does not require the fitting of dedicated slop tanks, slop tanks may be needed for certain washing procedures. Cargo tanks may be used as slop tanks.

Chapter 5 – Operational discharges of residues of noxious liquid substances

Regulation 13
Control of discharges of residues of noxious liquid substances

Subject to the provisions of regulation 3 of this Annex, the control of discharges of residues of noxious liquid substances or ballast water, tank washings or other mixtures containing such substances shall be in compliance with the following requirements.

1 *Discharge provisions*

1.1 The discharge into the sea of residues of substances assigned to category X, Y or Z or of those provisionally assessed as such or ballast water, tank washings or other mixtures containing such substances shall be prohibited unless such discharges are made in full compliance with the applicable operational requirements contained in this Annex.

1.2 Before any prewash or discharge procedure is carried out in accordance with this regulation, the relevant tank shall be emptied to the maximum extent in accordance with the procedures prescribed in the Manual.

1.3 The carriage of substances which have not been categorized, provisionally assessed or evaluated as referred to in regulation 6 of this Annex or of ballast water, tank washings or other mixtures containing such residues shall be prohibited along with any consequential discharge of such substances into the sea.

2 *Discharge standards*

2.1 Where the provisions in this regulation allow the discharge into the sea of residues of substances in category X, Y or Z or of those provisionally assessed as such or ballast water, tank washings or other mixtures containing such substances, the following discharge standards shall apply:

.1 the ship is proceeding en route at a speed of at least 7 knots in the case of self-propelled ships or at least 4 knots in the case of ships which are not self-propelled;

.2 the discharge is made below the waterline through the underwater discharge outlet(s) not exceeding the maximum rate for which the underwater discharge outlet(s) is (are) designed; and

.3 the discharge is made at a distance of not less than 12 nautical miles from the nearest land in a depth of water of not less than 25 m.

2.2 For ships constructed before 1 January 2007 the discharge into the sea of residues of substances in category Z or of those provisionally assessed as such or ballast water, tank washings or other mixtures containing such substances below the waterline is not mandatory.

2.3 The Administration may waive the requirements of paragraph 2.1.3 for substances in category Z, regarding the distance of not less than 12 nautical miles from the nearest land for ships solely engaged in voyages within waters subject to the sovereignty or jurisdiction of the State the flag of which the ship is entitled to fly. In addition, the Administration may waive the same requirement regarding the discharge distance of not less than 12 nautical miles from the nearest land for a particular ship entitled to fly the flag of their State, when engaged in voyages within waters subject to the sovereignty or jurisdiction of one adjacent State after the establishment of an agreement, in writing, of a waiver between the two coastal States involved provided that no third party will be affected. Information on such agreement shall be communicated to the Organization within 30 days for further circulation to the Parties to the Convention for their information and appropriate action if any.

3 *Ventilation of cargo residues*

Ventilation procedures approved by the Administration may be used to remove cargo residues from a tank. Such procedures shall be in accordance with appendix VII of this Annex. Any water subsequently introduced into the tank shall be regarded as clean and shall not be subject to the discharge requirements in this Annex.

4 *Exemption for a prewash*

On request of the ship's master, an exemption for a prewash may be granted by the Government of the receiving Party, where it is satisfied that:

.1 the unloaded tank is to be reloaded with the same substance or another substance compatible with the previous one and that the tank will not be washed or ballasted prior to loading; or

.2 the unloaded tank is neither washed nor ballasted at sea. The prewash in accordance with the applicable paragraph of this regulation shall be carried out at another port provided that it has been confirmed in writing that a reception facility at that port is available and is adequate for such a purpose; or

.3 the cargo residues will be removed by a ventilation procedure approved by the Administration in accordance with appendix VII of this Annex.

5 *The use of cleaning agents or additives*

5.1 When a washing medium other than water, such as mineral oil or chlorinated solvent, is used instead of water to wash a tank, its discharge shall be governed by the provisions of either Annex I or Annex II which would apply to the medium had it been carried as cargo. Tank washing procedures involving the use of such a medium shall be set out in the Manual and be approved by the Administration.

5.2 When small amounts of cleaning additives (detergent products) are added to water in order to facilitate tank washing, no additives containing pollution category X components shall be used except those components that are readily biodegradable and present in a total concentration of less than 10% of the cleaning additive. No restrictions additional to those applicable to the tank due to the previous cargo shall apply.

6 *Discharge of residues of category X*

6.1 Subject to the provision of paragraph 1, the following provisions shall apply:

.1 A tank from which a substance in category X has been unloaded shall be prewashed before the ship leaves the port of unloading. The resulting residues shall be discharged to a reception facility until the concentration of the substance in the effluent to such facility, as indicated by analyses of samples of the effluent taken by the surveyor, is at or below 0.1% by weight. When the required concentration level has been achieved, remaining tank washings shall continue to be discharged to the reception facility until the tank is empty. Appropriate entries of these operations shall be made in the Cargo Record Book and endorsed by the surveyor referred to in regulation 16.1.

.2 Any water subsequently introduced into the tank may be discharged into the sea in accordance with the discharge standards in regulation 13.2.

.3 Where the Government of the receiving party is satisfied that it is impracticable to measure the concentration of the substance in the effluent without causing undue delay to the ship, that Party may accept an alternative procedure as being equivalent to obtain the required concentration in regulation 13.6.1.1 provided that:

.3.1 the tank is prewashed in accordance with a procedure approved by the Administration in compliance with appendix VI of this Annex; and

.3.2 appropriate entries shall be made in the Cargo Record Book and endorsed by the surveyor referred to in regulation 16.1.

ANNEX I ANNEX II ANNEX III ANNEX IV ANNEX V ANNEX VI UNIFIED INTERPRETATIONS

7 *Discharge of residues of category Y and Z*

7.1 Subject to the provision of paragraph 1, the following provisions shall apply:

.1 With respect to the residue discharge procedures for substances in category Y or Z, the discharge standards in regulation 13.2 shall apply.

.2 If the unloading of a substance of category Y or Z is not carried out in accordance with the Manual, a prewash shall be carried out before the ship leaves the port of unloading, unless alternative measures are taken to the satisfaction of the surveyor referred to in regulation 16.1 of this Annex to remove the cargo residues from the ship to quantities specified in this Annex. The resulting tank washings of the prewash shall be discharged to a reception facility at the port of unloading or another port with a suitable reception facility provided that it has been confirmed in writing that a reception facility at that port is available and is adequate for such a purpose.

.3 For high-viscosity or solidifying substances in category Y, the following shall apply:

.3.1 a prewash procedure as specified in appendix VI shall be applied;

.3.2 the residue/water mixture generated during the prewash shall be discharged to a reception facility until the tank is empty; and

.3.3 any water subsequently introduced into the tank may be discharged into the sea in accordance with the discharge standards in regulation 13.2.

7.2 Operational requirements for ballasting and deballasting

7.2.1 After unloading, and, if required, after a prewash, a cargo tank may be ballasted. Procedures for the discharge of such ballast are set out in regulation 13.2.

7.2.2 Ballast introduced into a cargo tank which has been washed to such an extent that the ballast contains less than 1 ppm of the substance previously carried may be discharged into the sea without regard to the discharge rate, ship's speed and discharge outlet location, provided that the ship is not less than 12 nautical miles from the nearest land and in water that is not less than 25 m deep. The required degree of cleanliness has been achieved when a prewash as specified in appendix VI has been carried out and the tank has been subsequently washed with a complete cycle of the cleaning machine for ships built before 1 July 1994 or with a water quantity not less than that calculated with $k = 1.0$.

7.2.3 The discharge into the sea of clean or segregated ballast shall not be subject to the requirements of this Annex.

8 *Discharges in the Antarctic Area*

8.1 *Antarctic Area* means the sea area south of latitude 60° S.

8.2 In the Antarctic Area any discharge into the sea of noxious liquid substances or mixtures containing such substances is prohibited.

Regulation 14
Procedures and Arrangements Manual

1 Every ship certified to carry substances of category X, Y or Z shall have on board a Manual approved by the Administration. The Manual shall have a standard format in compliance with appendix IV to this Annex. In the case of a ship engaged in international voyages on which the language used is not English, French or Spanish, the text shall include a translation into one of these languages.

2 The main purpose of the Manual is to identify for the ship's officers the physical arrangements and all the operational procedures with respect to cargo handling, tank cleaning, slops handling and cargo tank ballasting and deballasting which must be followed in order to comply with the requirements of this Annex.

Regulation 15
Cargo Record Book

1 Every ship to which this Annex applies shall be provided with a Cargo Record Book, whether as part of the ship's official logbook or otherwise, in the form specified in appendix II to this Annex.

2 After completion of any operation specified in appendix II to this Annex, the operation shall be promptly recorded in the Cargo Record Book.

3 In the event of an accidental discharge of a noxious liquid substance or a mixture containing such a substance or a discharge under the provisions of regulation 3 of this Annex, an entry shall be made in the Cargo Record Book stating the circumstances of, and the reason for, the discharge.

4 Each entry shall be signed by the officer or officers in charge of the operation concerned and each page shall be signed by the master of the ship. The entries in the Cargo Record Book, for ships holding an International Pollution Prevention Certificate for the Carriage of Noxious Liquid Substances in Bulk or a certificate referred to in regulation 7 of this Annex, shall be at least in English, French or Spanish. Where entries in an official national language of the State whose flag the ship is entitled to fly are also used, this shall prevail in case of a dispute or discrepancy.

5 The Cargo Record Book shall be kept in such a place as to be readily available for inspection and, except in the case of unmanned ships under tow, shall be kept on board the ship. It shall be retained for a period of three years after the last entry has been made.

6 The competent authority of the Government of a Party may inspect the Cargo Record Book on board any ship to which this Annex applies while the ship is in its port, and may make a copy of any entry in that book and may require the master of the ship to certify that the copy is a true copy of such entry. Any copy so made which has been certified by the master of the ship as a true copy of an entry in the ship's Cargo Record Book shall be made admissible in any judicial proceedings as evidence of the facts stated in the entry. The inspection of a Cargo Record Book and the taking of a certified copy by the competent authority under this paragraph shall be performed as expeditiously as possible without causing the ship to be unduly delayed.

ANNEX I ANNEX II ANNEX III ANNEX IV ANNEX V ANNEX VI UNIFIED INTERPRETATIONS

Chapter 6 – Measures of control by port States

Regulation 16
Measures of control

1 The Government of each Party to the Convention shall appoint or authorize surveyors for the purpose of implementing this regulation. The surveyors shall execute control in accordance with control procedures developed by the Organization.*

2 When a surveyor appointed or authorized by the Government of the Party to the Convention has verified that an operation has been carried out in accordance with the requirements of the Manual, or has granted an exemption for a prewash, then that surveyor shall make an appropriate entry in the Cargo Record Book.

3 The master of a ship certified to carry noxious liquid substances in bulk shall ensure that the provisions of regulation 13 and of this regulation and chapter 2 of part II-A of the Polar Code when the ship is operating in Arctic waters have been complied with and that the Cargo Record Book is completed in accordance with regulation 15 whenever operations as referred to in that regulation take place.

4 A tank which has carried a category X substance shall be prewashed in accordance with regulation 13.6. The appropriate entries of these operations shall be made in the Cargo Record Book and endorsed by the surveyor referred to under paragraph 1 of this regulation.

5 Where the Government of the receiving party is satisfied that it is impracticable to measure the concentration of the substance in the effluent without causing undue delay to the ship, that Party may accept the alternative procedure referred to in regulation 13.6.3 provided that the surveyor referred to under paragraph 1 of this regulation certifies in the Cargo Record Book that:

.1 the tank, its pump and piping systems have been emptied; and

.2 the prewash has been carried out in accordance with the provisions of appendix VI of this Annex; and

.3 the tank washings resulting from such prewash have been discharged to a reception facility and the tank is empty.

6 At the request of the ship's master, the Government of the receiving Party may exempt the ship from the requirements for a prewash referred to in the applicable paragraphs of regulation 13 when one of the conditions of regulation 13.4 is met.

7 An exemption referred to in paragraph 6 of this regulation may only be granted by the Government of the receiving Party to a ship engaged in voyages to ports or terminals under the jurisdiction of other States Parties to the present Convention. When such an exemption has been granted, the appropriate entry made in the Cargo Record Book shall be endorsed by the surveyor referred to in paragraph 1 of this regulation.

8 If the unloading is not carried out in accordance with the pumping conditions for the tank approved by the Administrations and based on appendix V of this Annex, alternative measures may be taken to the satisfaction of the surveyor referred to in paragraph 1 of this regulation to remove the cargo residues from the ship to quantities specified in regulation 12 as applicable. The appropriate entries shall be made in the Cargo Record Book.

* Refer to Procedures for port State control, 2011 (resolution A.1052(27)).

9 *Port State control on operational requirements*[*]

9.1 A ship when in a port of another Party is subject to inspection by officers duly authorized by such Party concerning operational requirements under this Annex, where there are clear grounds for believing that the master or crew are not familiar with essential shipboard procedures relating to the prevention of pollution by noxious liquid substances.

9.2 In the circumstances given in paragraph 9.1 of this regulation, the Party shall take such steps as will ensure that the ship shall not sail until the situation has been brought to order in accordance with the requirements of this Annex.

9.3 Procedures relating to the port State control prescribed in article 5 of the present Convention shall apply to this regulation.

9.4 Nothing in this regulation shall be construed to limit the rights and obligations of a Party carrying out control over operational requirements specifically provided for in the present Convention.

[*] Refer to Procedures for port State control, 2011 (resolution A.1052(27)).

Chapter 7 – Prevention of pollution arising from an incident involving noxious liquid substances

Regulation 17
Shipboard marine pollution emergency plan for noxious liquid substances

1 Every ship of 150 gross tonnage and above certified to carry noxious liquid substances in bulk shall carry on board a shipboard marine pollution emergency plan for noxious liquid substances approved by the Administration.

2 Such a plan shall be based on the Guidelines* developed by the Organization and written in a working language or languages understood by the master and officers. The plan shall consist at least of:

.1 the procedure to be followed by the master or other persons having charge of the ship to report a noxious liquid substances pollution incident, as required in article 8 and Protocol I of the present Convention, based on the Guidelines developed by the Organization;†

.2 the list of authorities or persons to be contacted in the event of a noxious liquid substances pollution incident;

.3 a detailed description of the action to be taken immediately by persons on board to reduce or control the discharge of noxious liquid substances following the incident; and

.4 the procedures and point of contact on the ship for coordinating shipboard action with national and local authorities in combating the pollution.

3 In the case of ships to which regulation 37 of Annex I of the Convention also applies, such a plan may be combined with the shipboard oil pollution emergency plan required under regulation 37 of Annex I of the Convention. In this case, the title of such a plan shall be "Shipboard marine pollution emergency plan".

---

* Refer to Guidelines for the development of shipboard marine pollution emergency plans for oil and/or noxious liquid substances (resolution MEPC.85(44), as amended by resolution MEPC.137(53)).

† Refer to General principles for ship reporting systems and ship reporting requirements, including guidelines for reporting incidents involving dangerous goods, harmful substances and/or marine pollutants (resolution A.851(20), as amended by resolution MEPC.138(53)).

Chapter 8 – Reception facilities

Regulation 18

Reception facilities and cargo unloading terminal arrangements

1 The Government of each Party to the Convention undertakes to ensure the provision of reception facilities according to the needs of ships using its ports, terminals or repair ports as follows:

- **.1** ports and terminals involved in ships' cargo handling shall have adequate facilities for the reception of residues and mixtures containing such residues of noxious liquid substances resulting from compliance with this Annex, without undue delay for the ships involved.
- **.2** ship repair ports undertaking repairs to NLS tankers shall provide facilities adequate for the reception of residues and mixtures containing noxious liquid substances for ships calling at that port.

2 The Government of each Party shall determine the types of facilities provided for the purpose of paragraph 1 of this regulation at each cargo loading and unloading port, terminal and ship repair port in its territories and notify the Organization thereof.

3 Small Island Developing States may satisfy the requirements in paragraphs 1, 2 and 6 of this regulation through regional arrangements when, because of those States' unique circumstances, such arrangements are the only practical means to satisfy these requirements. Parties participating in a regional arrangement shall develop a Regional Reception Facilities Plan, taking into account the guidelines developed by the Organization.

The Government of each Party participating in the arrangement shall consult with the Organization for circulation to the Parties of the present Convention:

- **.1** how the Regional Reception Facilities Plan takes into account the Guidelines;
- **.2** particulars of the identified Regional Ships Waste Reception Centres; and
- **.3** particulars of those ports with only limited facilities.

4 Where regulation 13 of this annex requires a prewash and the Regional Reception Facility Plan is applicable to the port of unloading, the prewash and subsequent discharge to a reception facility shall be carried out as prescribed in regulation 13 of this annex or at a Regional Ship Waste Reception Centre specified in the applicable Regional Reception Facility Plan.

5 The Governments of Parties to the Convention, the coastlines of which border on any given special area, shall collectively agree and establish a date by which time the requirement of paragraph 1 of this regulation will be fulfilled and from which the requirements of the applicable paragraphs of regulation 13 in respect of that area shall take effect and notify the Organization of the date so established at least six months in advance of that date. The Organization shall then promptly notify all Parties of that date.

6 The Government of each Party to the Convention shall undertake to ensure that cargo unloading terminals shall provide arrangements to facilitate stripping of cargo tanks of ships unloading noxious liquid substances at these terminals. Cargo hoses and piping systems of the terminal, containing noxious liquid substances received from ships unloading these substances at the terminal, shall not be drained back to the ship.

7 Each Party shall notify the Organization, for transmission to the Parties concerned, of any case where facilities required under paragraph 1 or arrangements required under paragraph 6 of this regulation are alleged to be inadequate.

Chapter 9 – Verification of compliance with the provisions of this Convention

Regulation 19
Application

Parties shall use the provisions of the Code for Implementation in the execution of their obligations and responsibilities contained in this Annex.

Regulation 20
Verification of compliance

1 Every Party shall be subject to periodic audits by the Organization in accordance with the audit standard to verify compliance with and implementation of this Annex.

2 The Secretary-General of the Organization shall have responsibility for administering the Audit Scheme, based on the guidelines developed by the Organization.

3 Every Party shall have responsibility for facilitating the conduct of the audit and implementation of a programme of actions to address the findings, based on the guidelines adopted by the Organization.*

4 Audit of all Parties shall be:

.1 based on an overall schedule developed by the Secretary-General of the Organization, taking into account the guidelines developed by the Organization; and

.2 conducted at periodic intervals, taking into account the guidelines developed by the Organization.

---

* Refer to the Framework and Procedures for the IMO Member State Audit Scheme (resolution A.1067(28)).

Chapter 10 – International Code for Ships Operating in Polar Waters

Regulation 21
Definitions

For the purpose of this Annex,

1 *Polar Code* means the International Code for Ships Operating in Polar Waters, consisting of an introduction, part I-A and part II-A and parts I-B and II-B, as adopted by resolutions MSC.385(94) and MEPC.264(68), as may be amended, provided that:

- **.1** amendments to the environment-related provisions of the introduction and chapter 2 of part II-A of the Polar Code are adopted, brought into force and take effect in accordance with the provisions of article 16 of the present Convention concerning the amendment procedures applicable to an appendix to an annex; and
- **.2** amendments to part II-B of the Polar Code are adopted by the Marine Environment Protection Committee in accordance with its Rules of Procedure.

2 *Arctic waters* means those waters which are located north of a line from the latitude 58°00′.0 N and longitude 042°00′.0 W to latitude 64°37′.0 N, longitude 035°27′.0 W and thence by a rhumb line to latitude 67°03′.9 N, longitude 026°33′.4 W and thence by a rhumb line to the latitude 70°49′.56 N and longitude 008°59′.61 W (Sørkapp, Jan Mayen) and by the southern shore of Jan Mayen to 73°31′.6 N and 019°01′.0 E by the Island of Bjørnøya, and thence by a great circle line to the latitude 68°38′.29 N and longitude 043°23′.08 E (Cap Kanin Nos) and thence by the northern shore of the Asian Continent eastward to the Bering Strait and thence from the Bering Strait westward to latitude 60° N as far as Il'pyrskiy and following the 60th North parallel eastward as far as and including Etolin Strait and thence by the northern shore of the North American continent as far south as latitude 60° N and thence eastward along parallel of latitude 60° N, to longitude 056°37′.1 W and thence to the latitude 58°00′.0 N, longitude 042°00′.0 W.

3 *Polar waters* means Arctic waters and/or the Antarctic area.

Regulation 22
Application and requirements

1 This chapter applies to all ships certified to carry noxious liquid substances in bulk, operating in polar waters.

2 Unless expressly provided otherwise, any ship covered by paragraph 1 of this regulation shall comply with the environment-related provisions of the introduction and with chapter 2 of part II-A of the Polar Code, in addition to any other applicable requirements of this Annex.

3 In applying chapter 2 of part II-A of the Polar Code, consideration should be given to the additional guidance in part II-B of the Polar Code.

Appendices to Annex II

Appendix I

Guidelines for the categorization of noxious liquid substances*

Products are assigned to pollution categories based on an evaluation of their properties as reflected in the resultant GESAMP Hazard Profile as shown in the table below:

Rule	A1 Bio-accumulation	A2 Bio-degradation	B1 Acute toxicity	B2 Chronic toxicity	D3 Long-term health effects	E2 Effects on marine wildlife and on benthic habitats	Cat
1			≥ 5				X
2	≥ 4		4				
3		NR	4				
4	≥ 4	NR			CMRTNI		
5			4				Y
6			3				
7			2				
8	≥ 4	NR		Not 0			
9				≥ 1			
10						Fp, F or S if not inorganic	
11					CMRTNI		
12	Any product not meeting the criteria of rules 1 to 11 and 13						Z
13	All products identified as: ≤ 2 in column A1; R in column A2; blank in column D3; not Fp, F or S (if not organic) in column E2; and 0 (zero) in all other columns of the GESAMP Hazard Profile						OS

* Refer to Revised Guidelines for the provisional assessment of liquid substances transported in bulk (MEPC.1/Circ.512).

Abbreviated legend to the revised GESAMP Hazard Evaluation Procedure

Columns A and B – Aquatic environment					
Numerical rating	A			B	
	Bioaccumulation and biodegradation			Aquatic toxicity	
	A1* Bioaccumulation		A2* Biodegradation	B1* Acute toxicity	B2* Chronic toxicity
	log P_{OW}	BCF		$LC/EC/IC_{50}$ (mg/L)	NOEC (mg/L)
0	< 1 or > ca. 7	not measurable	R: readily biodegradable	> 1,000	> 1
1	≥ 1 – < 2	≥ 1 – < 10		> 100 – ≤ 1,000	> 0.1 – ≤ 1
2	≥ 2 – < 3	≥ 10 – < 100	NR: not readily biodegradable	> 10 – ≤ 100	> 0.01 – ≤ 0.1
3	≥ 3 – < 4	≥ 100 – < 500		> 1 – ≤ 10	> 0.001 – ≤ 0.01
4	≥ 4 – < 5	≥ 500 – < 4,000	inorg: inorganic substance	> 0.1 – ≤ 1	≤ 0.001
5	≥ 5 – < ca. 7	≥ 4,000		> 0.01 – ≤ 0.1	
6				≤ 0.01	

Columns C and D – Human health (Toxic effects to mammals)						
Numerical rating	C			D		
	Acute mammalian toxicity			Irritation, corrosion and long-term health effects		
	C1 Oral toxicity LD_{50} (mg/kg)	C2 Percutaneous toxicity LD_{50} (mg/kg)	C3 Inhalation toxicity LC_{50} (mg/L)	D1 Skin irritation and corrosion	D2 Eye irritation and corrosion	D3* Long-term health effects
0	> 2,000	> 2,000	> 20	not irritating	not irritating	C – Carcinogen
1	> 300 – ≤ 2,000	> 1,000 – ≤ 2,000	> 10 – ≤ 20	mildly irritating	mildly irritating	M – Mutagenic
2	> 50 – ≤ 300	> 200 – ≤ 1,000	> 2 – ≤ 10	irritating	irritating	R – Reprotoxic
3	> 5 – ≤ 50	> 50 – ≤ 200	> 0.5 – ≤ 2	severely irritating or corrosive 3A Corr. (≤ 4 h) 3B Corr. (≤ 1 h) 3C Corr. (≤ 3 min)	severely irritating	S – Sensitizing A – Aspiration hazard T – Target organ systemic toxicity L – Lung injury N – Neurotoxic I – Immunotoxic
4	≤ 5	≤ 50	≤ 0.5			

* These columns are used to define pollution categories.

Abbreviated legend to the revised GESAMP Hazard Evaluation Procedure *(continued)*

Column E – Interferences with other uses of the sea			
E1 **Tainting**	**E2*** **Physical effects on wildlife and benthic habitats**	**E3** **Interference with coastal amenities**	
		Numerical rating	**Description and action**
NT: not tainting (tested) T: tainting test positive	Fp: Persistent floater F: Floater S: Sinking substances	0	no interference **no warning**
		1	slightly objectionable **warning, no closure of amenity**
		2	moderately objectionable **possible closure of amenity**
		3	highly objectionable **closure of amenity**

* These columns are used to define pollution categories.

Appendix II

Form of Cargo Record Book for ships carrying noxious liquid substances in bulk

CARGO RECORD BOOK FOR SHIPS CARRYING
NOXIOUS LIQUID SUBSTANCES IN BULK

Name of ship . . . .

Distinctive number or letters . . . .

IMO Number . . . .

Gross tonnage . . . .

Period from: . . . . to . . . .

Name of ship . . . .

Distinctive number or letters . . . .

PLAN VIEW OF CARGO AND SLOP TANKS
(to be completed on board)

Identification of the tanks	Capacity

(Give the capacity of each tank in cubic metres)

Introduction

The following pages show a comprehensive list of items of cargo and ballast operations which are, when appropriate, to be recorded in the Cargo Record Book on a tank-to-tank basis in accordance with regulation 15.2 of Annex II of the International Convention for the Prevention of Pollution from Ships, 1973, as modified by the Protocol of 1978 relating thereto, as amended. The items have been grouped into operational sections, each of which is denoted by a letter.

When making entries in the Cargo Record Book, the date, operational code and item number shall be inserted in the appropriate columns and the required particulars shall be recorded chronologically in the blank spaces.

Each completed operation shall be signed for and dated by the officer or officers in charge and, if applicable, by a surveyor authorized by the competent authority of the State in which the ship is unloading. Each completed page shall be countersigned by the master of the ship.

LIST OF ITEMS TO BE RECORDED

Entries are required only for operations involving all categories of substances.

(A) Loading of cargo

1 Place of loading.

2 Identify tank(s), name of substance(s) and category(ies).

(B) Internal transfer of cargo

3 Name and category of cargo(es) transferred.

4 Identity of tanks:

.1 from:

.2 to:

5 Was (were) tank(s) in 4.1 emptied?

6 If not, quantity remaining in tank(s).

(C) Unloading of cargo

7 Place of unloading.

8 Identity of tank(s) unloaded.

9 Was (were) tank(s) emptied?

.1 If yes, confirm that the procedure for emptying and stripping has been performed in accordance with the ship's Procedures and Arrangements Manual (i.e. list, trim, stripping temperature).

.2 If not, quantity remaining in tank(s).

10 Does the ship's Procedures and Arrangements Manual require a prewash with subsequent disposal to reception facilities?

11 Failure of pumping and/or stripping system:

.1 time and nature of failure;

.2 reasons for failure;

.3 time when system has been made operational.

(D) Mandatory prewash in accordance with the ship's Procedures and Arrangements Manual

12 Identify tank(s), substance(s) and category(ies).

13 Washing method:

.1 number of cleaning machines per tank;

.2 duration of wash/washing cycles;

.3 hot/cold wash.

14 Prewash slops transferred to:

.1 reception facility in unloading port (identify port);*

.2 reception facility otherwise (identify port).*

* Ship's masters should obtain from the operator of the reception facilities, which include barges and tank trucks, a receipt or certificate specifying the quantity of tank washings transferred, together with the time and date of the transfer. The receipt or certificate should be kept together with the Cargo Record Book.

(E) Cleaning of cargo tanks except mandatory prewash (other prewash operations, final wash, ventilation, etc.)

15 State time, identify tank(s), substance(s) and category(ies) and state:

.1 washing procedure used;

.2 cleaning agent(s) (identify agent(s) and quantities);

.3 ventilation procedure used (state number of fans used, duration of ventilation).

16 Tank washings transferred:

.1 into the sea;

.2 to reception facility (identify port);*

.3 to slops collecting tank (identify tank).

(F) Discharge into the sea of tank washings

17 Identify tank(s):

.1 Were tank washings discharged during cleaning of tank(s)? If so, at what rate?

.2 Were tank washing(s) discharged from a slops collecting tank? If so, state quantity and rate of discharge.

18 Time pumping commenced and stopped.

19 Ship's speed during discharge.

(G) Ballasting of cargo tanks

20 Identity of tank(s) ballasted.

21 Time at start of ballasting.

(H) Discharge of ballast water from cargo tanks

22 Identity of tank(s).

23 Discharge of ballast:

.1 into the sea;

.2 to reception facilities (identify port).*

24 Time ballast discharge commenced and stopped.

25 Ship's speed during discharge.

(I) Accidental or other exceptional discharge

26 Time of occurrence.

27 Approximate quantity, substance(s) and category(ies).

28 Circumstances of discharge or escape and general remarks.

(J) Control by authorized surveyors

29 Identify port.

30 Identify tank(s), substance(s), category(ies) discharged ashore.

31 Have tank(s), pump(s), and piping system(s) been emptied?

32 Has a prewash in accordance with the ship's Procedures and Arrangements Manual been carried out?

33 Have tank washings resulting from the prewash been discharged ashore and is the tank empty?

34 An exemption has been granted from mandatory prewash.

35 Reasons for exemption.

36 Name and signature of authorized surveyor.

37 Organization, company, government agency for which surveyor works.

---

* Ship's masters should obtain from the operator of the reception facilities, which include barges and tank trucks, a receipt or certificate specifying the quantity of tank washings transferred, together with the time and date of the transfer. The receipt or certificate should be kept together with the Cargo Record Book.

(K) **Additional operational procedures and remarks**

Name of ship . . . .

Distinctive number or letters . . . .

IMO Number . . . .

CARGO/BALLAST OPERATIONS

Date	Code (letter)	Item (number)	Record of operations/signature of officer in charge/ name of and signature of authorized surveyor

Signature of master . . . .

Appendix III

Form of International Pollution Prevention Certificate for the Carriage of Noxious Liquid Substances in Bulk

INTERNATIONAL POLLUTION PREVENTION CERTIFICATE

FOR THE CARRIAGE OF NOXIOUS LIQUID SUBSTANCES IN BULK*

Issued under the provisions of the International Convention for the Prevention of Pollution from Ships, 1973, as modified by the Protocol of 1978 relating thereto, and as amended (hereinafter referred to as "the Convention") under the authority of the Government of:

. . . .

(full designation of the country)

by. . . . .

(full designation of the competent person or organization authorized under the provisions of the Convention)

Particulars of ship

Name of ship. . . . .

Distinctive number or letters. . . . .

IMO Number† . . . .

Port of registry . . . .

Gross tonnage. . . . .

THIS IS TO CERTIFY:

1 That the ship has been surveyed in accordance with regulation 8 of Annex II of the Convention.

2 That the survey showed that the structure, equipment, systems, fitting, arrangements and material of the ship and the condition thereof are in all respects satisfactory and that the ship complies with the applicable requirements of Annex II of the Convention.

3 That the ship has been provided with a Procedures and Arrangements Manual as required by regulation 14 of Annex II of the Convention, and that the arrangements and equipment of the ship prescribed in the Manual are in all respects satisfactory.

4 That the ship complies with the requirements of Annex II to MARPOL for the carriage in bulk of the following noxious liquid substances, provided that all relevant provisions of Annex II are observed.

Noxious liquid substances	Conditions of carriage (tank numbers etc.)	Pollution category
Continued on additional signed and dated sheets		

This certificate is valid until (dd/mm/yyyy). . . . .
subject to surveys in accordance with regulation 8 of Annex II of the Convention.

Completion date of the survey on which this certificate is based (dd/mm/yyyy) . . . .

* The NLS Certificate shall be at least in English, French or Spanish. Where entries in an official national language of the State whose flag the ship is entitled to fly are also used, this shall prevail in case of a dispute or discrepancy.

† Refer to the IMO Ship Identification Number Scheme (resolution A.1078(28)).

Issued at ....
(place of issue of certificate)

Date (dd/mm/yyyy) ....
(date of issue) *(signature of duly authorized official issuing the certificate)*

(seal or stamp of the authority, as appropriate)

ENDORSEMENT FOR ANNUAL AND INTERMEDIATE SURVEYS

THIS IS TO CERTIFY that, at a survey required by regulation 8 of Annex II of the Convention, the ship was found to comply with the relevant provisions of the Convention:

Annual survey

Signed. . . . .
(signature of duly authorized official)

Place . . . .

Date (dd/mm/yyyy) . . . .

(seal or stamp of the authority, as appropriate)

Annual/Intermediate* survey

Signed. . . . .
(signature of duly authorized official)

Place . . . .

Date (dd/mm/yyyy) . . . .

(seal or stamp of the authority, as appropriate)

Annual/Intermediate* survey

Signed. . . . .
(signature of duly authorized official)

Place . . . .

Date (dd/mm/yyyy) . . . .

(seal or stamp of the authority, as appropriate)

Annual survey

Signed. . . . .
(signature of duly authorized official)

Place . . . .

Date (dd/mm/yyyy) . . . .

(seal or stamp of the authority, as appropriate)

ANNUAL/INTERMEDIATE SURVEY IN ACCORDANCE WITH REGULATION 10.8.3

THIS IS TO CERTIFY that, at an annual/intermediate* survey in accordance with regulation 10.8.3 of Annex II of the Convention, the ship was found to comply with the relevant provisions of the Convention:

Signed. . . . .
(signature of duly authorized official)

Place . . . .

Date (dd/mm/yyyy) . . . .

(seal or stamp of the authority, as appropriate)

ENDORSEMENT TO EXTEND THE CERTIFICATE IF VALID FOR LESS THAN 5 YEARS WHERE REGULATION 10.3 APPLIES

The ship complies with the relevant provisions of the Convention, and this Certificate shall, in accordance with regulation 10.3 of Annex II of the Convention, be accepted as valid until (dd/mm/yyyy) . . . .

Signed. . . . .
(signature of duly authorized official)

Place . . . .

Date (dd/mm/yyyy) . . . .

(seal or stamp of the authority, as appropriate)

* Delete as appropriate.

ENDORSEMENT WHERE THE RENEWAL SURVEY HAS BEEN COMPLETED AND REGULATION 10.4 APPLIES

The ship complies with the relevant provisions of the Convention, and this Certificate shall, in accordance with regulation 10.4 of Annex II of the Convention, be accepted as valid until (dd/mm/yyyy) . . . .

Signed. . . . .

(signature of duly authorized official)

Place . . . .

Date (dd/mm/yyyy) . . . .

(seal or stamp of the authority, as appropriate)

ENDORSEMENT TO EXTEND THE VALIDITY OF THE CERTIFICATE UNTIL REACHING THE PORT OF SURVEY OR FOR A PERIOD OF GRACE WHERE REGULATION 10.5 OR 10.6 APPLIES

This Certificate shall, in accordance with regulation 10.5 or 10.6* of Annex II of the Convention, be accepted as valid until (dd/mm/yyyy) . . . .

Signed. . . . .

(signature of duly authorized official)

Place . . . .

Date (dd/mm/yyyy) . . . .

(seal or stamp of the authority, as appropriate)

ENDORSEMENT FOR ADVANCEMENT OF ANNIVERSARY DATE WHERE REGULATION 10.8 APPLIES

In accordance with regulation 10.8 of Annex II of the Convention, the new anniversary date is (dd/mm/yyyy). . . . .

Signed. . . . .

(signature of duly authorized official)

Place . . . .

Date (dd/mm/yyyy) . . . .

(seal or stamp of the authority, as appropriate)

In accordance with regulation 10.8 of Annex II of the Convention, the new anniversary date is (dd/mm/yyyy). . . . .

Signed. . . . .

(signature of duly authorized official)

Place . . . .

Date (dd/mm/yyyy) . . . .

(seal or stamp of the authority, as appropriate)

* Delete as appropriate.

Appendix IV
Standard format for the Procedures and Arrangements Manual

Note 1: The format consists of a standardized introduction and index of the leading paragraphs to each section. This standardized part shall be reproduced in the Manual of each ship. It shall be followed by the contents of each section as prepared for the particular ship. When a section is not applicable, "NA" shall be entered, so as not to lead to any disruption of the numbering as required by the standard format. Where the paragraphs of the standard format are printed in *italics*, the required information shall be described for that particular ship. The contents will vary from ship to ship because of design, trade and intended cargoes. Where the text is not in italics, that text of the standard format shall be copied into the Manual without any modification.

Note 2: If the Administration requires or accepts information and operational instructions in addition to those outlined in this Standard Format, they shall be included in Addendum D of the Manual.

Standard format

MARPOL ANNEX II
PROCEDURES AND ARRANGEMENTS MANUAL

Name of ship. . . . .

Distinctive number or letters. . . . .

IMO Number . . . .

Port of registry . . . .

Approval stamp of Administration:

Introduction

1 The International Convention for the Prevention of Pollution from Ships, 1973, as modified by the Protocol of 1978 relating thereto (hereinafter referred to as MARPOL) was established in order to prevent the pollution of the marine environment by discharges into the sea from ships of harmful substances or effluents containing such substances. In order to achieve its aim, MARPOL contains six Annexes in which detailed regulations are given with respect to the handling on board ships and the discharge into the sea or release into the atmosphere of six main groups of harmful substances, i.e. Annex I (Mineral oils), Annex II (Noxious liquid substances carried in bulk), Annex III (Harmful substances carried in packaged form), Annex IV (Sewage), Annex V (Garbage) and Annex VI (Air pollution).

2 Regulation 13 of Annex II of MARPOL (hereinafter referred to as "Annex II") prohibits the discharge into the sea of noxious liquid substances of categories X, Y or Z or of ballast water, tank washings or other residues or mixtures containing such substances, except in compliance with specified conditions including procedures and arrangements based upon standards developed by the International Maritime Organization (IMO) to ensure that the criteria specified for each category will be met.

3 Annex II requires that each ship which is certified for the carriage of noxious liquid substances in bulk shall be provided with a Procedures and Arrangements Manual, hereinafter referred to as the "Manual".

4 This Manual has been written in accordance with regulation 14 of Annex II and is concerned with the marine environmental aspects of the cleaning of cargo tanks and the discharge of residues and mixtures from these operations. The Manual is not a safety guide and reference shall be made to other publications specifically to evaluate safety hazards.

5 The purpose of the Manual is to identify the arrangements and equipment required to enable compliance with Annex II and to identify for the ship's officers all operational procedures with respect to cargo handling, tank cleaning, slops handling, residue discharging, ballasting and deballasting which must be followed in order to comply with the requirements of Annex II.

6 In addition, this Manual, together with the ship's Cargo Record Book and the Certificate issued under Annex II*, will be used by Administrations for control purposes in order to ensure full compliance with the requirements of Annex II by this ship.

7 The master shall ensure that no discharges into the sea of cargo residues or residue/water mixtures containing category X, Y or Z substances shall take place, unless such discharges are made in full compliance with the operational procedures contained in this Manual.

8 This Manual has been approved by the Administration and no alteration or revision shall be made to any part of it without the prior approval of the Administration.

Index of sections

Section 1 – Main features of MARPOL Annex II

1.1 The requirements of Annex II apply to all ships carrying noxious liquid substances in bulk. Substances posing a threat of harm to the marine environment are divided into three categories, X, Y and Z. Category X substances are those posing the greatest threat to the marine environment, whilst category Z substances are those posing the smallest threat.

1.2 Annex II prohibits the discharge into the sea of any effluent containing substances falling under these categories, except when the discharge is made under conditions which are specified in detail for each category. These conditions include, where applicable, such parameters as:

.1 the maximum quantity of substances per tank which may be discharged into the sea;

.2 the speed of the ship during the discharge;

.3 the minimum distance from the nearest land during discharge;

.4 the minimum depth of water at sea during discharge; and

.5 the need to effect the discharge below the waterline.

1.3 For certain sea areas identified as "special area" more stringent discharge criteria apply. Under Annex II the special area is the Antarctic area. In addition, under paragraph 2 of part II-A of the Polar Code, more stringent discharge criteria apply in Arctic waters.

1.4 Annex II requires that every ship is provided with pumping and piping arrangements to ensure that each tank designated for the carriage of category X, Y and Z substances does not retain after unloading a quantity of residue in excess of the quantity given in the Annex. For each tank intended for the carriage of such substances an assessment of the residue quantity has to be made. Only when the residue quantity as assessed is less than the quantity prescribed by the Annex may a tank be approved for the carriage of a category X, Y or Z substance.

1.5 In addition to the conditions referred to above, an important requirement contained in Annex II is that the discharge operations of certain cargo residues and certain tank cleaning and ventilation operations may only be carried out in accordance with approved procedures and arrangements.

1.6 To enable the requirement of paragraph 1.5 to be met, this Manual contains in section 2 all particulars of the ship's equipment and arrangements, in section 3 operational procedures for cargo unloading and tank stripping and in section 4 procedures for discharge of cargo residues, tank washing, slops collection, ballasting and deballasting as may be applicable to the substances the ship is certified to carry.

1.7 By following the procedures as set out in this Manual, it will be ensured that the ship complies with all relevant requirements of Annex II to MARPOL.

* Include only the Certificate issued to the particular ship: i.e. The International Pollution Prevention Certificate for the Carriage of Noxious Liquid Substances in Bulk or the Certificate of Fitness for the Carriage of Dangerous Chemicals in Bulk or the International Certificate of Fitness for the Carriage of Dangerous Chemicals in Bulk.

Section 2 – Description of the ship's equipment and arrangements

2.1 This section contains all particulars of the ship's equipment and arrangements necessary to enable the crew to follow the operational procedures set out in sections 3 and 4.

2.2 General arrangement of ship and description of cargo tanks

This section shall contain a brief description of the cargo area of the ship with the main features of the cargo tanks and their positions.

Line or schematic drawings showing the general arrangement of the ship and indicating the position and numbering of the cargo tanks and heating arrangements shall be included.

2.3 Description of cargo pumping and piping arrangements and stripping system

This section shall contain a description of the cargo pumping and piping arrangements and of the stripping system.

Line or schematic drawings shall be provided showing the following and be supported by textual explanation where necessary:

.1 *cargo piping arrangements with diameters;*

.2 *cargo pumping arrangements with pump capacities;*

.3 *piping arrangements of stripping system with diameters;*

.4 *pumping arrangements of stripping system with pump capacities;*

.5 *location of suction points of cargo lines and stripping lines inside every cargo tank;*

.6 *if a suction well is fitted, the location and cubic capacity thereof;*

.7 *line draining and stripping or blowing arrangements; and*

.8 *quantity and pressure of nitrogen or air required for line blowing if applicable.*

2.4 Description of ballast tanks and ballast pumping and piping arrangements

This section shall contain a description of the ballast tanks and ballast pumping and piping arrangements.

Line or schematic drawings and tables shall be provided showing the following:

.1 *a general arrangement showing the segregated ballast tanks and cargo tanks to be used as ballast tanks together with their capacities (cubic metres);*

.2 *ballast piping arrangement;*

.3 *pumping capacity for those cargo tanks which may also be used as ballast tanks; and*

.4 *any interconnection between the ballast piping arrangements and the underwater outlet system.*

2.5 Description of dedicated slop tanks with associated pumping and piping arrangements

This section shall contain a description of the dedicated slop tank(s), if any, with the associated pumping and piping arrangements.

Line or schematic drawings shall be provided showing the following:

.1 *which dedicated slop tanks are provided together with the capacities of such tanks;*

.2 *pumping and piping arrangements of dedicated slop tanks with piping diameters and their connection with the underwater discharge outlet.*

2.6 Description of underwater discharge outlet for effluents containing noxious liquid substances

This section shall contain information on position and maximum flow capacity of the underwater discharge outlet (or outlets) and the connections to this outlet from the cargo tanks and slop tanks.

Line or schematic drawings shall be provided showing the following:

.1 *location and number of underwater discharge outlets;*

.2 *connections to underwater discharge outlets;*

.3 *location of all seawater intakes in relation to underwater discharge outlets.*

2.7 Description of flow rate indicating and recording devices

[Deleted]

2.8 Description of cargo tank ventilation system

This section shall contain a description of the cargo tank ventilation system.

Line or schematic drawings and tables shall be provided showing the following and supported by textual explanation if necessary:

.1 the noxious liquid substances the ship is certified fit to carry having a vapour pressure over 5 kPa at 20°C suitable for cleaning by ventilation to be listed in paragraph 4.4.10 of the Manual;

.2 ventilation piping and fans;

.3 positions of the ventilation openings;

.4 the minimum flow rate of the ventilation system to adequately ventilate the bottom and all parts of the cargo tank;

.5 the location of structures inside the tank affecting ventilation;

.6 the method of ventilating the cargo pipeline system, pumps, filters, etc.; and

.7 means for ensuring that the tank is dry.

2.9 Description of tank washing arrangements and wash water heating system

This section shall contain a description of the cargo tank washing arrangements, wash water heating system and all necessary tank washing equipment.

Line or schematic drawings and tables or charts shall be provided showing the following:

.1 arrangements of piping dedicated for tank washing with pipeline diameters;

.2 type of tank cleaning machines with capacities and pressure rating;

.3 maximum number of tank cleaning machines which can operate simultaneously;

.4 position of deck openings for cargo tank washing;

.5 the number of cleaning machines and their location required for ensuring complete coverage of the cargo tank walls;

.6 maximum capacity of wash water which can be heated to 60°C by the installed heating equipment; and

.7 maximum number of tank cleaning machines which can be operated simultaneously at 60°C.

Section 3 – Cargo unloading procedures and tank stripping

3.1 This section contains operational procedures in respect of cargo unloading and tank stripping which must be followed in order to ensure compliance with the requirements of Annex II.

3.2 Cargo unloading

This section shall contain procedures to be followed including the pump and cargo unloading and suction line to be used for each tank. Alternative methods may be given.

The method of operation of the pump or pumps and the sequence of operation of all valves shall be given.

The basic requirement is to unload the cargo to the maximum extent.

3.3 Cargo tank stripping

This section shall contain procedures to be followed during the stripping of each cargo tank.

The procedures shall include the following:

.1 operation of stripping system;

.2 list and trim requirements;

.3 line draining and stripping or blowing arrangements if applicable; and

.4 duration of the stripping time of the water test.

3.4 Cargo temperature

This section shall contain information on the heating requirements of cargoes which have been identified as being required to be at a certain minimum temperature during unloading.

Information shall be given on control of the heating system and the method of temperature measurement.

3.5 Procedures to be followed when a cargo tank cannot be unloaded in accordance with the required procedures

This section shall contain information on the procedures to be followed in the event that the requirements contained in sections 3.3 and/or 3.4 cannot be met due to circumstances such as the following:

.1 failure of cargo tank stripping system; and

.2 failure of cargo tank heating system.

3.6 Cargo Record Book

The Cargo Record Book shall be completed in the appropriate places on completion of any cargo operation.

Section 4 – Procedures relating to the cleaning of cargo tanks, the discharge of residues, ballasting and deballasting

4.1 This section contains operational procedures in respect of tank cleaning, ballast and slops handling which must be followed in order to ensure compliance with the requirements of Annex II.

4.2 The following paragraphs outline the sequence of actions to be taken and contain the information essential to ensure that noxious liquid substances are discharged without posing a threat of harm to the marine environment.

4.3 [Deleted]

4.4 The information necessary to establish the procedures for discharging the residue of the cargo, cleaning, ballasting and deballasting the tank shall take into account the following:

.1 **Category of substance**

The category of the substance should be obtained from the relevant Certificate.

.2 **Stripping efficiency of tank pumping system**

The contents of this section will depend on the design of the ship and whether it is a new ship or existing ship (See flow diagram and pumping/stripping requirements).

.3 **Vessel within or outside special area**

This section shall contain instructions on whether the tank washings can be discharged into the sea within a special area (as defined in section 1.3) or outside a special area. The different requirements shall be made clear and will depend on the design and trade of the ship.

No discharges into the sea of residues of noxious liquid substances, or mixtures containing such substances, are allowed within the polar waters.

.4 **Solidifying or high-viscosity substance**

The properties of the substance should be obtained from the shipping document.

.5 **Miscibility with water**

[Deleted]

.6 **Compatibility with slops containing other substances**

This section shall contain instructions on the permissible and non-permissible mixing of cargo slops. Reference should be made to compatibility guides.

.7 **Discharge to reception facility**

This section shall identify those substances the residues of which are required to be prewashed and discharged to a reception facility.

.8 **Discharging into the sea**

This section shall contain information on the factors to be considered in order to identify whether the residue/water mixtures are permitted to be discharged into the sea.

.9 **Use of cleaning agents or additives**

This section shall contain information on the use and disposal of cleaning agents (e.g. solvents used for tank cleaning) and additives to tank washing water (e.g. detergents).

.10 **Use of ventilation procedures for tank cleaning**

This section shall make reference to all substances suitable for the use of ventilation procedures.

4.5 Having assessed the above information, the correct operational procedures to be followed should be identified using the instructions and flow diagram of section 5. Appropriate entries shall be made in the Cargo Record Book indicating the procedure adopted.

Section 5 – Information and procedures

This section shall contain procedures, which will depend on the age of the ship and pumping efficiency. Examples of flow diagram referred to in this section are given at addendum A and incorporate comprehensive requirements applicable to both new and existing ships. The Manual for a particular ship shall only contain those requirements specifically applicable to that ship.

Information relating to melting point and viscosity, for those substances which have a melting point equal to or greater than 0°C or a viscosity equal or greater than 50 mPa·s at 20°C, should be obtained from the shipping document.

For substances allowed to be carried, reference is made to the relevant Certificate.

The Manual shall contain:

Table 1	[Deleted]
Table 2	Cargo tank information
Addendum A	Flow diagram
Addendum B	Prewash procedures
Addendum C	Ventilation procedures
Addendum D	Additional information and operational instructions when required or accepted by the Administration

Outlines of the above table and addenda are shown below.

Table 2 – Cargo tank information

Tank no.*	Capacity (m^3)	Stripping quantity (litres)

* Tank numbers should be identical to those in the ship's Certificate of Fitness.

Addendum A

Flow diagrams – Cleaning of cargo tanks and disposal of tank washings/ballast containing residues of category X, Y, and Z substances

Note 1: This flow diagram shows the basic requirements applicable to all age groups of ships and is for guidance only.

Note 2: All discharges into the sea are regulated by Annex II.

Note 3: Within the Antarctic area, any discharge into the sea of noxious liquid substances or mixtures containing such substances is prohibited.

Ship details	Stripping requirements (in L)		
	Category X	Category Y	Category Z
New ships: keel laid after 1 January 2007	75	75	75
IBC ships until 1 January 2007	100 + 50 tolerance	100 + 50 tolerance	300 + 50 tolerance
BCH ships	300 + 50 tolerance	300 + 50 tolerance	900 + 50 tolerance
Other ships: keel laid before 1 January 2007	N/A	N/A	Empty to the most possible extent

Cleaning and disposal procedures (CDP)

(Start at the top of the column under the CDP number specified and complete each item procedure in the sequence where marked)

No.	Operation	Procedure number				
		1(a)	1(b)	2(a)	2(b)	3
1	Strip tank and piping to maximum extent, at least in compliance with the procedures in section 3 of this Manual	X	X	X	X	X
2	Apply prewash in accordance with addendum B of this Manual and discharge residue to reception facility	X	X			
3	Apply subsequent wash, additional to the prewash, with: a complete cycle of the cleaning machine(s) (for ships built before 1 July 1994) a water quantity not less than calculated with "k" = 1.0 (for ships built on or after 1 July 1994)		X			
4	Apply ventilation procedure in accordance with addendum C of this Manual					X
5	Ballast tanks or wash tank to commercial standards	X		X	X	X
6	Ballast added to tank		X			
7	Conditions for discharge of ballast/residue/water mixtures other than prewash:					
	.1 distance from land > 12 nautical miles	X		X	X	
	.2 ship's speed > 7 knots	X		X	X	
	.3 water depth > 25 m	X		X	X	
	.4 Using underwater discharge (not exceeding permissible discharge rate)	X		X		
8	Conditions for discharge of ballast:					
	.1 distance from land > 12 nautical miles		X			
	.2 water depth > 25 m		X			
9	Any water subsequently introduced into a tank may be discharged into the sea without restrictions	X	X	X	X	X

Addendum B

Prewash procedures

This addendum to the Manual shall contain prewash procedures based on appendix VI of Annex II. These procedures shall contain specific requirements for the use of the tank washing arrangements and equipment provided on the particular ship and include the following:

.1 cleaning machine positions to be used;

.2 slops pumping out procedure;

.3 requirements for hot washing;

.4 number of cycles of cleaning machine (or time); and

.5 minimum operating pressures.

Addendum C

Ventilation procedures

This addendum to the Manual shall contain ventilation procedures based on appendix 7 of Annex II. The procedures shall contain specific requirements for the use of the cargo tank ventilation system, or equipment, fitted on the particular ship and shall include the following:

.1 ventilation positions to be used;

.2 minimum flow or speed of fans;

.3 procedures for ventilating cargo pipeline, pumps, filters, etc.; and

.4 procedures for ensuring that tanks are dry on completion.

Addendum D

Additional information and operational instructions required or accepted by the Administration

This addendum to the Manual shall contain additional information and operational instructions required or accepted by the Administration.

Appendix V

Assessment of residue quantities in cargo tanks, pumps and associated piping

1 Introduction

1.1 Purpose

1.1.1 The purpose of this appendix is to provide the procedure for testing the efficiency of cargo pumping systems.

1.2 Background

1.2.1 The ability of the pumping system of a tank to comply with regulation 12.1, 12.2 or 12.3 is determined by performing a test in accordance with the procedure set out in section 3 of this appendix. The quantity measured is termed the "stripping quantity". The stripping quantity of each tank shall be recorded in the ship's Manual.

1.2.2 After having determined the stripping quantity of one tank, the Administration may use the determined quantities for a similar tank, provided the Administration is satisfied that the pumping system in that tank is similar and operating properly.

2 Design criteria and performance test

2.1 The cargo pumping systems should be designed to meet the required maximum amount of residue per tank and associated piping as specified in regulation 12 of Annex II to the satisfaction of the Administration.

2.2 In accordance with regulation 12.5 the cargo pumping systems shall be tested with water to prove their performance. Such water tests shall, by measurement, show that the system meets the requirements of regulation 12. In respect of regulations 12.1 and 12.2 a tolerance of 50 L per tank is acceptable.

3 Water performance test

3.1 Test condition

3.1.1 The ship's trim and list shall be such as to provide favourable drainage to the suction point. During the water test the ship's trim shall not exceed 3° by the stern, and the ship's list shall not exceed 1°.

3.1.2 The trim and list chosen for the water test shall be recorded. This shall be the minimum favourable trim and list used during the water test.

3.1.3 During the water test, means shall be provided to maintain a backpressure of not less than 100 kPa at the cargo tank's unloading manifold (see figures 5-1 and 5-2).

3.1.4 The time taken to complete the water test shall be recorded for each tank, recognizing that this may need to be amended as a result of subsequent tests.

3.2 Test procedure

3.2.1 Ensure that the cargo tank to be tested and its associated piping have been cleaned and that the cargo tank is safe for entry.

3.2.2 Fill the cargo tank with water to a depth necessary to carry out normal end of unloading procedures.

3.2.3 Discharge and strip water from the cargo tank and its associated piping in accordance with the proposed procedures.

3.2.4 Collect all water remaining in the cargo tank and its associated piping into a calibrated container for measurement. Water residues shall be collected, inter alia, from the following points:

.1 the cargo tank suction and its vicinity;

.2 any entrapped areas on the cargo tank bottom;

.3 the low point drain of the cargo pump; and

.4 all low point drains of piping associated with the cargo tank up to the manifold valve.

3.2.5 The total water volumes collected above determine the stripping quantity for the cargo tank.

3.2.6 Where a group of tanks is served by a common pump or piping, the water test residues associated with the common system(s) may be apportioned equally among the tanks provided that the following operational restriction is included in the ship's approved Manual: "For sequential unloading of tanks in this group, the pump or piping is not to be washed until all tanks in the group have been unloaded."

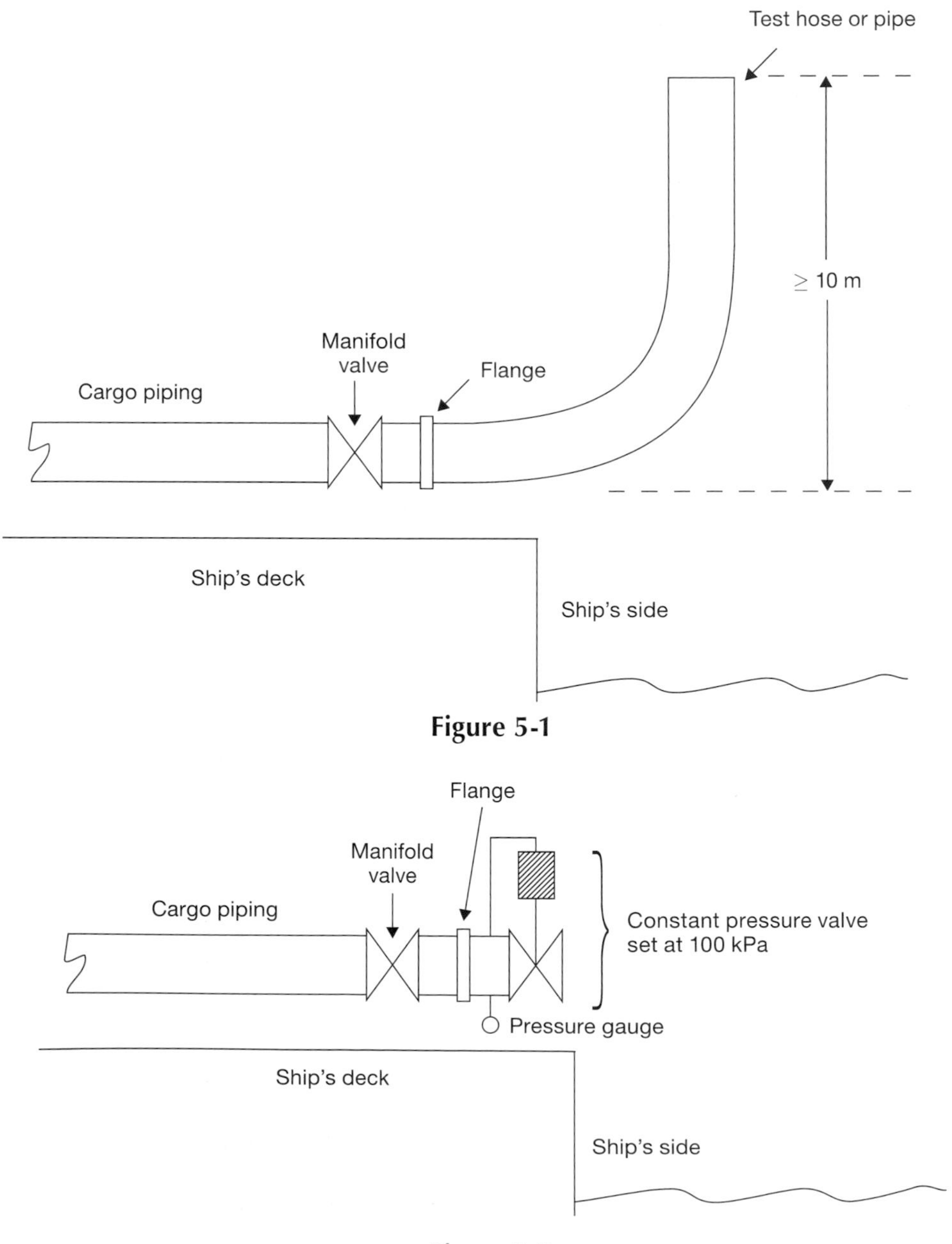

Figure 5-1

Figure 5-2

The above figures illustrate test arrangements that would provide a backpressure of not less than 100 kPa at the cargo tank's unloading manifold.

Appendix VI
Prewash procedures

A For ships built before 1 July 1994

A prewash procedure is required in order to meet certain Annex II requirements. This appendix explains how these prewash procedures shall be performed.

Prewash procedures for non-solidifying substances

1 Tanks shall be washed by means of a rotary water jet, operated at sufficiently high water pressure. In the case of category X substances, cleaning machines shall be operated in such locations that all tank surfaces are washed. In the case of category Y substances, only one location need be used.

2 During washing, the amount of water in the tank shall be minimized by continuously pumping out slops and promoting flow to the suction point (positive list and trim). If this condition cannot be met, the washing procedure shall be repeated three times, with thorough stripping of the tank between washings.

3 Those substances which have a viscosity equal to or greater than 50 mPa·s at 20°C shall be washed with hot water (temperature at least 60°C), unless the properties of such substances make the washing less effective.

4 The number of cycles of the cleaning machine used shall not be less than that specified in table 6-1. A cleaning machine cycle is defined as the period between two consecutive identical orientations of the tank cleaning machine (rotation through 360°).

5 After washing, the tank cleaning machine(s) shall be kept operating long enough to flush the pipeline, pump and filter, and discharge to shore reception facilities shall be continued until the tank is empty.

Prewash procedures for solidifying substances

1 Tanks shall be washed as soon as possible after unloading. If possible, tanks shall be heated prior to washing.

2 Residues in hatches and manholes shall preferably be removed prior to the prewash.

3 Tanks shall be washed by means of a rotary water jet operated at sufficiently high water pressure and in locations to ensure that all tank surfaces are washed.

4 During washing, the amount of water in the tank shall be minimized by pumping out slops continuously and promoting flow to the suction point (positive list and trim). If this condition cannot be met, the washing procedure shall be repeated three times with thorough stripping of the tank between washings.

5 Tanks shall be washed with hot water (temperature at least 60°C) unless the properties of such substances make the washing less effective.

6 The number of cycles of the cleaning machine used shall not be less than that specified in table 6-1. A cleaning machine cycle is defined as the period between two consecutive identical orientations of the machine (rotation through 360°).

7 After washing, the cleaning machine(s) shall be kept operating long enough to flush the pipeline, pump and filter, and discharge to shore reception facilities shall be continued until the tank is empty.

Table 6-1 – Number of cleaning machine cycles to be used in each location

Category of substance	Number of cleaning machine cycles	
	Non-solidifying substances	Solidifying substances
Category X	1	2
Category Y	$\frac{1}{2}$	1

B For ships built on or after 1 July 1994 and recommendatory for ships built before 1 July 1994

A prewash procedure is required in order to meet certain Annex II requirements. This appendix explains how these prewash procedures shall be performed and how the minimum volumes of washing media to be used shall be determined. Smaller volumes of washing media may be used based on actual verification testing to the satisfaction of the Administration. Where reduced volumes are approved, an entry to that effect must be recorded in the Manual.

If a medium other than water is used for the prewash, the provisions of regulation 13.5.1 apply.

Prewash procedures for non-solidifying substances without recycling

1 Tanks shall be washed by means of a rotary jet(s), operated at sufficiently high water pressure. In the case of category X substances, cleaning machines shall be operated in such locations that all tank surfaces are washed. In the case of category Y substances, only one location need be used.

2 During washing, the amount of liquid in the tank shall be minimized by continuously pumping out slops and promoting flow to the suction point. If this condition cannot be met, the washing procedure shall be repeated three times, with thorough stripping of the tank between washings.

3 Those substances which have a viscosity equal to or greater than 50 mPa·s at 20°C shall be washed with hot water (temperature at least 60°C), unless the properties of such substances make the washing less effective.

4 The quantities of wash water used shall not be less than those specified in paragraph 20 or determined according to paragraph 21.

5 After prewashing, the tanks and lines shall be thoroughly stripped.

Prewash procedures for solidifying substances without recycling

6 Tanks shall be washed as soon as possible after unloading. If possible, tanks should be heated prior to washing.

7 Residues in hatches and manholes should preferably be removed prior to the prewash.

8 Tanks shall be washed by means of a rotary jet(s) operated at sufficiently high water pressure and in locations to ensure that all tank surfaces are washed.

9 During washing, the amount of liquid in the tank shall be minimized by pumping out slops continuously and promoting flow to the suction point. If this condition cannot be met, the washing procedure shall be repeated three times with thorough stripping of the tank between washings.

10 Tanks shall be washed with hot water (temperature at least 60°C), unless the properties of such substances make the washing less effective.

11 The quantities of wash water used shall not be less than those specified in paragraph 20 or determined according to paragraph 21.

12 After prewashing, the tanks and lines shall be thoroughly stripped.

Prewash procedures with recycling of washing medium

13 Washing with a recycled washing medium may be adopted for the purpose of washing more than one cargo tank. In determining the quantity, due regard must be given to the expected amount of residues in the tanks and the properties of the washing medium and whether any initial rinse or flushing is employed. Unless sufficient data are provided, the calculated end concentration of cargo residues in the washing medium shall not exceed 5% based on nominal stripping quantities.

14 The recycled washing medium shall only be used for washing tanks having contained the same or similar substance.

15 A quantity of washing medium sufficient to allow continuous washing shall be added to the tank or tanks to be washed.

16 All tank surfaces shall be washed by means of a rotary jet(s) operated at sufficiently high pressure. The recycling of the washing medium may either be within the tank to be washed or via another tank, e.g. a slop tank.

17 The washing shall be continued until the accumulated throughput is not less than that corresponding to the relevant quantities given in paragraph 20 or determined according to paragraph 21.

18 Solidifying substances and substances with a viscosity equal to or greater than 50 mPa·s at 20°C shall be washed with hot water (temperature at least 60°C) when water is used as the washing medium, unless the properties of such substances make the washing less effective.

19 After completing the tank washing with recycling to the extent specified in paragraph 17, the washing medium shall be discharged and the tank thoroughly stripped. Thereafter, the tank shall be subjected to a rinse, using clean washing medium, with continuous drainage and discharged to a reception facility. The rinse shall as a minimum cover the tank bottom and be sufficient to flush the pipelines, pump and filter.

Minimum quantity of water to be used in a prewash

20 The minimum quantity of water to be used in a prewash is determined by the residual quantity of noxious liquid substance in the tank, the tank size, the cargo properties, the permitted concentration in any subsequent wash water effluent, and the area of operation. The minimum quantity is given by the following formula:

$$Q = k(15r^{0.8} + 5r^{0.7} \times V/1{,}000)$$

where

Q = the required minimum quantity in cubic metres

r = the residual quantity per tank in cubic metres. The value of r shall be the value demonstrated in the actual stripping efficiency test, but shall not be taken lower than 0.100 m^3 for a tank volume of 500 m^3 and above and 0.040 m^3 for a tank volume of 100 m^3 and below. For tank sizes between 100 m^3 and 500 m^3 the minimum value of r allowed to be used in the calculations is obtained by linear interpolation.

For category X substances the value of r shall either be determined based on stripping tests according to the Manual, observing the lower limits as given above, or be taken to be 0.9 m^3.

V = tank volume in cubic metres

k = a factor having values as follows:

Category X, non-solidifying, low-viscosity substance, $k = 1.2$

Category X, solidifying or high-viscosity substance, $k = 2.4$

Category Y, non-solidifying, low-viscosity substance, $k = 0.5$

Category Y, solidifying or high-viscosity substance, $k = 1.0$

The table below is calculated using the formula with a *k* factor of 1 and may be used as an easy reference.

Stripping quantity (m^3)	Tank volume (m^3)		
	100	500	3000
≤ 0.04	1.2	2.9	5.4
0.10	2.5	2.9	5.4
0.30	5.9	6.8	12.2
0.90	14.3	16.1	27.7

21 Verification testing for approval of prewash volumes lower than those given in paragraph 20 may be carried out to the satisfaction of the Administration to prove that the requirements of regulation 13 are met, taking into account the substances the ship is certified to carry. The prewash volume so verified shall be adjusted for other prewash conditions by application of the factor *k* as defined in paragraph 20.

Appendix VII

Ventilation procedures

1 Cargo residues of substances with a vapour pressure greater than 5 kPa at 20°C may be removed from a cargo tank by ventilation.

2 Before residues of noxious liquid substances are ventilated from a tank, the safety hazards relating to cargo flammability and toxicity shall be considered. With regard to safety aspects, the operational requirements for openings in cargo tanks in SOLAS 74, as amended, the International Bulk Chemical Code, the Bulk Chemical Code, and the ventilation procedures in the International Chamber of Shipping (ICS) *Tanker Safety Guide (Chemicals)* should be consulted.

3 Port authorities may also have regulations on cargo tank ventilation.

4 The procedures for ventilation of cargo residues from a tank are as follows:

.1 the pipelines shall be drained and further cleared of liquid by means of ventilation equipment;

.2 the list and trim shall be adjusted to the minimum levels possible so that evaporation of residues in the tank is enhanced;

.3 ventilation equipment producing an airjet which can reach the tank bottom shall be used. Figure 7-1 could be used to evaluate the adequacy of ventilation equipment used for ventilating a tank of a given depth;

.4 ventilation equipment shall be placed in the tank opening closest to the tank sump or suction point;

.5 ventilation equipment shall, when practicable, be positioned so that the airjet is directed at the tank sump or suction point and impingement of the airjet on tank structural members is to be avoided as much as possible; and

.6 ventilation shall continue until no visible remains of liquid can be observed in the tank. This shall be verified by a visual examination or an equivalent method.

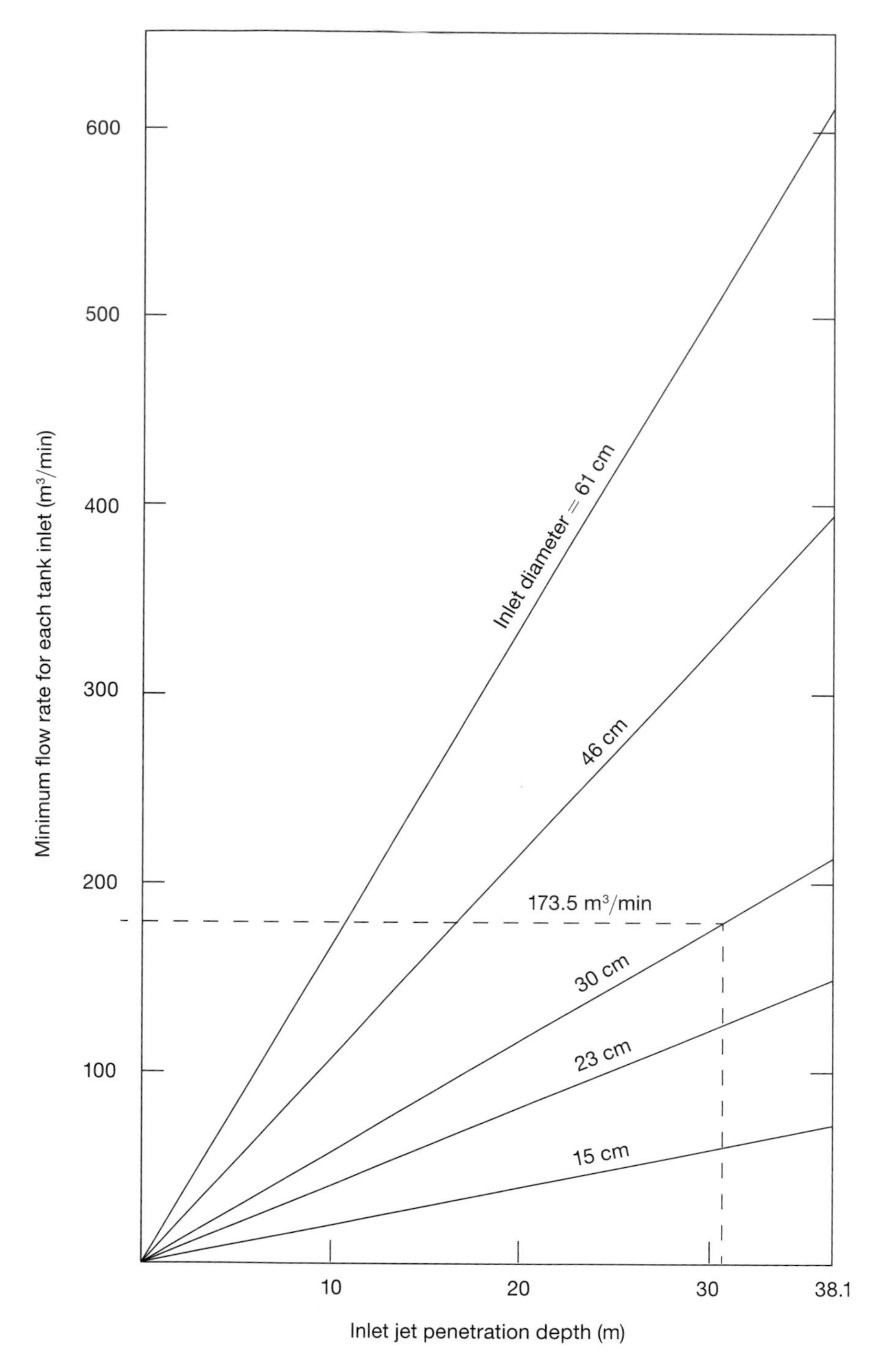

Figure 7-1 – *Minimum flow rate as a function of jet penetration depth. Jet penetration depth shall be compared against tank height.*

MARPOL Annex III

Regulations for the prevention of pollution by harmful substances carried by sea in packaged form

MARPOL Annex III

Regulations for the prevention of pollution by harmful substances carried by sea in packaged form

Chapter 1 – General

Regulation 1
Definitions

For the purposes of this Annex:

1 *Harmful substances* are those substances which are identified as marine pollutants in the International Maritime Dangerous Goods Code (IMDG Code)* or which meet the criteria in the appendix of this Annex.

2 *Packaged form* is defined as the forms of containment specified for harmful substances in the IMDG Code.

3 *Audit* means a systematic, independent and documented process for obtaining audit evidence and evaluating it objectively to determine the extent to which audit criteria are fulfilled.

4 *Audit Scheme* means the IMO Member State Audit Scheme established by the Organization and taking into account the guidelines developed by the Organization.†

5 *Code for Implementation* means the IMO Instruments Implementation Code (III Code) adopted by the Organization by resolution A.1070(28).

6 *Audit Standard* means the Code for Implementation.

Regulation 2
Application

1 The carriage of harmful substances is prohibited, except in accordance with the provisions of this Annex.

2 To supplement the provisions of this Annex, the Government of each Party to the Convention shall issue, or cause to be issued, detailed requirements on packing, marking, labelling, documentation, stowage, quantity limitations and exceptions for preventing or minimizing pollution of the marine environment by harmful substances.

3 For the purposes of this Annex, empty packagings which have been used previously for the carriage of harmful substances shall themselves be treated as harmful substances unless adequate precautions have been taken to ensure that they contain no residue that is harmful to the marine environment.

4 The requirements of this Annex do not apply to ship's stores and equipment.

---

* Refer to the IMDG Code (resolution MSC.122(75), as amended).

† Refer to the Framework and Procedures for the IMO Member State Audit Scheme (resolution A.1067(28)).

Regulation 3
Packing

Packages shall be adequate to minimize the hazard to the marine environment, having regard to their specific contents.

Regulation 4
Marking and labelling

1 Packages containing a harmful substance shall be durably marked or labelled to indicate that the substance is a harmful substance in accordance with the relevant provisions of the IMDG Code.

2 The method of affixing marks or labels on packages containing a harmful substance shall be in accordance with the relevant provisions of the IMDG Code.

Regulation 5*
Documentation

1 Transport information relating to the carriage of harmful substances shall be in accordance with the relevant provisions of the IMDG Code and shall be made available to the person or organization designated by the port State authority.

2 Each ship carrying harmful substances shall have a special list, manifest or stowage plan setting forth, in accordance with the relevant provisions of the IMDG Code, the harmful substances on board and the location thereof. A copy of one of these documents shall be made available before departure to the person or organization designated by the port State authority.

Regulation 6
Stowage

Harmful substances shall be properly stowed and secured so as to minimize the hazards to the marine environment without impairing the safety of the ship and persons on board.

Regulation 7
Quantity limitations

Certain harmful substances may, for sound scientific and technical reasons, need to be prohibited for carriage or be limited as to the quantity which may be carried aboard any one ship. In limiting the quantity, due consideration shall be given to size, construction and equipment of the ship, as well as the packaging and the inherent nature of the substances.

Regulation 8
Exceptions

1 Jettisoning of harmful substances carried in packaged form shall be prohibited, except where necessary for the purpose of securing the safety of the ship or saving life at sea.

* Reference to "documents" in this regulation does not preclude the use of electronic data processing (EDP) and electronic data interchange (EDI) transmission techniques as an aid to paper documentation.

2 Subject to the provisions of the present Convention, appropriate measures based on the physical, chemical and biological properties of harmful substances shall be taken to regulate the washing of leakages overboard, provided that compliance with such measures would not impair the safety of the ship and persons on board.

Regulation 9

*Port State control on operational requirements**

1 A ship when in a port or an offshore terminal of another Party is subject to inspection by officers duly authorized by such Party concerning operational requirements under this Annex.

2 Where there are clear grounds for believing that the master or crew are not familiar with essential shipboard procedures relating to the prevention of pollution by harmful substances, the Party shall take such steps, including carrying out detailed inspection and, if required, will ensure that the ship shall not sail until the situation has been brought to order in accordance with the requirements of this Annex.

3 Procedures relating to the port State control prescribed in article 5 of the present Convention shall apply to this regulation.

4 Nothing in this regulation shall be construed to limit the rights and obligations of a Party carrying out control over operational requirements specifically provided for in the present Convention.

* Refer to Procedures for port State control, 2011 (resolution A.1052(27)).

ANNEX I ANNEX II ANNEX III ANNEX IV ANNEX V ANNEX VI UNIFIED INTERPRETATIONS

Chapter 2 – Verification of compliance with the provisions of this Annex

Regulation 10
Application

Parties shall use the provisions of the Code for Implementation in the execution of their obligations and responsibilities contained in this Annex.

Regulation 11
Verification of compliance

1 Every Party shall be subject to periodic audits by the Organization in accordance with the audit standard to verify compliance with and implementation of this Annex.

2 The Secretary-General of the Organization shall have responsibility for administering the Audit Scheme, based on the guidelines developed by the Organization.

3 Every Party shall have responsibility for facilitating the conduct of the audit and implementation of a programme of actions to address the findings, based on the guidelines developed by the Organization.*

4 Audit of all Parties shall be:

.1 based on an overall schedule developed by the Secretary General of the Organization, taking into account the guidelines developed by the Organization; and

.2 conducted at periodic intervals, taking into account the guidelines developed by the Organization

.

* Refer to the Framework and Procedures for the IMO Member State Audit Scheme (resolution A.1067(28)).

Appendix to Annex III

Criteria for the identification of harmful substances in packaged form

For the purpose of this Annex, substances, other than radioactive materials,* identified by any one of the following criteria are harmful substances:†

(a) Acute (short-term) aquatic hazard

Category: Acute 1	
96 hr LC_{50} (for fish)	$\leq$ 1 mg/L and/or
48 hr EC_{50} (for crustacea)	$\leq$ 1 mg/L and/or
72 or 96 hr ErC_{50} (for algae or other aquatic plants)	$\leq$ 1 mg/L

(b) Long-term aquatic hazard

(i) Non-rapidly degradable substances for which there are adequate chronic toxicity data available

Category: Chronic 1	
Chronic NOEC or EC_x (for fish)	$\leq$ 0.1 mg/L and/or
Chronic NOEC or EC_x (for crustacea)	$\leq$ 0.1 mg/L and/or
Chronic NOEC or EC_x (for algae or other aquatic plants)	$\leq$ 0.1 mg/L
Category: Chronic 2	
Chronic NOEC or EC_x (for fish)	$\leq$ 1 mg/L and/or
Chronic NOEC or EC_x (for crustacea)	$\leq$ 1 mg/L and/or
Chronic NOEC or EC_x (for algae or other aquatic plants)	$\leq$ 1 mg/L

(ii) Rapidly degradable substances for which there are adequate chronic toxicity data available

Category: Chronic 1	
Chronic NOEC or EC_x (for fish)	$\leq$ 0.01 mg/L and/or
Chronic NOEC or EC_x (for crustacea)	$\leq$ 0.01 mg/L and/or
Chronic NOEC or EC_x (for algae or other aquatic plants)	$\leq$ 0.01 mg/L
Category: Chronic 2	
Chronic NOEC or EC_x (for fish)	$\leq$ 0.1 mg/L and/or
Chronic NOEC or EC_x (for crustacea)	$\leq$ 0.1 mg/L and/or
Chronic NOEC or EC_x (for algae or other aquatic plants)	$\leq$ 0.1 mg/L

* Refer to class 7, as defined in chapter 2.7 of the IMDG Code.

† The criteria are based on those developed by the United Nations Globally Harmonized System of Classification and Labelling of Chemicals (GHS), as amended. For definitions of acronyms or terms used in this appendix, refer to the relevant paragraphs of the IMDG Code.

(iii) Substances for which adequate chronic toxicity data are not available

Category: Chronic 1	
96 hr LC_{50} (for fish)	$\leq$ 1 mg/L and/or
48 hr EC_{50} (for crustacea)	$\leq$ 1 mg/L and/or
72 or 96 hr ErC_{50} (for algae or other aquatic plants)	$\leq$ 1 mg/L
and the substance is not rapidly degradable and/or the experimentally determined BCF is $\geq$ 500 (or, if absent, the log $K_{ow} \geq 4$).	
Category: Chronic 2	
96 hr LC_{50} (for fish)	$>$ 1 mg/L but $\leq$ 10 mg/L and/or
48 hr EC_{50} (for crustacea)	$>$ 1 mg/L but $\leq$ 10 mg/L and/or
72 or 96 hr ErC_{50} (for algae or other aquatic plants)	$>$ 1 mg/L but $\leq$ 10 mg/L
and the substance is not rapidly degradable and/or the experimentally determined BCF is $\geq$ 500 (or, if absent, the log $K_{ow} \geq 4$).	

Additional guidance on the classification process for substances and mixtures is included in the IMDG Code.

MARPOL Annex IV

Regulations for the prevention of pollution by sewage from ships

MARPOL Annex IV

Regulations for the prevention of pollution by sewage from ships

Chapter 1 – General

Regulation 1
Definitions

For the purposes of this Annex:

1 *New ship* means a ship:

.1 for which the building contract is placed, or in the absence of a building contract, the keel of which is laid, or which is at a similar stage of construction, on or after the date of entry into force of this Annex;* or

SEE INTERPRETATION 1

.2 the delivery of which is three years or more after the date of entry into force of this Annex.

SEE INTERPRETATION 2

2 *Existing ship* means a ship which is not a new ship.

3 *Sewage* means:

.1 drainage and other wastes from any form of toilets and urinals;

.2 drainage from medical premises (dispensary, sick bay, etc.) via wash basins, wash tubs and scuppers located in such premises;

.3 drainage from spaces containing living animals; or

.4 other waste waters when mixed with the drainages defined above.

4 *Holding tank* means a tank used for the collection and storage of sewage.

5 *Nearest land*. The term "from the nearest land" means from the baseline from which the territorial sea of the territory in question is established in accordance with international law except that, for the purposes of the present Convention, "from the nearest land" off the north-eastern coast of Australia shall mean from a line drawn from a point on the coast of Australia in:

latitude 11°00′ S, longitude 142°08′ E
to a point in latitude 10°35′ S, longitude 141°55′ E,
thence to a point latitude 10°00′ S, longitude 142°00′ E,
thence to a point latitude 09°10′ S, longitude 143°52′ E,

* Annex IV entered into force on 27 September 2003.

thence to a point latitude 09°00′ S, longitude 144°30′ E,
thence to a point latitude 10°41′ S, longitude 145°00′ E,
thence to a point latitude 13°00′ S, longitude 145°00′ E,
thence to a point latitude 15°00′ S, longitude 146°00′ E,
thence to a point latitude 17°30′ S, longitude 147°00′ E,
thence to a point latitude 21°00′ S, longitude 152°55′ E,
thence to a point latitude 24°30′ S, longitude 154°00′ E,
thence to a point on the coast of Australia in
latitude 24°42′ S, longitude 153°15′ E.

6 *Special area* means a sea area where for recognized technical reasons in relation to its oceanographical and ecological condition and to the particular character of its traffic the adoption of special mandatory methods for the prevention of sea pollution by sewage is required.

The special areas are:

.1 the Baltic Sea area as defined in regulation 1.11.2 of Annex I; and

.2 any other sea area designated by the Organization in accordance with criteria and procedures for designation of special areas with respect to prevention of pollution by sewage from ships.*

7 *International voyage* means a voyage from a country to which the present Convention applies to a port outside such country, or conversely.

8 *Person* means member of the crew and passengers.

9 A *passenger* means every person other than:

.1 the master and the members of the crew or other persons employed or engaged in any capacity on board a ship on the business of that ship; and

.2 a child under one year of age.

10 A *passenger ship* means a ship which carries more than 12 passengers.

For the application of regulation 11.3, a *new passenger ship* is a passenger ship:

.1 for which the building contract is placed, or in the absence of a building contract, the keel of which is laid, or which is in a similar stage of construction, on or after 1 January 2016; or

.2 the delivery of which is two years or more after 1 January 2016.

An *existing passenger ship* is a passenger ship which is not a new passenger ship.

11 *Anniversary date* means the day and the month of each year which will correspond to the date of expiry of the International Sewage Pollution Prevention Certificate.

12 *Audit* means a systematic, independent and documented process for obtaining audit evidence and evaluating it objectively to determine the extent to which audit criteria are fulfilled.

13 *Audit Scheme* means the IMO Member State Audit Scheme established by the Organization and taking into account the guidelines developed by the Organization.†

14 *Code for Implementation* means the IMO Instruments Implementation Code (III Code) adopted by the Organization by resolution A.1070(28).

15 *Audit Standard* means the Code for Implementation.

* Refer to Guidelines for the designation of special areas under MARPOL and guidelines for the identification and designation of particularly sensitive sea areas (resolution A.927(22)).

† Refer to the Framework and Procedures for the IMO Member State Audit Scheme (resolution A.1067(28)).

Regulation 2
*Application**

1 The provisions of this Annex shall apply to the following ships engaged in international voyages:

.1 new ships of 400 gross tonnage and above; and

.2 new ships of less than 400 gross tonnage which are certified to carry more than 15 persons; and

.3 existing ships of 400 gross tonnage and above, five years after the date of entry into force of this Annex; and

.4 existing ships of less than 400 gross tonnage which are certified to carry more than 15 persons, five years after the date of entry into force of this Annex.

2 The Administration shall ensure that existing ships, according to subparagraphs 1.3 and 1.4 of this regulation, the keels of which are laid or which are of a similar stage of construction before 2 October 1983 shall be equipped, as far as practicable, to discharge sewage in accordance with the requirements of regulation 11 of the Annex.

Regulation 3
Exceptions

1 Regulation 11 of this Annex and section 4.2 of chapter 4 of part II-A of the Polar Code shall not apply to:

.1 the discharge of sewage from a ship necessary for the purpose of securing the safety of a ship and those on board or saving life at sea; or

.2 the discharge of sewage resulting from damage to a ship or its equipment if all reasonable precautions have been taken before and after the occurrence of the damage, for the purpose of preventing or minimizing the discharge.

* MEPC 52 (11 to 15 October 2004) confirmed that 27 September 2003 was the one and only entry into force date of MARPOL Annex IV (see document MEPC 52/24, paragraphs 6.16 to 6.19).

Chapter 2 – Surveys and certification*

Regulation 4
Surveys

1 Every ship which, in accordance with regulation 2, is required to comply with the provisions of this Annex shall be subject to the surveys specified below:

.1 An initial survey before the ship is put in service or before the Certificate required under regulation 5 of this Annex is issued for the first time, which shall include a complete survey of its structure, equipment, systems, fittings, arrangements and material in so far as the ship is covered by this Annex. This survey shall be such as to ensure that the structure, equipment, systems, fittings, arrangements and materials fully comply with the applicable requirements of this Annex.

.2 A renewal survey at intervals specified by the Administration, but not exceeding five years, except where regulation 8.2, 8.5, 8.6 or 8.7 of this Annex is applicable. The renewal survey shall be such as to ensure that the structure, equipment, systems, fittings, arrangements and materials fully comply with applicable requirements of this Annex.

.3 An additional survey, either general or partial, according to the circumstances, shall be made after a repair resulting from investigations prescribed in paragraph 9 of this regulation, or whenever any important repairs or renewals are made. The survey shall be such as to ensure that the necessary repairs or renewals have been effectively made, that the material and workmanship of such repairs or renewals are in all respects satisfactory and that the ship complies in all respects with the requirements of this Annex.

2 The Administration shall establish appropriate measures for ships which are not subject to the provisions of paragraph 1 of this regulation in order to ensure that the applicable provisions of this Annex are complied with.

3 Surveys of ships as regards the enforcement of the provisions of this Annex shall be carried out by officers of the Administration. The Administration may, however, entrust the surveys either to surveyors nominated for the purpose or to organizations recognized by it.

4 An Administration nominating surveyors or recognizing organizations to conduct surveys as set forth in paragraph 3 of this regulation shall, as a minimum, empower any nominated surveyor or recognized organization to:

.1 require repairs to a ship; and

.2 carry out surveys if requested by the appropriate authorities of a Port State.

The Administration shall notify the Organization of the specific responsibilities and conditions of the authority delegated to the nominated surveyors or recognized organizations, for circulation to Parties to the present Convention for the information of their officers.

5 When a nominated surveyor or recognized organization determines that the condition of the ship or its equipment does not correspond substantially with the particulars of the Certificate or is such that the ship is not fit to proceed to sea without presenting an unreasonable threat of harm to the marine environment, such surveyor or organization shall immediately ensure that corrective action is taken and shall in due course

* Refer to Global and uniform implementation of the harmonized system of survey and certification (HSSC) (resolution A.883(21)), Survey guidelines under the harmonized system of survey and certification, 2007 (resolution A.997(25)), Communication of information on the authorization of recognized organizations (ROs) (MSC/Circ.1010-MEPC/Circ.382), and the information collected via the Global Integrated Shipping Information System (GISIS).

notify the Administration. If such corrective action is not taken, the Certificate should be withdrawn and the Administration shall be notified immediately and if the ship is in a port of another Party, the appropriate authorities of the Port State shall also be notified immediately. When an officer of the Administration, a nominated surveyor or recognized organization has notified the appropriate authorities of the Port State, the Government of the Port State concerned shall give such officer, surveyor or organization any necessary assistance to carry out their obligations under this regulation. When applicable, the Government of the Port State concerned shall take such steps as will ensure that the ship shall not sail until it can proceed to sea or leave the port for the purpose of proceeding to the nearest appropriate repair yard available without presenting an unreasonable threat of harm to the marine environment.

6 In every case, the Administration concerned shall fully guarantee the completeness and efficiency of the survey and shall undertake to ensure the necessary arrangements to satisfy this obligation.

7 The condition of the ship and its equipment shall be maintained to conform with the provisions of the present Convention to ensure that the ship in all respects will remain fit to proceed to sea without presenting an unreasonable threat of harm to the marine environment.

8 After any survey of the ship under paragraph 1 of this regulation has been completed, no change shall be made in the structure, equipment, systems, fittings, arrangements or materials covered by the survey, without the sanction of the Administration, except the direct replacement of such equipment and fittings.

9 Whenever an accident occurs to a ship or a defect is discovered which substantially affects the integrity of the ship or the efficiency or completeness of its equipment covered by this Annex, the master or owner of the ship shall report at the earliest opportunity to the Administration, the recognized organization or the nominated surveyor responsible for issuing the relevant Certificate, who shall cause investigations to be initiated to determine whether a survey as required by paragraph 1 of this regulation is necessary. If the ship is in a port of another Party, the master or owner shall also report immediately to the appropriate authorities of the Port State and the nominated surveyor or recognized organization shall ascertain that such report has been made.

Regulation 5
Issue or endorsement of Certificate

1 An International Sewage Pollution Prevention Certificate shall be issued, after an initial or renewal survey in accordance with the provisions of regulation 4 of this Annex, to any ship which is engaged in voyages to ports or offshore terminals under the jurisdiction of other Parties to the Convention. In the case of existing ships this requirement shall apply five years after the date of entry into force of this Annex.

2 Such Certificate shall be issued or endorsed either by the Administration or by any persons or organization* duly authorized by it. In every case, the Administration assumes full responsibility for the Certificate.

Regulation 6
Issue or endorsement of a Certificate by another Government

1 The Government of a Party to the Convention may, at the request of the Administration, cause a ship to be surveyed and, if satisfied that the provisions of this Annex are complied with, shall issue or authorize the issue of an International Sewage Pollution Prevention Certificate to the ship, and where appropriate, endorse or authorize the endorsement of that Certificate on the ship in accordance with this Annex.

2 A copy of the Certificate and a copy of the survey report shall be transmitted as soon as possible to the Administration requesting the survey.

* Refer to Guidelines for the authorization of organizations acting on behalf of the Administration (resolution A.739(18), as amended by resolution MSC.208(81)), and Specifications on the survey and certification functions of Recognized Organizations acting on behalf of the Administration (resolution A.789(19), as may be amended).

3 A Certificate so issued shall contain a statement to the effect that it has been issued at the request of the Administration and it shall have the same force and receive the same recognition as the Certificate issued under regulation 5 of this Annex.

4 No International Sewage Pollution Prevention Certificate shall be issued to a ship which is entitled to fly the flag of a State which is not a Party.

Regulation 7
Form of Certificate

The International Sewage Pollution Prevention Certificate shall be drawn up in the form corresponding to the model given in the appendix to this Annex and shall be at least in English, French or Spanish. If an official language of the issuing country is also used, this shall prevail in case of a dispute or discrepancy.

Regulation 8
*Duration and validity of Certificate**

1 An International Sewage Pollution Prevention Certificate shall be issued for a period specified by the Administration which shall not exceed five years.

2.1 Notwithstanding the requirements of paragraph 1 of this regulation, when the renewal survey is completed within three months before the expiry date of the existing Certificate, the new Certificate shall be valid from the date of completion of the renewal survey to a date not exceeding five years from the date of expiry of the existing Certificate.

2.2 When the renewal survey is completed after the expiry date of the existing Certificate, the new Certificate shall be valid from the date of completion of the renewal survey to a date not exceeding five years from the date of expiry of the existing Certificate.

2.3 When the renewal survey is completed more than three months before the expiry date of the existing Certificate, the new Certificate shall be valid from the date of completion of the renewal survey to a date not exceeding five years from the date of completion of the renewal survey.

3 If a Certificate is issued for a period of less than five years, the Administration may extend the validity of the Certificate beyond the expiry date to the maximum period specified in paragraph 1 of this regulation.

4 If a renewal survey has been completed and a new Certificate cannot be issued or placed on board the ship before the expiry date of the existing Certificate, the person or organization authorized by the Administration may endorse the existing Certificate and such a Certificate shall be accepted as valid for a further period which shall not exceed five months from the expiry date.

5 If a ship at the time when a Certificate expires is not in a port in which it is to be surveyed, the Administration may extend the period of validity of the Certificate but this extension shall be granted only for the purpose of allowing the ship to complete its voyage to the port in which it is to be surveyed and then only in cases where it appears proper and reasonable to do so. No Certificate shall be extended for a period longer than three months, and a ship to which an extension is granted shall not, on its arrival in the port in which it is to be surveyed, be entitled by virtue of such extension to leave that port without having a new Certificate. When the renewal survey is completed, the new Certificate shall be valid to a date not exceeding five years from the date of expiry of the existing Certificate before the extension was granted.

* Refer to Guidance on the timing of replacement of existing certificates issued after the entry into force of amendments to certificates in IMO instruments (MSC-MEPC.5/Circ.6).

6 A Certificate issued to a ship engaged on short voyages which has not been extended under the fore-going provisions of this regulation may be extended by the Administration for a period of grace of up to one month from the date of expiry stated on it. When the renewal survey is completed, the new Certificate shall be valid to a date not exceeding five years from the date of expiry of the existing Certificate before the extension was granted.

7 In special circumstances, as determined by the Administration, a new Certificate need not be dated from the date of expiry of the existing Certificate as required by paragraph 2.2, 5 or 6 of this regulation. In these special circumstances, the new Certificate shall be valid to a date not exceeding five years from the date of completion of the renewal survey.

8 A Certificate issued under regulation 5 or 6 of this Annex shall cease to be valid in any of the following cases:

.1 if the relevant surveys are not completed within the periods specified under regulation 4.1 of this Annex; or

.2 upon transfer of the ship to the flag of another State. A new Certificate shall only be issued when the Government issuing the new Certificate is fully satisfied that the ship is in compliance with the requirements of regulations 4.7 and 4.8 of this Annex. In the case of a transfer between Parties, if requested within 3 months after the transfer has taken place, the Government of the Party whose flag the ship was formerly entitled to fly shall, as soon as possible, transmit to the Administration copies of the Certificate carried by the ship before the transfer and, if available, copies of the relevant survey reports.

ANNEX I ANNEX II ANNEX III ANNEX IV ANNEX V ANNEX VI UNIFIED INTERPRETATIONS

Chapter 3 – Equipment and control of discharge

Regulation 9
Sewage systems

1 Every ship which, in accordance with regulation 2, is required to comply with the provisions of this Annex shall be equipped with one of the following sewage systems:

.1 a sewage treatment plant which shall be of a type approved by the Administration, taking into account the standards and test methods developed by the Organization,[*] or

SEE INTERPRETATION 3

.2 a sewage comminuting and disinfecting system approved by the Administration. Such system shall be fitted with facilities to the satisfaction of the Administration, for the temporary storage of sewage when the ship is less than 3 nautical miles from the nearest land, or

.3 a holding tank of the capacity to the satisfaction of the Administration for the retention of all sewage, having regard to the operation of the ship, the number of persons on board and other relevant factors. The holding tank shall be constructed to the satisfaction of the Administration and shall have a means to indicate visually the amount of its contents.

2 By derogation from paragraph 1, every passenger ship which, in accordance with regulation 2, is required to comply with the provisions of this Annex, and for which regulation 11.3 applies while in a special area, shall be equipped with one of the following sewage systems:

.1 a sewage treatment plant which shall be of a type approved by the Administration, taking into account the standards and test methods developed by the Organization,[†] or

.2 a holding tank of the capacity to the satisfaction of the Administration for the retention of all sewage, having regard to the operation of the ship, the number of persons on board and other relevant factors. The holding tank shall be constructed to the satisfaction of the Administration and shall have a means to indicate visually the amount of its contents.

Regulation 10
Standard discharge connections

1 To enable pipes of reception facilities to be connected with the ship's discharge pipeline, both lines shall be fitted with a standard discharge connection in accordance with the following table:

SEE INTERPRETATION 4

[*] Refer to the Recommendation on international effluent standards and guidelines for performance tests for sewage treatment plants (resolution MEPC.2(VI)), or Revised guidelines on implementation of effluent standards and performance tests for sewage treatment plants (resolution MEPC.159(55)) (see Unified Interpretation 3).

[†] Refer to the Recommendation on international effluent standards and guidelines for performance tests for sewage treatment plants (resolution MEPC.2(VI)), Revised guidelines on implementation of effluent standards and performance tests for sewage treatment plants (resolution MEPC.159(55)) (see Unified Interpretation 3), or 2012 Guidelines on implementation of effluent standards and performance tests for sewage treatment plants (resolution MEPC.227(64)).

Standard dimensions of flanges for discharge connections

Description	Dimension
Outside diameter	210 mm
Inner diameter	According to pipe outside diameter
Bolt circle diameter	170 mm
Slots in flange	4 holes, 18 mm in diameter, equidistantly placed on a bolt circle of the above diameter, slotted to the flange periphery. The slot width to be 18 mm
Flange thickness	16 mm
Bolts and nuts: quantity and diameter	4, each of 16 mm in diameter and of suitable length
The flange is designed to accept pipes up to a maximum internal diameter of 100 mm and shall be of steel or other equivalent material having a flat face. This flange, together with a suitable gasket, shall be suitable for a service pressure of 600 kPa.For ships having a moulded depth of 5 m and less, the inner diameter of the discharge connection may be 38 mm.	

2 For ships in dedicated trades, i.e. passenger ferries, alternatively the ship's discharge pipeline may be fitted with a discharge connection which can be accepted by the Administration, such as quick-connection couplings.

Regulation 11
Discharge of sewage

A Discharge of sewage from ships other than passenger ships in all areas and discharge of sewage from passenger ships outside special areas

1 Subject to the provisions of regulation 3 of this Annex, the discharge of sewage into the sea is prohibited, except when:

.1 the ship is discharging comminuted and disinfected sewage using a system approved by the Administration in accordance with regulation 9.1.2 of this Annex at a distance of more than 3 nautical miles from the nearest land, or sewage which is not comminuted or disinfected, at a distance of more than 12 nautical miles from the nearest land, provided that, in any case, the sewage that has been stored in holding tanks, or sewage originating from spaces containing living animals, shall not be discharged instantaneously but at a moderate rate when the ship is en route and proceeding at not less than 4 knots; the rate of discharge shall be approved by the Administration based upon standards developed by the Organization;* or

.2 the ship has in operation an approved sewage treatment plant which has been certified by the Administration to meet the operational requirements referred to in regulation 9.1.1 of this Annex, and the effluent shall not produce visible floating solids nor cause discoloration of the surrounding water.

2 The provisions of paragraph 1 shall not apply to ships operating in the waters under the jurisdiction of a State and visiting ships from other States while they are in these waters and are discharging sewage in accordance with such less stringent requirements as may be imposed by such State.

B Discharge of sewage from passenger ships within a special area

3 Subject to the provisions of regulation 3 of this Annex, the discharge of sewage from a passenger ship within a special area shall be prohibited:

.1 for new passenger ships on, or after 1 January 2016, subject to paragraph 2 of regulation 13; and

* Refer to the Recommendation on standards for the rate of discharge of untreated sewage from ships (resolution MEPC.157(55)).

.2 for existing passenger ships on, or after 1 January 2018, subject to paragraph 2 of regulation 13,

except when the following conditions are satisfied:

the ship has in operation an approved sewage treatment plant which has been certified by the Administration to meet the operational requirements referred to in regulation 9.2.1 of this Annex, and the effluent shall not produce visible floating solids nor cause discoloration of the surrounding water.

C *General requirements*

4 When the sewage is mixed with wastes or waste water covered by other Annexes of the present Convention, the requirements of those Annexes shall be complied with in addition to the requirements of this Annex.

Chapter 4 – Reception facilities

Regulation 12
Reception facilities[*]

1 The Government of each Party to the Convention, which requires ships operating in waters under its jurisdiction and visiting ships while in its waters to comply with the requirements of regulation 11.1, undertakes to ensure the provision of facilities at ports and terminals for the reception of sewage, without causing delay to ships, adequate to meet the needs of the ships using them.

2 Small Island Developing States may satisfy the requirements in paragraphs 1 to 3 of this regulation through regional arrangements when, because of those States' unique circumstances, such arrangements are the only practical means to satisfy these requirements. Parties participating in a regional arrangement shall develop a Regional Reception Facilities Plan, taking into account the guidelines developed by the Organization.

The Government of each Party participating in the arrangement shall consult with the Organization, for circulation to the Parties of the present Convention:

- **.1** how the Regional Reception Facilities Plan takes into account the Guidelines;
- **.2** particulars of the identified Regional Ships Waste Reception Centres; and
- **.3** particulars of those ports with only limited facilities.

3 The Government of each Party shall notify the Organization, for transmission to the Contracting Governments concerned, of all cases where the facilities provided under this regulation are alleged to be inadequate.

Regulation 13
Reception facilities for passenger ships in special areas

1 Each Party, the coastline of which borders a special area, undertakes to ensure that:

- **.1** facilities for the reception of sewage are provided in ports and terminals which are in a special area and which are used by passenger ships;
- **.2** the facilities are adequate to meet the needs of those passenger ships; and
- **.3** the facilities are operated so as not to cause undue delay to those passenger ships.

2 The Government of each Party concerned shall notify the Organization of the measures taken pursuant to paragraph 1 of this regulation. Upon receipt of sufficient notifications in accordance with paragraph 1 of this regulation, the Organization shall establish a date from which the requirements of regulation 11.3 in respect of the area in question shall take effect. The Organization shall notify all Parties of the date so established no less than 12 months in advance of that date. Until the date so established, ships while navigating in the special area shall comply with the requirements of regulation 11.1 of this Annex.

[*] Refer to Consolidated guidance for port reception facility providers and users (MEPC.1/Circ.834).

Chapter 5 – Port State control

Regulation 14

*Port State control on operational requirements**

1 A ship when in a port or an offshore terminal of another Party is subject to inspection by officers duly authorized by such Party concerning operational requirements under this Annex, where there are clear grounds for believing that the master or crew are not familiar with essential shipboard procedures relating to the prevention of pollution by sewage.

2 In the circumstances given in paragraph 1 of this regulation, the Party shall take such steps as will ensure that the ship shall not sail until the situation has been brought to order in accordance with the requirements of this Annex.

3 Procedures relating to the port State control prescribed in article 5 of the present Convention shall apply to this regulation.

4 Nothing in this regulation shall be construed to limit the rights and obligations of a Party carrying out control over operational requirements specifically provided for in the present Convention.

* Refer to Procedures for port State control, 2011 (resolution A.1052(27)).

Chapter 6 – Verification of compliance with the provisions of this Annex

Regulation 15
Application

Parties shall use the provisions of the Code for Implementation in the execution of their obligations and responsibilities contained in this Annex.

Regulation 16
Verification of compliance

1 Every Party shall be subject to periodic audits by the Organization in accordance with the audit standard to verify compliance with and implementation of this Annex.

2 The Secretary-General of the Organization shall have responsibility for administering the Audit Scheme, based on the guidelines developed by the Organization.

3 Every Party shall have responsibility for facilitating the conduct of the audit and implementation of a programme of actions to address the findings, based on the guidelines developed by the Organization.

4 Audit of all Parties shall be:

.1 based on an overall schedule developed by the Secretary-General of the Organization, taking into account the guidelines developed by the Organization; and

.2 conducted at periodic intervals, taking into account the guidelines developed by the Organization.*

* Refer to the Framework and Procedures for the IMO Member State Audit Scheme (resolution A.1067(28)).

Chapter 7 – International Code for Ships Operating in Polar Waters

Regulation 17
Definitions

For the purpose of this Annex,

1 *Polar Code* means the International Code for ships operating in polar waters, consisting of an introduction, part I-A and part II-A and parts I-B and II-B, as adopted by resolutions MSC.385(94) and MEPC.264(68), as may be amended, provided that:

.1 amendments to the environment-related provisions of the introduction and chapter 4 of part II-A of the Polar Code are adopted, brought into force and take effect in accordance with the provisions of article 16 of the present Convention concerning the amendment procedures applicable to an appendix to an annex; and

.2 amendments to part II-B of the Polar Code are adopted by the Marine Environment Protection Committee in accordance with its Rules of Procedure.

2 *Antarctic area* means the sea area south of latitude 60°S.

3 *Arctic waters* means those waters which are located north of a line from the latitude 58°00′.0 N and longitude 042°00′.0 W to latitude 64°37′.0 N, longitude 035°27′.0 W and thence by a rhumb line to latitude 67°03′.9 N, longitude 026°33′.4 W and thence by a rhumb line to the latitude 70°49′.56 N and longitude 008°59′.61 W (Sørkapp, Jan Mayen) and by the southern shore of Jan Mayen to 73°31′.6 N and 019°01′.0 E by the Island of Bjørnøya, and thence by a great circle line to the latitude 68°38′.29 N and longitude 043°23′.08 E (Cap Kanin Nos) and thence by the northern shore of the Asian Continent eastward to the Bering Strait and thence from the Bering Strait westward to latitude 60° N as far as Il′pyrskiy and following the 60th North parallel eastward as far as and including Etolin Strait and thence by the northern shore of the North American continent as far south as latitude 60° N and thence eastward along parallel of latitude 60° N, to longitude 056°37′.1 W and thence to the latitude 58°00′.0 N, longitude 042°00′.0 W.

4 *Polar waters* means Arctic waters and/or the Antarctic area.

Regulation 18
Application and requirements

1 This chapter applies to all ships certified in accordance with this Annex operating in polar waters.

2 Unless expressly provided otherwise, any ship covered by paragraph 1 of this regulation shall comply with the environment-related provisions of the introduction and with chapter 4 of part II-A of the Polar Code, in addition to any other applicable requirements of this Annex.

Appendix to Annex IV

Form of International Sewage Pollution Prevention Certificate

INTERNATIONAL SEWAGE POLLUTION
PREVENTION CERTIFICATE

Issued under the provisions of the International Convention for the Prevention of Pollution from Ships, 1973, as modified by the Protocol of 1978 relating thereto, as amended, (hereinafter referred to as "the Convention") under the authority of the Government of:

. . . .

(full designation of the country)

by. . . . .

(full designation of the competent person or organization authorized under the provisions of the Convention)

Particulars of ship*

Name of ship. . . . .

Distinctive number or letters. . . . .

Port of registry . . . .

Gross tonnage. . . . .

Number of persons which the ship is certified to carry. . . . .

IMO Number† . . . .

New/existing ship‡

Date on which keel was laid or ship was at a similar stage of construction or, where applicable, date on which work for a conversion or an alteration or modification of a major character was commenced . . . .

THIS IS TO CERTIFY:

1 That the ship is equipped with a sewage treatment plant/comminuter/holding tank‡ and a discharge pipeline in compliance with regulations 9 and 10 of Annex IV of the Convention as follows:

‡1.1 Description of the sewage treatment plant:

Type of sewage treatment plant . . . .

Name of manufacturer . . . .

The sewage treatment plant is certified by the Administration to meet the effluent standards as provided for in resolution MEPC.2(VI).

‡1.2 Description of comminuter:

Type of comminuter . . . .

Name of manufacturer . . . .

Standard of sewage after disinfection . . . .

* Alternatively, the particulars of the ship may be placed horizontally in boxes.

† Refer to the IMO Ship Identification Number Scheme (resolution A.1078(28)).

‡ Delete as appropriate.

*1.3 Description of holding tank:
Total capacity of the holding tank . . . . m^3
Location . . . .

1.4 A pipeline for the discharge of sewage to a reception facility, fitted with a standard shore connection.

2 That the ship has been surveyed in accordance with regulation 4 of Annex IV of the Convention.

3 That the survey shows that the structure, equipment, systems, fittings, arrangements and material of the ship and the condition thereof are in all respects satisfactory and that the ship complies with the applicable requirements of Annex IV of the Convention.

This Certificate is valid until (dd/mm/yyyy). . . . .†
subject to surveys in accordance with regulation 4 of Annex IV of the Convention.

Completion date of the survey on which this Certificate is based (dd/mm/yyyy). . . . .

Issued at . . . .
(place of issue of Certificate)

Date (dd/mm/yyyy) . . . .
(date of issue) *(signature of duly authorized official issuing the Certificate)*

(seal or stamp of the authority, as appropriate)

* Delete as appropriate.

† Insert the date of expiry as specified by the Administration in accordance with regulation 8.1 of Annex IV of the Convention. The day and the month of this date correspond to the anniversary date as defined in regulation 1.8 of Annex IV of the Convention.

ENDORSEMENT TO EXTEND THE CERTIFICATE IF VALID FOR LESS THAN 5 YEARS WHERE REGULATION 8.3 APPLIES

The ship complies with the relevant provisions of the Convention, and this Certificate shall, in accordance with regulation 8.3 of Annex IV of the Convention, be accepted as valid until (dd/mm/yyyy) ....

Signed...
(signature of duly authorized official)

Place...

Date (dd/mm/yyyy)...............................

(seal or stamp of the authority, as appropriate)

ENDORSEMENT WHERE THE RENEWAL SURVEY HAS BEEN COMPLETED AND REGULATION 8.4 APPLIES

The ship complies with the relevant provisions of the Convention, and this Certificate shall, in accordance with regulation 8.4 of Annex IV of the Convention, be accepted as valid until (dd/mm/yyyy) ....

Signed...
(signature of duly authorized official)

Place...

Date (dd/mm/yyyy)...............................

(seal or stamp of the authority, as appropriate)

ENDORSEMENT TO EXTEND THE VALIDITY OF THE CERTIFICATE UNTIL REACHING THE PORT OF SURVEY OR FOR A PERIOD OF GRACE WHERE REGULATION 8.5 OR 8.6 APPLIES

This Certificate shall, in accordance with regulation 8.5 or 8.6* of Annex IV of the Convention, be accepted as valid until (dd/mm/yyyy) ....

Signed...
(signature of duly authorized official)

Place...

Date (dd/mm/yyyy)...............................

(seal or stamp of the authority, as appropriate)

* Delete as appropriate.

Unified Interpretations of Annex IV

1 Definition of "a similar stage of construction"

Reg. 1.1.1

"*A similar stage of construction*" means the stage at which:

.1 construction identifiable with a specific ship begins; and

.2 assembly of that ship has commenced comprising at least 50 tonnes or one per cent of the estimated mass of all structural material, whichever is less.

2 Building contract date, keel-laying date and delivery date

Reg. 1.1.2

1 Under certain provisions of the SOLAS and MARPOL Conventions, the application of regulations to a ship is governed by the dates:

.1 for which the building contract is placed on or after dd/mm/yyyy; or

.2 in the absence of a building contract, the keel of which is laid or which is at a similar stage of construction on or after dd/mm/yyyy; or

.3 the delivery of which is on or after dd/mm/yyyy.

2 For the application of such provisions, the date on which the building contract is placed for optional ships should be interpreted to be the date on which the original building contract to construct the series of ships is signed between the shipowner and the shipbuilder provided:

.1 the option for construction of the optional ship(s) is ultimately exercised within the period of one year after the date of the original building contract for the series of ships; and

.2 the optional ships are of the same design plans and constructed by the same shipbuilder as that for the series of ships.

3 The application of regulations governed as described in paragraph 1, above, is to be applied as follows:

.1 if a building contract signing date occurs on or after the contract date specified for a particular set of regulation amendments, then, that set of regulation amendments applies;

.2 only in the absence of a building contract does the keel laying date criteria apply and, if a ship's keel laying date occurs on or after the keel laying date specified for a particular set of regulation amendments, then, that set of regulation amendments applies; and

.3 regardless of the building contract signing date or keel laying date, if a ship's delivery date occurs on or after the delivery date specified for a particular set of regulation amendments, then, that set of regulation amendments applies except in the case where the Administration has accepted that the delivery of the ships was delayed due to unforeseen circumstances beyond the control of the shipbuilder and the owner.*

3 Installed on board a ship on or after 1 January 2010

Reg. 9.1.1

For application of resolution MEPC.159(55), the phrase "*installed on board a ship on or after 1 January 2010*" shall be interpreted as follows:

.1 For new ships, installations on board ships the keels of which are laid or which are at a similar stage of construction on or after 1 January 2010.

.2 For existing ships, new installations with a contractual delivery date to the ship on or after 1 January 2010 or, in the absence of a contractual delivery date, the actual delivery of the equipment to the ship on or after 1 January 2010.

* Refer to Unified Interpretation of "Unforeseen delay in the delivery of ships" (MSC.1/Circ.1247 and MARPOL Annex I, Unified Interpretation 6).

4 Standard discharge connections

Reg. 10.1

All ships subject to Annex IV, irrespective of their size and of the presence of a sewage treatment plant or sewage holding tank, shall be provided with a pipeline and the relevant shore connection flange for discharging sewage to port sewage treatment facility.

MARPOL Annex V

Regulations for the prevention of pollution by garbage from ships

MARPOL Annex V

Regulations for the prevention of pollution by garbage from ships*

Chapter 1 – General

Regulation 1
Definitions

For the purposes of this Annex:

1 *Animal carcasses* means the bodies of any animals that are carried on board as cargo and that die or are euthanized during the voyage.

2 *Cargo residues* means the remnants of any cargo which are not covered by other Annexes to the present Convention and which remain on the deck or in holds following loading or unloading, including loading and unloading excess or spillage, whether in wet or dry condition or entrained in wash water but does not include cargo dust remaining on the deck after sweeping or dust on the external surfaces of the ship.

3 *Cooking oil* means any type of edible oil or animal fat used or intended to be used for the preparation or cooking of food, but does not include the food itself that is prepared using these oils.

4 *Domestic wastes* means all types of wastes not covered by other Annexes that are generated in the accommodation spaces on board the ship. Domestic wastes does not include grey water.

5 *En route* means that the ship is underway at sea on a course or courses, including deviation from the shortest direct route, which as far as practicable for navigational purposes, will cause any discharge to be spread over as great an area of the sea as is reasonable and practicable.

6 *Fishing gear* means any physical device or part thereof or combination of items that may be placed on or in the water or on the sea-bed with the intended purpose of capturing, or controlling for subsequent capture or harvesting, marine or fresh water organisms.

7 *Fixed or floating platforms* means fixed or floating structures located at sea which are engaged in the exploration, exploitation or associated offshore processing of sea-bed mineral resources.

8 *Food wastes* means any spoiled or unspoiled food substances and includes fruits, vegetables, dairy products, poultry, meat products and food scraps generated aboard ship.

9 *Garbage* means all kinds of food wastes, domestic wastes and operational wastes, all plastics, cargo residues, incinerator ashes, cooking oil, fishing gear, and animal carcasses generated during the normal operation of the ship and liable to be disposed of continuously or periodically except those substances which are defined or listed in other Annexes to the present Convention. Garbage does not include fresh fish and parts thereof generated as a result of fishing activities undertaken during the voyage, or as a result of aquaculture activities which involve the transport of fish including shellfish for placement in the aquaculture facility and the transport of harvested fish including shellfish from such facilities to shore for processing.

* Refer to 2017 Guidelines for implementation of MARPOL Annex V (resolution MEPC.295(71)).

10 *Incinerator ashes* means ash and clinkers resulting from shipboard incinerators used for the incineration of garbage.

11 *Nearest land.* The term "from the nearest land" means from the baseline from which the territorial sea of the territory in question is established in accordance with international law, except that, for the purposes of the present Annex, "from the nearest land" off the north-eastern coast of Australia shall mean from a line drawn from a point on the coast of Australia in:

latitude 11°00′ S, longitude 142°08′ E
to a point in latitude 10°35′ S, longitude 141°55′ E,
thence to a point latitude 10°00′ S, longitude 142°00′ E,
thence to a point latitude 09°10′ S, longitude 143°52′ E,
thence to a point latitude 09°00′ S, longitude 144°30′ E,
thence to a point latitude 10°41′ S, longitude 145°00′ E,
thence to a point latitude 13°00′ S, longitude 145°00′ E,
thence to a point latitude 15°00′ S, longitude 146°00′ E,
thence to a point latitude 17°30′ S, longitude 147°00′ E,
thence to a point latitude 21°00′ S, longitude 152°55′ E,
thence to a point latitude 24°30′ S, longitude 154°00′ E,
thence to a point on the coast of Australia in
latitude 24°42′ S, longitude 153°15′ E.

12 *Operational wastes* means all solid wastes (including slurries) not covered by other Annexes that are collected on board during normal maintenance or operations of a ship, or used for cargo stowage and handling. Operational wastes also includes cleaning agents and additives contained in cargo hold and external wash water. Operational wastes does not include grey water, bilge water, or other similar discharges essential to the operation of a ship, taking into account the guidelines developed by the Organization.

13 *Plastic* means a solid material which contains as an essential ingredient one or more high molecular mass polymers and which is formed (shaped) during either manufacture of the polymer or the fabrication into a finished product by heat and/or pressure. Plastics have material properties ranging from hard and brittle to soft and elastic. For the purposes of this annex, "all plastics" means all garbage that consists of or includes plastic in any form, including synthetic ropes, synthetic fishing nets, plastic garbage bags and incinerator ashes from plastic products.

14 *Special area* means a sea area where for recognized technical reasons in relation to its oceanographic and ecological condition and to the particular character of its traffic the adoption of special mandatory methods for the prevention of sea pollution by garbage is required.

For the purposes of this Annex the special areas are the Mediterranean Sea area, the Baltic Sea area, the Black Sea area, the Red Sea area, the Gulfs area, the North Sea area, the Antarctic area and the Wider Caribbean Region, which are defined as follows:

.1 The Mediterranean Sea area means the Mediterranean Sea proper including the gulfs and seas therein with the boundary between the Mediterranean and the Black Sea constituted by the 41° N parallel and bounded to the west by the Straits of Gibraltar at the meridian 5°36′ W.

.2 The Baltic Sea area means the Baltic Sea proper with the Gulf of Bothnia and the Gulf of Finland and the entrance to the Baltic Sea bounded by the parallel of the Skaw in the Skagerrak at 57°44.8′ N.

.3 The Black Sea area means the Black Sea proper with the boundary between the Mediterranean and the Black Sea constituted by the parallel 41° N.

.4 The Red Sea area means the Red Sea proper including the Gulfs of Suez and Aqaba bounded at the south by the rhumb line between Ras si Ane (12°28.5′ N, 43°19.6′ E) and Husn Murad (12°40.4′ N, 43°30.2′ E).

.5 The Gulfs area means the sea area located north-west of the rhumb line between Ras al Hadd (22°30′ N, 59°48′ E) and Ras al Fasteh (25°04′ N, 61°25′ E).

.6 The North Sea area means the North Sea proper including seas therein with the boundary between:

.1 the North Sea southwards of latitude 62°N and eastwards of longitude 4° W;

.2 the Skagerrak, the southern limit of which is determined east of the Skaw by latitude 57°44.8′ N; and

.3 the English Channel and its approaches eastwards of longitude 5° W and northwards of latitude 48°30′ N.

.7 The Antarctic area means the sea area south of latitude 60° S.

.8 The Wider Caribbean Region means the Gulf of Mexico and Caribbean Sea proper including the bays and seas therein and that portion of the Atlantic Ocean within the boundary constituted by the 30° N parallel from Florida eastward to 77°30′ W meridian, thence a rhumb line to the intersection of 20° N parallel and 59° W meridian, thence a rhumb line to the intersection of 7°20′ N parallel and 50° W meridian, thence a rhumb line drawn southwesterly to the eastern boundary of French Guiana.

15 *Audit* means a systematic, independent and documented process for obtaining audit evidence and evaluating it objectively to determine the extent to which audit criteria are fulfilled.

16 *Audit Scheme* means the IMO Member State Audit Scheme established by the Organization and taking into account the guidelines developed by the Organization.*

17 *Code for Implementation* means the IMO Instruments Implementation Code (III Code) adopted by the Organization by resolution A.1070(28).

18 *Audit Standard* means the Code for Implementation.

Regulation 2
Application

Unless expressly provided otherwise, the provisions of this Annex shall apply to all ships.

Regulation 3
General prohibition on discharge of garbage into the sea

1 Discharge of all garbage into the sea is prohibited, except as provided otherwise in regulations 4, 5, 6 and 7 of this Annex and section 5.2 of part II-A of the Polar Code, as defined in regulation 13.1 of this Annex.

2 Except as provided in regulation 7 of this Annex, discharge into the sea of all plastics, including but not limited to synthetic ropes, synthetic fishing nets, plastic garbage bags and incinerator ashes from plastic products is prohibited.

3 Except as provided in regulation 7 of this Annex, the discharge into the sea of cooking oil is prohibited.

Regulation 4
Discharge of garbage outside special areas

1 Discharge of the following garbage into the sea outside special areas shall only be permitted while the ship is en route and as far as practicable from the nearest land, but in any case not less than:

.1 3 nautical miles from the nearest land for food wastes which have been passed through a comminuter or grinder. Such comminuted or ground food wastes shall be capable of passing through a screen with openings no greater than 25 mm.

.2 12 nautical miles from the nearest land for food wastes that have not been treated in accordance with subparagraph .1 above.

* Refer to the Framework and Procedures for the IMO Member State Audit Scheme (resolution A.1067(28)).

.3 12 nautical miles from the nearest land for cargo residues that cannot be recovered using commonly available methods for unloading. These cargo residues shall not contain any substances classified as harmful to the marine environment, taking into account guidelines developed by the Organization.

.4 For animal carcasses, discharge shall occur as far from the nearest land as possible, taking into account the guidelines developed by the Organization.

2 Cleaning agents or additives contained in cargo hold, deck and external surfaces wash water may be discharged into the sea, but these substances must not be harmful to the marine environment, taking into account guidelines developed by the Organization.

3 When garbage is mixed with or contaminated by other substances prohibited from discharge or having different discharge requirements, the more stringent requirements shall apply.

Regulation 5
Special requirements for discharge of garbage from fixed or floating platforms

1 Subject to the provisions of paragraph 2 of this regulation, the discharge into the sea of any garbage is prohibited from fixed or floating platforms and from all other ships when alongside or within 500 m of such platforms.

2 Food wastes may be discharged into the sea from fixed or floating platforms located more than 12 nautical miles from the nearest land and from all other ships when alongside or within 500 m of such platforms, but only when the wastes have been passed through a comminuter or grinder. Such comminuted or ground food wastes shall be capable of passing through a screen with openings no greater than 25 mm.

Regulation 6
Discharge of garbage within special areas

1 Discharge of the following garbage into the sea within special areas shall only be permitted while the ship is en route and as follows:

.1 Discharge into the sea of food wastes as far as practicable from the nearest land, but not less than 12 nautical miles from the nearest land or the nearest ice shelf. Food wastes shall be comminuted or ground and shall be capable of passing through a screen with openings no greater than 25 mm. Food wastes shall not be contaminated by any other garbage type. Discharge of introduced avian products, including poultry and poultry parts, is not permitted in the Antarctic area unless it has been treated to be made sterile.

.2 Discharge of cargo residues that cannot be recovered using commonly available methods for unloading, where all the following conditions are satisfied:

.1 cargo residues, cleaning agents or additives, contained in hold washing water do not include any substances classified as harmful to the marine environment, taking into account guidelines developed by the Organization;

.2 both the port of departure and the next port of destination are within the special area and the ship will not transit outside the special area between those ports;

.3 no adequate reception facilities are available at those ports taking into account guidelines developed by the Organization; and

.4 where the conditions of subparagraphs 2.1, 2.2 and 2.3 of this paragraph have been fulfilled, discharge of cargo hold washing water containing residues shall be made as far as practicable from the nearest land or the nearest ice shelf and not less than 12 nautical miles from the nearest land or the nearest ice shelf.

2 Cleaning agents or additives contained in deck and external surfaces wash water may be discharged into the sea, but only if these substances are not harmful to the marine environment, taking into account guidelines developed by the Organization.

3 The following rules (in addition to the rules in paragraph 1 of this regulation) apply with respect to the Antarctic area:

- **.1** Each Party at whose ports ships depart en route to or arrive from the Antarctic area undertakes to ensure that as soon as practicable adequate facilities are provided for the reception of all garbage from all ships, without causing undue delay, and according to the needs of the ships using them.
- **.2** Each Party shall ensure that all ships entitled to fly its flag, before entering the Antarctic area, have sufficient capacity on board for the retention of all garbage, while operating in the area and have concluded arrangements to discharge such garbage at a reception facility after leaving the area.

4 When garbage is mixed with or contaminated by other substances prohibited from discharge or having different discharge requirements, the more stringent requirements shall apply.

Regulation 7
Exceptions

1 Regulations 3, 4, 5 and 6 of this Annex and section 5.2 of chapter 5 of part II-A of the Polar Code shall not apply to:

- **.1** the discharge of garbage from a ship necessary for the purpose of securing the safety of a ship and those on board or saving life at sea; or
- **.2** the accidental loss of garbage resulting from damage to a ship or its equipment, provided that all reasonable precautions have been taken before and after the occurrence of the damage, to prevent or minimize the accidental loss; or
- **.3** the accidental loss of fishing gear from a ship provided that all reasonable precautions have been taken to prevent such loss; or
- **.4** the discharge of fishing gear from a ship for the protection of the marine environment or for the safety of that ship or its crew.

2 Exception of en route:

- **.1** The en route requirements of regulations 4 and 6 of this Annex and chapter 5 of part II-A of the Polar Code shall not apply to the discharge of food wastes where it is clear the retention on board of these food wastes presents an imminent health risk to the people on board.

Regulation 8
*Reception facilities**

1 Each Party undertakes to ensure the provision of adequate facilities at ports and terminals for the reception of garbage without causing undue delay to ships, and according to the needs of the ships using them.

2 Reception facilities within special areas

- **.1** Each Party, the coastline of which borders a special area, undertakes to ensure that as soon as possible, in all ports and terminals within the special area, adequate reception facilities are provided, taking into account the needs of ships operating in these areas.
- **.2** Each Party concerned shall notify the Organization of the measures taken pursuant to paragraph 2.1 of this regulation. Upon receipt of sufficient notifications the Organization shall establish a date from which the requirements of regulation 6 of this Annex in respect of the area in question are to take effect. The Organization shall notify all Parties of the date so established no less than 12 months in advance of that date. Until the date so established, ships that are navigating in a special area shall comply with the requirements of regulation 4 of this Annex as regards discharges outside special areas.

* Refer to Consolidated guidance for port reception facility providers and users (MEPC.1/Circ.834).

3 Small Island Developing States may satisfy the requirements in paragraphs 1 and 2.1 of this regulation through regional arrangements when, because of those States' unique circumstances, such arrangements are the only practical means to satisfy these requirements. Parties participating in a regional arrangement shall develop a Regional Reception Facilities Plan, taking into account the guidelines developed by the Organization.*

The Government of each Party participating in the Arrangement shall consult with the Organization for circulation to the Parties of the present Convention:

.1 how the Regional Reception Facilities Plan takes into account the Guidelines;

.2 particulars of the identified Regional Ships Waste Reception Centres; and

.3 particulars of those ports with only limited facilities.

4 Each Party shall notify the Organization for transmission to the Contracting Parties concerned of all cases where the facilities provided under this regulation are alleged to be inadequate.

Regulation 9
Port State control on operational requirements†

1 A ship when in a port or an offshore terminal of another Party is subject to inspection by officers duly authorized by such Party concerning operational requirements under this Annex, where there are clear grounds for believing that the master or crew are not familiar with essential shipboard procedures relating to the prevention of pollution by garbage.

2 In the circumstances given in paragraph 1 of this regulation, the Party shall take such steps as will ensure that the ship shall not sail until the situation has been brought to order in accordance with the requirements of this Annex.

3 Procedures relating to the port State control prescribed in article 5 of the present Convention shall apply to this regulation.

4 Nothing in this regulation shall be construed to limit the rights and obligations of a Party carrying out control over operational requirements specifically provided for in the present Convention.

Regulation 10
Placards, garbage management plans‡ *and garbage record-keeping*

1 **.1** Every ship of 12 m or more in length overall and fixed or floating platforms shall display placards which notify the crew and passengers of the discharge requirements of regulations 3, 4, 5 and 6 of this Annex and section 5.2 of part II-A of the Polar Code, as applicable.

.2 The placards shall be written in the working language of the ship's crew and, for ships engaged in voyages to ports or offshore terminals under the jurisdiction of other Parties to the Convention, shall also be in English, French or Spanish.

2 Every ship of 100 gross tonnage and above, and every ship which is certified to carry 15 or more persons, and fixed or floating platforms shall carry a garbage management plan which the crew shall follow. This plan shall provide written procedures for minimizing, collecting, storing, processing and disposing of garbage, including the use of the equipment on board. It shall also designate the person or persons in charge of carrying out the plan. Such a plan shall be based on the guidelines developed by the Organization‡ and written in the working language of the crew.

---

* Refer to the 2012 Guidelines for the development of a regional reception facilities plan (resolution MEPC.221(63)).

† Refer to Procedures for port State control, 2011 (resolution A.1052(27)).

‡ Refer to 2012 Guidelines for the development of garbage management plans (resolution MEPC.220(63), as amended).

3 Every ship of 400 gross tonnage and above and every ship which is certified to carry 15 or more persons engaged in voyages to ports or offshore terminals under the jurisdiction of another Party to the Convention and every fixed or floating platform shall be provided with a Garbage Record Book. The Garbage Record Book, whether as a part of the ship's official logbook or otherwise, shall be in the form specified in the appendix to this Annex:

.1 Each discharge into the sea or to a reception facility, or a completed incineration, shall be promptly recorded in the Garbage Record Book and signed for on the date of the discharge or incineration by the officer in charge. Each completed page of the Garbage Record Book shall be signed by the master of the ship. The entries in the Garbage Record Book shall be at least in English, French or Spanish. Where the entries are also made in an official language of the State whose flag the ship is entitled to fly, the entries in that language shall prevail in case of a dispute or discrepancy.

.2 The entry for each discharge or incineration shall include date and time, position of the ship, category of the garbage and the estimated amount discharged or incinerated.

.3 The Garbage Record Book shall be kept on board the ship or the fixed or floating platform, and in such a place as to be readily available for inspection at all reasonable times. This document shall be preserved for a period of at least two years from the date of the last entry made in it.

.4 In the event of any discharge or accidental loss referred to in regulation 7 of this Annex an entry shall be made in the Garbage Record Book, or in the case of any ship of less than 400 gross tonnage, an entry shall be made in the ship's official logbook, of the location, circumstances of, and the reasons for the discharge or loss, details of the items discharged or lost, and the reasonable precautions taken to prevent or minimize such discharge or accidental loss.

4 The Administration may waive the requirements for Garbage Record Books for:

.1 any ship engaged on voyages of 1 h or less in duration which is certified to carry 15 or more persons; or

.2 fixed or floating platforms.

5 The competent authority of the Government of a Party to the Convention may inspect the Garbage Record Books or ship's official logbook on board any ship to which this regulation applies while the ship is in its ports or offshore terminals and may make a copy of any entry in those books, and may require the master of the ship to certify that the copy is a true copy of such an entry. Any copy so made, which has been certified by the master of the ship as a true copy of an entry in the ship's Garbage Record Book or ship's official logbook, shall be admissible in any judicial proceedings as evidence of the facts stated in the entry. The inspection of a Garbage Record Book or ship's official logbook and the taking of a certified copy by the competent authority under this paragraph shall be performed as expeditiously as possible without causing the ship to be unduly delayed.

6 The accidental loss or discharge of fishing gear as provided for in regulations 7.1.3 and 7.1.4 which poses a significant threat to the marine environment or navigation shall be reported to the State whose flag the ship is entitled to fly, and, where the loss or discharge occurs within waters subject to the jurisdiction of a coastal State, also to that coastal State.

Chapter 2 – Verification of compliance with the provisions of this Annex

Regulation 11
Application

Parties shall use the provisions of the Code for Implementation in the execution of their obligations and responsibilities contained in this Annex.

Regulation 12
Verification of compliance

1 Every Party shall be subject to periodic audits by the Organization in accordance with the audit standard to verify compliance with and implementation of this Annex.

2 The Secretary-General of the Organization shall have responsibility for administering the Audit Scheme, based on the guidelines developed by the Organization.*

3 Every Party shall have responsibility for facilitating the conduct of the audit and implementation of a programme of actions to address the findings, based on the guidelines developed by the Organization.*

4 Audit of all Parties shall be:

.1 based on an overall schedule developed by the Secretary-General of the Organization, taking into account the guidelines developed by the Organization; and

.2 conducted at periodic intervals, taking into account the guidelines developed by the Organization.

* Refer to the Framework and Procedures for the IMO Member State Audit Scheme (resolution A.1067(28)).

Chapter 3 – International Code for Ships Operating in Polar Waters

Regulation 13
Definitions

For the purpose of this Annex,

1 *Polar Code* means the International Code for Ships Operating in Polar Waters, consisting of an introduction, part I-A and part II-A and parts I-B and II-B, as adopted by resolutions MSC.385(94) and MEPC.264(68), as may be amended, provided that:

.1 amendments to the environment-related provisions of the introduction and chapter 5 of part II-A of the Polar Code are adopted, brought into force and take effect in accordance with the provisions of article 16 of the present Convention concerning the amendment procedures applicable to an appendix to an annex; and

.2 amendments to part II-B of the Polar Code are adopted by the Marine Environment Protection Committee in accordance with its Rules of Procedure.

2 *Arctic waters* means those waters which are located north of a line from the latitude 58°00′.0 N and longitude 042°00′.0 W to latitude 64°37′.0 N, longitude 035°27′.0 W and thence by a rhumb line to latitude 67°03′.9 N, longitude 026°33′.4 W and thence by a rhumb line to the latitude 70°49′.56 N and longitude 008°59′.61 W (Sørkapp, Jan Mayen) and by the southern shore of Jan Mayen to 73°31′.6 N and 019°01′.0 E by the Island of Bjørnøya, and thence by a great circle line to the latitude 68°38′.29 N and longitude 043°23′.08 E (Cap Kanin Nos) and thence by the northern shore of the Asian Continent eastward to the Bering Strait and thence from the Bering Strait westward to latitude 60° N as far as Il'pyrskiy and following the 60th North parallel eastward as far as and including Etolin Strait and thence by the northern shore of the North American continent as far south as latitude 60° N and thence eastward along parallel of latitude 60° N, to longitude 056°37′.1 W and thence to the latitude 58°00′.0 N, longitude 042°00′.0 W.

3 *Polar waters* means Arctic waters and/or the Antarctic area.

Regulation 14
Application and requirements

1 This chapter applies to all ships to which this Annex applies, operating in polar waters.

2 Unless expressly provided otherwise, any ship covered by paragraph 1 of this regulation shall comply with the environment-related provisions of the introduction and with chapter 5 of part II-A of the Polar Code, in addition to any other applicable requirements of this Annex.

3 In applying chapter 5 of part II-A of the Polar Code, consideration should be given to the additional guidance in part II-B of the Polar Code.

Appendix to Annex V

Form of Garbage Record Book

GARBAGE RECORD BOOK

Name of ship . . . .

Distinctive number or letters. . . . .

IMO Number . . . .

Period . . . . from: . . . . to. . . . .

1 Introduction

In accordance with regulation 10 of Annex V of the International Convention for the Prevention of Pollution from Ships, 1973, as modified by the Protocol of 1978 (MARPOL), a record is to be kept of each discharge operation or completed incineration. This includes discharges into the sea, to reception facilities, or to other ships, as well as the accidental loss of garbage.

2 Garbage and garbage management

Garbage means all kinds of food wastes, domestic wastes and operational wastes, all plastics, cargo residues, incinerator ashes, cooking oil, fishing gear, and animal carcasses generated during the normal operation of the ship and liable to be disposed of continuously or periodically except those substances which are defined or listed in other Annexes to the present Convention. Garbage does not include fresh fish and parts thereof generated as a result of fishing activities undertaken during the voyage, or as a result of aquaculture activities which involve the transport of fish including shellfish for placement in the aquaculture facility and the transport of harvested fish including shellfish from such facilities to shore for processing.

The Guidelines for the Implementation of MARPOL Annex V should also be referred to for relevant information.

3 Description of the garbage

Garbage is to be grouped into categories for the purposes of the Garbage Record Book (or ship's official logbook) as follows:

A Plastics

B Food wastes

C Domestic Wastes

D Cooking Oil

E Incinerator ashes

F Operational wastes

G Cargo residues

H Animal Carcass(es)

I Fishing Gear

4 Entries in the Garbage Record Book

4.1 Entries in the Garbage Record Book shall be made on each of the following occasions:

4.1.1 When garbage is discharged to a reception facility* ashore or to other ships:

.1 Date and time of discharge

.2 Port or facility, or name of ship

.3 Categories of garbage discharged

.4 Estimated amount discharged for each category in cubic metres

.5 Signature of officer in charge of the operation.

4.1.2 When garbage is incinerated:

.1 Date and time of start and stop of incineration

.2 Position of the ship (latitude and longitude) at the start and stop of incineration

.3 Categories of garbage incinerated

.4 Estimated amount incinerated in cubic metres

.5 Signature of the officer in charge of the operation.

4.1.3 When garbage is discharged into the sea in accordance with regulations 4, 5 or 6 of MARPOL Annex V or chapter 5 of part II-A of the Polar Code:

.1 Date and time of discharge

.2 Position of the ship (latitude and longitude). Note: for cargo residue discharges, include discharge start and stop positions.

.3 Category of garbage discharged

.4 Estimated amount discharged for each category in cubic metres

.5 Signature of the officer in charge of the operation.

4.1.4 Accidental or other exceptional discharges or loss of garbage into the sea, including in accordance with regulation 7 of MARPOL Annex V:

.1 Date and time of occurrence

.2 Port or position of the ship at time of occurrence (latitude, longitude and water depth if known)

.3 Categories of garbage discharged or lost

.4 Estimated amount for each category in cubic metres

.5 The reason for the discharge or loss and general remarks.

4.2 Amount of garbage

The amount of garbage on board should be estimated in cubic metres, if possible separately according to category. The Garbage Record Book contains many references to estimated amount of garbage. It is recognized that the accuracy of estimating amounts of garbage is left to interpretation. Volume estimates will differ before and after processing. Some processing procedures may not allow for a usable estimate of volume, e.g. the continuous processing of food waste. Such factors should be taken into consideration when making and interpreting entries made in a record..

* In line with the standard format for waste delivery receipt, MEPC.1/Circ.834, ships' masters should obtain from the operator of the reception facilities, which includes barges and trucks, a receipt or certificate specifying the estimated amount of garbage transferred. The receipts or certificates must be kept together with the Garbage Record Book.

RECORD OF GARBAGE DISCHARGES

Ship's name ....

Distinctive number or letters ....

IMO No. ....

Garbage categories:

A. Plastics
B. Food wastes
C. Domestic wastes (e.g. paper products, rags, glass, metal, bottles, crockery, etc.)
D. Cooking oil
E. Incinerator Ashes
F. Operational wastes
G. Cargo residues
H. Animal Carcass(es)
I. Fishing gear

Date/ time	Position of the ship/ remarks (e.g. accidental loss)	Category	Estimated amount discharged/ incinerated	To sea	To reception facility	Incineration	Certification/ signature

Master's signature .... Date ....

MARPOL Annex VI

Regulations for the prevention of air pollution from ships

MARPOL Annex VI*

Regulations for the prevention of air pollution from ships

Chapter 1 – General

Regulation 1
Application

The provisions of this Annex shall apply to all ships, except where expressly provided otherwise in regulations 3, 5, 6, 13, 15, 16, 18, 19, 20, 21 and 22 of this Annex.

Regulation 2
Definitions

For the purpose of this Annex:

1 *Annex* means Annex VI to the International Convention for the Prevention of Pollution from Ships, 1973 (MARPOL), as modified by the Protocol of 1978 relating thereto, and as modified by the Protocol of 1997, as amended by the Organization, provided that such amendments are adopted and brought into force in accordance with the provisions of article 16 of the present Convention.

2 *A similar stage of construction* means the stage at which:

.1 construction identifiable with a specific ship begins; and

.2 assembly of that ship has commenced comprising at least 50 tonnes or one per cent of the estimated mass of all structural material, whichever is less.

3 *Anniversary date* means the day and the month of each year that will correspond to the date of expiry of the International Air Pollution Prevention Certificate.

4 *Auxiliary control device* means a system, function or control strategy installed on a marine diesel engine that is used to protect the engine and/or its ancillary equipment against operating conditions that could result in damage or failure, or that is used to facilitate the starting of the engine. An auxiliary control device may also be a strategy or measure that has been satisfactorily demonstrated not to be a defeat device.

5 *Continuous feeding* is defined as the process whereby waste is fed into a combustion chamber without human assistance while the incinerator is in normal operating conditions with the combustion chamber operative temperature between 850°C and 1,200°C.

6 *Defeat device* means a device that measures, senses or responds to operating variables (e.g. engine speed, temperature, intake pressure or any other parameter) for the purpose of activating, modulating, delaying or deactivating the operation of any component or the function of the emission control system such that the effectiveness of the emission control system is reduced under conditions encountered during normal operation, unless the use of such a device is substantially included in the applied emission certification test procedures.

* The original MARPOL Annex VI entered into force on 19 May 2005. The revised MARPOL Annex VI adopted by resolution MEPC. 176(58) entered into force on 1 July 2010. The amendments thereto, adopted by resolutions MEPC.190(60), MEPC.194(61), MEPC.202(62), MEPC.203(62), MEPC.217(63), MEPC.247(66), MEPC.251(66) and MEPC.258(67), have entered into force.

7 *Emission* means any release of substances, subject to control by this Annex, from ships into the atmosphere or sea.

8 *Emission control area* means an area where the adoption of special mandatory measures for emissions from ships is required to prevent, reduce and control air pollution from NO_x or SO_x and particulate matter or all three types of emissions and their attendant adverse impacts on human health and the environment. Emission control areas shall include those listed in, or designated under, regulations 13 and 14 of this Annex.

9 *Fuel oil* means any fuel delivered to and intended for combustion purposes for propulsion or operation on board a ship, including gas, distillate and residual fuels.

10 *Gross tonnage* means the gross tonnage calculated in accordance with the tonnage measurement regulations contained in Annex I to the International Convention on Tonnage Measurements of Ships, 1969, or any successor Convention.

11 *Installations* in relation to regulation 12 of this Annex means the installation of systems, equipment, including portable fire-extinguishing units, insulation, or other material on a ship, but excludes the repair or recharge of previously installed systems, equipment, insulation or other material, or the recharge of portable fire-extinguishing units.

12 *Installed* means a marine diesel engine that is or is intended to be fitted on a ship, including a portable auxiliary marine diesel engine, only if its fuelling, cooling or exhaust system is an integral part of the ship. A fuelling system is considered integral to the ship only if it is permanently affixed to the ship. This definition includes a marine diesel engine that is used to supplement or augment the installed power capacity of the ship and is intended to be an integral part of the ship.

13 *Irrational emission control strategy* means any strategy or measure that, when the ship is operated under normal conditions of use, reduces the effectiveness of an emission control system to a level below that expected on the applicable emission test procedures.

14 *Marine diesel engine* means any reciprocating internal combustion engine operating on liquid or dual fuel, to which regulation 13 of this Annex applies, including booster/compound systems if applied. In addition, a gas-fuelled engine installed on a ship constructed on or after 1 March 2016 or a gas-fuelled additional or non-identical replacement engine installed on or after that date is also considered as a marine diesel engine.

15 *NO_x Technical Code* means the Technical Code on Control of Emission of Nitrogen Oxides from Marine Diesel Engines adopted by resolution 2 of the 1997 MARPOL Conference, as amended by the Organization, provided that such amendments are adopted and brought into force in accordance with the provisions of article 16 of the present Convention.

16 *Ozone-depleting substances* means controlled substances defined in paragraph (4) of article 1 of the Montreal Protocol on Substances that Deplete the Ozone Layer, 1987, listed in Annexes A, B, C or E to the said Protocol in force at the time of application or interpretation of this Annex.

Ozone-depleting substances that may be found on board ship include, but are not limited to:

Halon 1211	Bromochlorodifluoromethane
Halon 1301	Bromotrifluoromethane
Halon 2402	1,2-Dibromo-1,1,2,2-tetraflouroethane (also known as Halon 114B2)
CFC-11	Trichlorofluoromethane
CFC-12	Dichlorodifluoromethane
CFC-113	1,1,2-Trichloro-1,2,2-trifluoroethane
CFC-114	1,2-Dichloro-1,1,2,2-tetrafluoroethane
CFC-115	Chloropentafluoroethane

17 *Shipboard incineration* means the incineration of wastes or other matter on board a ship, if such wastes or other matter were generated during the normal operation of that ship.

18 *Shipboard incinerator* means a shipboard facility designed for the primary purpose of incineration.

19 *Ships constructed* means ships the keels of which are laid or that are at a similar stage of construction.

20 *Sludge oil* means sludge from the fuel oil or lubricating oil separators, waste lubricating oil from main or auxiliary machinery, or waste oil from bilge water separators, oil filtering equipment or drip trays.

21 *Tanker* in relation to regulation 15 of this Annex means an oil tanker as defined in regulation 1 of Annex I of the present Convention or a chemical tanker as defined in regulation 1 of Annex II of the present Convention.

For the purpose of chapter 4:

22 *Existing ship* means a ship which is not a new ship.

23 *New ship* means a ship:

.1 for which the building contract is placed on or after 1 January 2013; or

.2 in the absence of a building contract, the keel of which is laid or which is at a similar stage of construction on or after 1 July 2013; or

.3 the delivery of which is on or after 1 July 2015.

SEE INTERPRETATION 1

24 *Major conversion* means in relation to chapter 4 of this Annex a conversion of a ship:

.1 which substantially alters the dimensions, carrying capacity or engine power of the ship; or

.2 which changes the type of the ship; or

.3 the intent of which in the opinion of the Administration is substantially to prolong the life of the ship; or

.4 which otherwise so alters the ship that, if it were a new ship, it would become subject to relevant provisions of the present Convention not applicable to it as an existing ship; or

.5 which substantially alters the energy efficiency of the ship and includes any modifications that could cause the ship to exceed the applicable required EEDI as set out in regulation 21 of this Annex.

SEE INTERPRETATION 2

25 *Bulk carrier* means a ship which is intended primarily to carry dry cargo in bulk, including such types as ore carriers as defined in regulation 1 of chapter XII of SOLAS 74 (as amended) but excluding combination carriers.

26 *Gas carrier* in relation to chapter 4 of this Annex means a cargo ship, other than an LNG carrier as defined in paragraph 38 of this regulation, constructed or adapted and used for the carriage in bulk of any liquefied gas.

27 *Tanker* in relation to chapter 4 of this Annex means an oil tanker as defined in regulation 1 of Annex I of the present Convention or a chemical tanker or an NLS tanker as defined in regulation 1 of Annex II of the present Convention.

28 *Containership* means a ship designed exclusively for the carriage of containers in holds and on deck.

29 *General cargo ship* means a ship with a multi-deck or single deck hull designed primarily for the carriage of general cargo. This definition excludes specialized dry cargo ships, which are not included in the calculation of reference lines for general cargo ships, namely livestock carrier, barge carrier, heavy load carrier, yacht carrier, nuclear fuel carrier.

30 *Refrigerated cargo carrier* means a ship designed exclusively for the carriage of refrigerated cargoes in holds.

SEE INTERPRETATION 3

31 *Combination carrier* means a ship designed to load 100% deadweight with both liquid and dry cargo in bulk.

32 *Passenger ship* means a ship which carries more than 12 passengers.

33 *Ro-ro cargo ship (vehicle carrier)* means a multi deck roll-on-roll-off cargo ship designed for the carriage of empty cars and trucks.

34 *Ro-ro cargo ship* means a ship designed for the carriage of roll-on-roll-off cargo transportation units.

35 *Ro-ro passenger ship* means a passenger ship with roll-on-roll-off cargo spaces.

36 *Attained EEDI* is the EEDI value achieved by an individual ship in accordance with regulation 20 of this Annex.

37 *Required EEDI* is the maximum value of attained EEDI that is allowed by regulation 21 of this Annex for the specific ship type and size.

38 *LNG carrier* in relation to chapter 4 of this Annex means a cargo ship constructed or adapted and used for the carriage in bulk of liquefied natural gas (LNG).

39 *Cruise passenger ship* in relation to chapter 4 of this Annex means a passenger ship not having a cargo deck, designed exclusively for commercial transportation of passengers in overnight accommodations on a sea voyage.

40 *Conventional propulsion* in relation to chapter 4 of this Annex means a method of propulsion where a main reciprocating internal combustion engine(s) is the prime mover and coupled to a propulsion shaft either directly or through a gear box.

41 *Non-conventional propulsion* in relation to chapter 4 of this Annex means a method of propulsion, other than conventional propulsion, including diesel-electric propulsion, turbine propulsion, and hybrid propulsion systems.

42 *Cargo ship having ice-breaking capability* in relation to chapter 4 of this Annex means a cargo ship which is designed to break level ice independently with a speed of at least 2 knots when the level ice thickness is 1.0 m or more having ice bending strength of at least 500 kPa.

43 *A ship delivered on or after 1 September 2019* means a ship:

.1 for which the building contract is placed on or after 1 September 2015; or

.2 in the absence of a building contract, the keel of which is laid, or which is at a similar stage of construction, on or after 1 March 2016; or

.3 the delivery of which is on or after 1 September 2019.

For the purposes of this Annex:

44 *Audit* means a systematic, independent and documented process for obtaining audit evidence and evaluating it objectively to determine the extent to which audit criteria are fulfilled.

45 *Audit Scheme* means the IMO Member State Audit Scheme established by the Organization and taking into account the guidelines developed by the Organization.*

46 *Code for Implementation* means the IMO Instruments Implementation Code (III Code) adopted by the Organization by resolution A.1070(28).

47 *Audit Standard* means the Code for Implementation.

Regulation 3
Exceptions and exemptions

General

1 Regulations of this Annex shall not apply to:

.1 any emission necessary for the purpose of securing the safety of a ship or saving life at sea; or

.2 any emission resulting from damage to a ship or its equipment:

.2.1 provided that all reasonable precautions have been taken after the occurrence of the damage or discovery of the emission for the purpose of preventing or minimizing the emission; and

.2.2 except if the owner or the master acted either with intent to cause damage, or recklessly and with knowledge that damage would probably result.

Trials for ship emission reduction and control technology research

2 The Administration of a Party may, in cooperation with other Administrations as appropriate, issue an exemption from specific provisions of this Annex for a ship to conduct trials for the development of ship emission reduction and control technologies and engine design programmes. Such an exemption shall only be provided if the applications of specific provisions of the Annex or the revised NO_x Technical Code 2008 could impede research into the development of such technologies or programmes. A permit for such an exemption shall only be provided to the minimum number of ships necessary and be subject to the following provisions:

.1 for marine diesel engines with a per cylinder displacement up to 30 L, the duration of the sea trial shall not exceed 18 months. If additional time is required, a permitting Administration or Administrations may permit a renewal for one additional 18-month period; or

.2 for marine diesel engines with a per cylinder displacement at or above 30 L, the duration of the ship trial shall not exceed five years and shall require a progress review by the permitting Administration or Administrations at each intermediate survey. A permit may be withdrawn based on this review if the testing has not adhered to the conditions of the permit or if it is determined that the technology or programme is not likely to produce effective results in the reduction and control of ship emissions. If the reviewing Administration or Administrations determine that additional time is required to conduct a test of a particular technology or programme, a permit may be renewed for an additional time period not to exceed five years.

Emissions from sea-bed mineral activities

3.1 Emissions directly arising from the exploration, exploitation and associated offshore processing of sea-bed mineral resources are, consistent with article 2(3)(b)(ii) of the present Convention, exempt from the provisions of this Annex. Such emissions include the following:

.1 emissions resulting from the incineration of substances that are solely and directly the result of exploration, exploitation and associated offshore processing of sea-bed mineral resources, including but not limited to the flaring of hydrocarbons and the burning of cuttings, muds, and/or stimulation fluids during well completion and testing operations, and flaring arising from upset conditions;

* Refer to the Framework and Procedures for the IMO Member State Audit Scheme (resolution A.1067(28)).

.2 the release of gases and volatile compounds entrained in drilling fluids and cuttings;

.3 emissions associated solely and directly with the treatment, handling or storage of sea-bed minerals; and

.4 emissions from marine diesel engines that are solely dedicated to the exploration, exploitation and associated offshore processing of sea-bed mineral resources.

3.2 The requirements of regulation 18 of this Annex shall not apply to the use of hydrocarbons that are produced and subsequently used on site as fuel, when approved by the Administration.

Regulation 4
Equivalents

1 The Administration of a Party may allow any fitting, material, appliance or apparatus to be fitted in a ship or other procedures, alternative fuel oils, or compliance methods used as an alternative to that required by this Annex if such fitting, material, appliance or apparatus or other procedures, alternative fuel oils, or compliance methods are at least as effective in terms of emissions reductions as that required by this Annex, including any of the standards set forth in regulations 13 and 14.

2 The Administration of a Party that allows a fitting, material, appliance or apparatus or other procedures, alternative fuel oils, or compliance methods used as an alternative to that required by this Annex shall communicate to the Organization for circulation to the Parties particulars thereof, for their information and appropriate action, if any.

3 The Administration of a Party should take into account any relevant guidelines developed by the Organization* pertaining to the equivalents provided for in this regulation.

4 The Administration of a Party that allows the use of an equivalent as set forth in paragraph 1 of this regulation shall endeavour not to impair or damage its environment, human health, property or resources or those of other States.

* Refer to 2015 Guidelines for exhaust gas cleaning systems (resolution MEPC.259(68)).

Chapter 2 – Survey, certification and means of control

Regulation 5
Surveys

1 Every ship of 400 gross tonnage and above and every fixed and floating drilling rig and other platforms shall, to ensure compliance with the requirements of chapter 3 of this Annex, be subject to the surveys specified below:

.1 An initial survey before the ship is put into service or before the certificate required under regulation 6 of this Annex is issued for the first time. This survey shall be such as to ensure that the equipment, systems, fittings, arrangements and material fully comply with the applicable requirements of chapter 3 of this Annex;

.2 A renewal survey at intervals specified by the Administration, but not exceeding five years, except where regulation 9.2, 9.5, 9.6 or 9.7 of this Annex is applicable. The renewal survey shall be such as to ensure that the equipment, systems, fittings, arrangements and material fully comply with applicable requirements of chapter 3 of this Annex;

.3 An intermediate survey within three months before or after the second anniversary date or within three months before or after the third anniversary date of the certificate which shall take the place of one of the annual surveys specified in paragraph 1.4 of this regulation. The intermediate survey shall be such as to ensure that the equipment and arrangements fully comply with the applicable requirements of chapter 3 of this Annex and are in good working order. Such intermediate surveys shall be endorsed on the IAPP Certificate issued under regulation 6 or 7 of this Annex;

.4 An annual survey within three months before or after each anniversary date of the certificate, including a general inspection of the equipment, systems, fittings, arrangements and material referred to in paragraph 1.1 of this regulation to ensure that they have been maintained in accordance with paragraph 5 of this regulation and that they remain satisfactory for the service for which the ship is intended. Such annual surveys shall be endorsed on the IAPP Certificate issued under regulation 6 or 7 of this Annex; and

.5 An additional survey either general or partial, according to the circumstances, shall be made whenever any important repairs or renewals are made as prescribed in paragraph 5 of this regulation or after a repair resulting from investigations prescribed in paragraph 6 of this regulation. The survey shall be such as to ensure that the necessary repairs or renewals have been effectively made, that the material and workmanship of such repairs or renewals are in all respects satisfactory and that the ship complies in all respects with the requirements of chapter 3 of this Annex.

2 In the case of ships of less than 400 gross tonnage, the Administration may establish appropriate measures in order to ensure that the applicable provisions of chapter 3 of this Annex are complied with.

3 Surveys of ships as regards the enforcement of the provisions of this Annex shall be carried out by officers of the Administration.

.1 The Administration may, however, entrust the surveys either to surveyors nominated for the purpose or to organizations recognized by it. Such organizations shall comply with the guidelines adopted by the Organization;*

* Refer to Guidelines for the authorization of organizations acting on behalf of the Administration (resolution A.739(18), as amended by resolution MSC.208(81)), and Specifications on the survey and certification functions of recognized organizations acting on behalf of the Administration (resolution A.789(19), as may be amended). Refer also to Survey Guidelines under the Harmonized System of Survey and Certification for the revised MARPOL Annex VI (resolution MEPC.180(59)).

.2 The survey of marine diesel engines and equipment for compliance with regulation 13 of this Annex shall be conducted in accordance with the revised NO_x Technical Code 2008;

.3 When a nominated surveyor or recognized organization determines that the condition of the equipment does not correspond substantially with the particulars of the certificate, it shall ensure that corrective action is taken and shall in due course notify the Administration. If such corrective action is not taken, the certificate shall be withdrawn by the Administration. If the ship is in a port of another Party, the appropriate authorities of the port State shall also be notified immediately. When an officer of the Administration, a nominated surveyor or recognized organization has notified the appropriate authorities of the port State, the Government of the port State concerned shall give such officer, surveyor or organization any necessary assistance to carry out their obligations under this regulation; and

.4 In every case, the Administration concerned shall fully guarantee the completeness and efficiency of the survey and shall undertake to ensure the necessary arrangements to satisfy this obligation.

4 Ships to which chapter 4 of this Annex applies shall also be subject to the surveys specified below, taking into account the guidelines adopted by the Organization:*

.1 An initial survey before a new ship is put in service and before the International Energy Efficiency Certificate is issued. The survey shall verify that the ship's attained EEDI is in accordance with the requirements in chapter 4 of this Annex, and that the SEEMP required by regulation 22 of this Annex is on board;

.2 A general or partial survey, according to the circumstances, after a major conversion of a new ship to which this regulation applies. The survey shall ensure that the attained EEDI is recalculated as necessary and meets the requirement of regulation 21 of this Annex, with the reduction factor applicable to the ship type and size of the converted ship in the phase corresponding to the date of contract or keel laying or delivery determined for the original ship in accordance with regulation 2.23 of this Annex;

.3 In cases where the major conversion of a new or existing ship is so extensive that the ship is regarded by the Administration as a newly constructed ship, the Administration shall determine the necessity of an initial survey on attained EEDI. Such a survey, if determined necessary, shall ensure that the attained EEDI is calculated and meets the requirement of regulation 21 of this Annex, with the reduction factor applicable corresponding to the ship type and size of the converted ship at the date of the contract of the conversion, or in the absence of a contract, the commencement date of the conversion. The survey shall also verify that the SEEMP required by regulation 22 of this Annex is on board; and

.4 For existing ships, the verification of the requirement to have a SEEMP on board according to regulation 22 of this Annex shall take place at the first intermediate or renewal survey identified in paragraph 1 of this regulation, whichever is the first, on or after 1 January 2013.

SEE INTERPRETATION 4

5 The equipment shall be maintained to conform with the provisions of this Annex and no changes shall be made in the equipment, systems, fittings, arrangements or material covered by the survey, without the express approval of the Administration. The direct replacement of such equipment and fittings with equipment and fittings that conform with the provisions of this Annex is permitted.

6 Whenever an accident occurs to a ship or a defect is discovered that substantially affects the efficiency or completeness of its equipment covered by this Annex, the master or owner of the ship shall report at the earliest opportunity to the Administration, a nominated surveyor or recognized organization responsible for issuing the relevant certificate.

* Refer to 2014 Guidelines on survey and certification of the Energy Efficiency Design Index (resolution MEPC.254(67), as amended by resolution MEPC.261(68)).

Regulation 6
Issue or endorsement of Certificates

International Air Pollution Prevention Certificate

1 An International Air Pollution Prevention Certificate shall be issued, after an initial or renewal survey in accordance with the provisions of regulation 5 of this Annex, to:

.1 any ship of 400 gross tonnage and above engaged in voyages to ports or offshore terminals under the jurisdiction of other Parties; and

.2 platforms and drilling rigs engaged in voyages to waters under the sovereignty or jurisdiction of other Parties.

2 A ship constructed before the date this Annex enters into force for that particular ship's Administration, shall be issued with an International Air Pollution Prevention Certificate in accordance with paragraph 1 of this regulation no later than the first scheduled dry-docking after the date of such entry into force, but in no case later than three years after this date.

3 Such certificate shall be issued or endorsed either by the Administration or by any person or organization duly authorized by it.* In every case, the Administration assumes full responsibility for the certificate.

International Energy Efficiency Certificate

4 An International Energy Efficiency Certificate for the ship shall be issued after a survey in accordance with the provisions of regulation 5.4 of this Annex to any ship of 400 gross tonnage and above before that ship may engage in voyages to ports or offshore terminals under the jurisdiction of other Parties.

SEE INTERPRETATION 4

5 The certificate shall be issued or endorsed either by the Administration or any organization duly authorized by it.* In every case, the Administration assumes full responsibility for the certificate.

Regulation 7
Issue of a Certificate by another Party

1 A Party may, at the request of the Administration, cause a ship to be surveyed and, if satisfied that the provisions of this Annex are complied with, shall issue or authorize the issuance of an International Air Pollution Prevention Certificate or an International Energy Efficiency Certificate to the ship, and where appropriate, endorse or authorize the endorsement of such certificates on the ship, in accordance with this Annex.

2 A copy of the certificate and a copy of the survey report shall be transmitted as soon as possible to the requesting Administration.

3 A certificate so issued shall contain a statement to the effect that it has been issued at the request of the Administration and it shall have the same force and receive the same recognition as a certificate issued under regulation 6 of this Annex.

4 No International Air Pollution Prevention Certificate or an International Energy Efficiency Certificate shall be issued to a ship which is entitled to fly the flag of a State which is not a Party.

* Refer to Guidelines for the authorization of organizations acting on behalf of the Administration (resolution A.739(18), as amended by resolution MSC.208(81)), and Specifications on the survey and certification functions of recognized organizations acting on behalf of the Administration (resolution A.789(19), as may be amended).

Regulation 8
Form of Certificates

International Air Pollution Prevention Certificate

1 The International Air Pollution Prevention (IAPP) Certificate shall be drawn up in a form corresponding to the model given in appendix I to this Annex and shall be at least in English, French or Spanish. If an official language of the issuing country is also used, this shall prevail in case of a dispute or discrepancy.

SEE INTERPRETATION 5

International Energy Efficiency Certificate

2 The International Energy Efficiency Certificate shall be drawn up in a form corresponding to the model given in appendix VIII to this Annex and shall be at least in English, French or Spanish. If an official language of the issuing Party is also used, this shall prevail in case of a dispute or discrepancy.

Regulation 9
Duration and validity of Certificates

International Air Pollution Prevention Certificate

1 An International Air Pollution Prevention (IAPP) Certificate shall be issued for a period specified by the Administration, which shall not exceed five years.

2 Notwithstanding the requirements of paragraph 1 of this regulation:

- **.1** when the renewal survey is completed within three months before the expiry date of the existing certificate, the new certificate shall be valid from the date of completion of the renewal survey to a date not exceeding five years from the date of expiry of the existing certificate;
- **.2** when the renewal survey is completed after the expiry date of the existing certificate, the new certificate shall be valid from the date of completion of the renewal survey to a date not exceeding five years from the date of expiry of the existing certificate; and
- **.3** when the renewal survey is completed more than three months before the expiry date of the existing certificate, the new certificate shall be valid from the date of completion of the renewal survey to a date not exceeding five years from the date of completion of the renewal survey.

3 If a certificate is issued for a period of less than five years, the Administration may extend the validity of the certificate beyond the expiry date to the maximum period specified in paragraph 1 of this regulation, provided that the surveys referred to in regulations 5.1.3 and 5.1.4 of this Annex applicable when a certificate is issued for a period of five years are carried out as appropriate.

4 If a renewal survey has been completed and a new certificate cannot be issued or placed on board the ship before the expiry date of the existing certificate, the person or organization authorized by the Administration may endorse the existing certificate and such a certificate shall be accepted as valid for a further period that shall not exceed five months from the expiry date.

5 If a ship, at the time when a certificate expires, is not in a port in which it is to be surveyed, the Administration may extend the period of validity of the certificate, but this extension shall be granted only for the purpose of allowing the ship to complete its voyage to the port in which it is to be surveyed, and then only in cases where it appears proper and reasonable to do so. No certificate shall be extended for a period longer than three months, and a ship to which an extension is granted shall not, on its arrival in the port in which it is to be surveyed, be entitled by virtue of such extension to leave that port without having a new certificate. When the renewal survey is completed, the new certificate shall be valid to a date not exceeding five years from the date of expiry of the existing certificate before the extension was granted.

6 A certificate issued to a ship engaged on short voyages that has not been extended under the foregoing provisions of this regulation may be extended by the Administration for a period of grace of up to one month from the date of expiry stated on it. When the renewal survey is completed, the new certificate shall be valid to a date not exceeding five years from the date of expiry of the existing certificate before the extension was granted.

7 In special circumstances, as determined by the Administration, a new certificate need not be dated from the date of expiry of the existing certificate as required by paragraph 2.1, 5 or 6 of this regulation. In these special circumstances, the new certificate shall be valid to a date not exceeding five years from the date of completion of the renewal survey.

8 If an annual or intermediate survey is completed before the period specified in regulation 5 of this Annex, then:

.1 the anniversary date shown on the certificate shall be amended by endorsement to a date that shall not be more than three months later than the date on which the survey was completed;

.2 the subsequent annual or intermediate survey required by regulation 5 of this Annex shall be completed at the intervals prescribed by that regulation using the new anniversary date; and

.3 the expiry date may remain unchanged, provided one or more annual or intermediate surveys, as appropriate, are carried out so that the maximum intervals between the surveys prescribed by regulation 5 of this Annex are not exceeded.

9 A certificate issued under regulation 6 or 7 of this Annex shall cease to be valid in any of the following cases:

.1 if the relevant surveys are not completed within the periods specified under regulation 5.1 of this Annex;

.2 if the certificate is not endorsed in accordance with regulation 5.1.3 or 5.1.4 of this Annex; and

.3 upon transfer of the ship to the flag of another State. A new certificate shall only be issued when the Government issuing the new certificate is fully satisfied that the ship is in compliance with the requirements of regulation 5.4 of this Annex. In the case of a transfer between Parties, if requested within three months after the transfer has taken place, the Government of the Party whose flag the ship was formerly entitled to fly shall, as soon as possible, transmit to the Administration copies of the certificate carried by the ship before the transfer and, if available, copies of the relevant survey reports.

International Energy Efficiency Certificate

10 The International Energy Efficiency Certificate shall be valid throughout the life of the ship subject to the provisions of paragraph 11 below.

11 An International Energy Efficiency Certificate issued under this Annex shall cease to be valid in any of the following cases:

.1 if the ship is withdrawn from service or if a new certificate is issued following major conversion of the ship; or

.2 upon transfer of the ship to the flag of another State. A new certificate shall only be issued when the Government issuing the new certificate is fully satisfied that the ship is in compliance with the requirements of chapter 4 of this Annex. In the case of a transfer between Parties, if requested within three months after the transfer has taken place, the Government of the Party whose flag the ship was formerly entitled to fly shall, as soon as possible, transmit to the Administration copies of the certificate carried by the ship before the transfer and, if available, copies of the relevant survey reports.

Regulation 10
Port State control on operational requirements

1 A ship, when in a port or an offshore terminal under the jurisdiction of another Party, is subject to inspection by officers duly authorized by such Party concerning operational requirements under this Annex,* where there are clear grounds for believing that the master or crew are not familiar with essential shipboard procedures relating to the prevention of air pollution from ships.

2 In the circumstances given in paragraph 1 of this regulation, the Party shall take such steps as to ensure that the ship shall not sail until the situation has been brought to order in accordance with the requirements of this Annex.

3 Procedures relating to the port State control prescribed in article 5 of the present Convention shall apply to this regulation.

4 Nothing in this regulation shall be construed to limit the rights and obligations of a Party carrying out control over operational requirements specifically provided for in the present Convention.

5 In relation to chapter 4 of this Annex, any port State inspection shall be limited to verifying, when appropriate, that there is a valid International Energy Efficiency Certificate on board, in accordance with article 5 of the Convention.

Regulation 11
Detection of violations and enforcement

1 Parties shall cooperate in the detection of violations and the enforcement of the provisions of this Annex, using all appropriate and practicable measures of detection and environmental monitoring, adequate procedures for reporting and accumulation of evidence.

2 A ship to which this Annex applies may, in any port or offshore terminal of a Party, be subject to inspection by officers appointed or authorized by that Party for the purpose of verifying whether the ship has emitted any of the substances covered by this Annex in violation of the provision of this Annex. If an inspection indicates a violation of this Annex, a report shall be forwarded to the Administration for any appropriate action.

3 Any Party shall furnish to the Administration evidence, if any, that the ship has emitted any of the substances covered by this Annex in violation of the provisions of this Annex. If it is practicable to do so, the competent authority of the former Party shall notify the master of the ship of the alleged violation.

4 Upon receiving such evidence, the Administration so informed shall investigate the matter, and may request the other Party to furnish further or better evidence of the alleged contravention. If the Administration is satisfied that sufficient evidence is available to enable proceedings to be brought in respect of the alleged violation, it shall cause such proceedings to be taken in accordance with its law as soon as possible. The Administration shall promptly inform the Party that has reported the alleged violation, as well as the Organization, of the action taken.

5 A Party may also inspect a ship to which this Annex applies when it enters the ports or offshore terminals under its jurisdiction, if a request for an investigation is received from any Party together with sufficient evidence that the ship has emitted any of the substances covered by the Annex in any place in violation of this Annex. The report of such investigation shall be sent to the Party requesting it and to the Administration so that the appropriate action may be taken under the present Convention.

6 The international law concerning the prevention, reduction and control of pollution of the marine environment from ships, including that law relating to enforcement and safeguards, in force at the time of application or interpretation of this Annex, applies, mutatis mutandis, to the rules and standards set forth in this Annex.

* Refer to Procedures for port State control (resolution A.1052(27)). Refer also to 2009 Guidelines for port State control under the revised MARPOL Annex VI (resolution MEPC.181(59)).

Chapter 3 – Requirements for control of emissions from ships

Regulation 12
Ozone-depleting substances

1 This regulation does not apply to permanently sealed equipment where there are no refrigerant charging connections or potentially removable components containing ozone-depleting substances.

2 Subject to the provisions of regulation 3.1, any deliberate emissions of ozone-depleting substances shall be prohibited. Deliberate emissions include emissions occurring in the course of maintaining, servicing, repairing or disposing of systems or equipment, except that deliberate emissions do not include minimal releases associated with the recapture or recycling of an ozone-depleting substance. Emissions arising from leaks of an ozone-depleting substance, whether or not the leaks are deliberate, may be regulated by Parties.

3.1 Installations that contain ozone-depleting substances, other than hydrochlorofluorocarbons, shall be prohibited:

.1 on ships constructed on or after 19 May 2005; or

.2 in the case of ships constructed before 19 May 2005, which have a contractual delivery date of the equipment to the ship on or after 19 May 2005 or, in the absence of a contractual delivery date, the actual delivery of the equipment to the ship on or after 19 May 2005.

3.2 Installations that contain hydrochlorofluorocarbons shall be prohibited:

.1 on ships constructed on or after 1 January 2020; or

.2 in the case of ships constructed before 1 January 2020, which have a contractual delivery date of the equipment to the ship on or after 1 January 2020 or, in the absence of a contractual delivery date, the actual delivery of the equipment to the ship on or after 1 January 2020.

4 The substances referred to in this regulation, and equipment containing such substances, shall be delivered to appropriate reception facilities when removed from ships.

5 Each ship subject to regulation 6.1 shall maintain a list of equipment containing ozone-depleting substances.*

6 Each ship subject to regulation 6.1 that has rechargeable systems that contain ozone-depleting substances shall maintain an *ozone-depleting substances record book*. This record book may form part of an existing logbook or electronic recording system as approved by the Administration.

7 Entries in the ozone-depleting substances record book shall be recorded in terms of mass (kg) of substance and shall be completed without delay on each occasion, in respect of the following:

.1 recharge, full or partial, of equipment containing ozone-depleting substances;

.2 repair or maintenance of equipment containing ozone-depleting substances;

.3 discharge of ozone-depleting substances to the atmosphere:

.3.1 deliberate; and

.3.2 non-deliberate;

.4 discharge of ozone-depleting substances to land-based reception facilities; and

.5 supply of ozone-depleting substances to the ship.

* See appendix I, Supplement to International Air Pollution Prevention Certificate (IAPP Certificate), section 2.1.

Regulation 13
Nitrogen oxides (NO_x)

Application

1.1 This regulation shall apply to:

.1 each marine diesel engine with a power output of more than 130 kW installed on a ship; and

.2 each marine diesel engine with a power output of more than 130 kW that undergoes a major conversion on or after 1 January 2000 except when demonstrated to the satisfaction of the Administration that such engine is an identical replacement to the engine that it is replacing and is otherwise not covered under paragraph 1.1.1 of this regulation.

SEE INTERPRETATION 6

1.2 This regulation does not apply to:

.1 a marine diesel engine intended to be used solely for emergencies, or solely to power any device or equipment intended to be used solely for emergencies on the ship on which it is installed, or a marine diesel engine installed in lifeboats intended to be used solely for emergencies; and

.2 a marine diesel engine installed on a ship solely engaged in voyages within waters subject to the sovereignty or jurisdiction of the State the flag of which the ship is entitled to fly, provided that such engine is subject to an alternative NO_x control measure established by the Administration.

1.3 Notwithstanding the provisions of paragraph 1.1 of this regulation, the Administration may provide an exclusion from the application of this regulation for any marine diesel engine that is installed on a ship constructed, or for any marine diesel engine that undergoes a major conversion, before 19 May 2005, provided that the ship on which the engine is installed is solely engaged in voyages to ports or offshore terminals within the State the flag of which the ship is entitled to fly.

Major conversion

2.1 For the purpose of this regulation, *major conversion* means a modification on or after 1 January 2000 of a marine diesel engine that has not already been certified to the standards set forth in paragraph 3, 4, or 5.1.1 of this regulation where:

.1 the engine is replaced by a marine diesel engine or an additional marine diesel engine is installed, or

.2 any substantial modification, as defined in the revised NO_x Technical Code 2008, is made to the engine, or

.3 the maximum continuous rating of the engine is increased by more than 10% compared to the maximum continuous rating of the original certification of the engine.

2.2 For a major conversion involving the replacement of a marine diesel engine with a non-identical marine diesel engine, or the installation of an additional marine diesel engine, the standards in this regulation at the time of the replacement or addition of the engine shall apply. In the case of replacement engines only, if it is not possible for such a replacement engine to meet the standards set forth in paragraph 5.1.1 of this regulation (Tier III, as applicable), then that replacement engine shall meet the standards set forth in paragraph 4 of this regulation (Tier II), taking into account guidelines developed by the Organization.*

SEE INTERPRETATIONS 6 AND 7

* Refer to 2013 Guidelines as required by regulation 13.2.2 of MARPOL Annex VI in respect of non-identical replacement engines not required to meet the Tier III limit (resolution MEPC.230(65)).

2.3 A marine diesel engine referred to in paragraph 2.1.2 or 2.1.3 of this regulation shall meet the following standards:

.1 for ships constructed prior to 1 January 2000, the standards set forth in paragraph 3 of this regulation shall apply; and

.2 for ships constructed on or after 1 January 2000, the standards in force at the time the ship was constructed shall apply.

Tier I*

3 Subject to regulation 3 of this Annex, the operation of a marine diesel engine that is installed on a ship constructed on or after 1 January 2000 and prior to 1 January 2011 is prohibited, except when the emission of nitrogen oxides (calculated as the total weighted emission of NO_2) from the engine is within the following limits, where n = rated engine speed (crankshaft revolutions per minute):

.1 17.0 g/kWh when n is less than 130 rpm;

.2 $45 \cdot n^{(-0.2)}$ g/kWh when n is 130 or more but less than 2,000 rpm;

.3 9.8 g/kWh when n is 2,000 rpm or more.

Tier II

4 Subject to regulation 3 of this Annex, the operation of a marine diesel engine that is installed on a ship constructed on or after 1 January 2011 is prohibited, except when the emission of nitrogen oxides (calculated as the total weighted emission of NO_2) from the engine is within the following limits, where n = rated engine speed (crankshaft revolutions per minute):

.1 14.4 g/kWh when n is less than 130 rpm;

.2 $44 \cdot n^{(-0.23)}$ g/kWh when n is 130 or more but less than 2,000 rpm;

.3 7.7 g/kWh when n is 2,000 rpm or more.

Tier III

5.1 Subject to regulation 3 of this Annex, in an emission control area designated for Tier III NO_x control under paragraph 6 of this regulation, the operation of a marine diesel engine that is installed on a ship:

.1 is prohibited except when the emission of nitrogen oxides (calculated as the total weighted emission of NO_x)† from the engine is within the following limits, where n = rated engine speed (crankshaft revolutions per minute):

.1.1 3.4 g/kWh when n is less than 130 rpm;

.1.2 $9 \cdot n^{(-0.2)}$ g/kWh when n is 130 or more but less than 2,000 rpm;

.1.3 2.0 g/kWh when n is 2,000 rpm or more;

when:

.2 that ship is constructed on or after 1 January 2016 and is operating in the North American Emission Control Area or the United States Caribbean Sea Emission Control Area;

when:

.3 that ship is operating in an emission control area designated for Tier III NO_x control under paragraph 6 of this regulation, other than an emission control area described in paragraph 5.1.2 of this regulation, and is constructed on or after the date of adoption of such an emission control area, or a later date as may be specified in the amendment designating the NO_x Tier III emission control area, whichever is later.

* Refer to Guidelines for the application of the NO_x Technical Code relative to certification and amendments of Tier I engines (MEPC.1/Circ.679).

† "NO_x" is corrected to "NO_2" by Amendments to regulation 13 of MARPOL Annex VI (resolution MEPC.271(69)); refer to Additional information.

5.2 The standards set forth in paragraph 5.1.1 of this regulation shall not apply to:

.1 a marine diesel engine installed on a ship with a length (*L*), as defined in regulation 1.19 of Annex I to the present Convention, of less than 24 m when it has been specifically designed, and is used solely, for recreational purposes; or

.2 a marine diesel engine installed on a ship with a combined nameplate diesel engine propulsion power of less than 750 kW if it is demonstrated, to the satisfaction of the Administration, that the ship cannot comply with the standards set forth in paragraph 5.1.1 of this regulation because of design or construction limitations of the ship; or

.3 a marine diesel engine installed on a ship constructed prior to 1 January 2021 of less than 500 gross tonnage, with a length (*L*), as defined in regulation 1.19 of Annex I to the present convention, of 24 m or over when it has been specifically designed, and is used solely, for recreational purposes.

Emission control area

6 For the purposes of this regulation, emission control areas shall be:

.1 the North American area, which means the area described by the coordinates provided in appendix VII to this Annex;

.2 the United States Caribbean sea area, which means the area described by the coordinates provided in appendix VII to this Annex; and

.3 any other sea area, including any port area, designated by the Organization in accordance with the criteria and procedures set forth in appendix III to this Annex.

Marine diesel engines installed on a ship constructed prior to 1 January 2000

7.1 Notwithstanding paragraph 1.1.1 of this regulation, a marine diesel engine with a power output of more than 5,000 kW and a per cylinder displacement at or above 90 L installed on a ship constructed on or after 1 January 1990 but prior to 1 January 2000 shall comply with the emission limits set forth in paragraph 7.4 of this regulation, provided that an approved method* for that engine has been certified by an Administration of a Party and notification of such certification has been submitted to the Organization by the certifying Administration.† Compliance with this paragraph shall be demonstrated through one of the following:

.1 installation of the certified approved method, as confirmed by a survey using the verification procedure specified in the approved method file, including appropriate notation on the ship's International Air Pollution Prevention Certificate of the presence of the approved method; or

.2 certification of the engine confirming that it operates within the limits set forth in paragraph 3, 4, or 5.1.1 of this regulation and an appropriate notation of the engine certification on the ship's International Air Pollution Prevention Certificate.

7.2 Paragraph 7.1 of this regulation shall apply no later than the first renewal survey that occurs 12 months or more after deposit of the notification in paragraph 7.1. If a shipowner of a ship on which an approved method is to be installed can demonstrate to the satisfaction of the Administration that the approved method was not commercially available despite best efforts to obtain it, then that approved method shall be installed on the ship no later than the next annual survey of that ship that falls after the approved method is commercially available.

7.3 With regard to a marine diesel engine with a power output of more than 5,000 kW and a per cylinder displacement at or above 90 L installed on a ship constructed on or after 1 January 1990, but prior to 1 January 2000, the International Air Pollution Prevention Certificate shall, for a marine diesel engine to which paragraph 7.1 of this regulation applies, indicate one of the following:

.1 an approved method has been applied pursuant to paragraph 7.1.1 of this regulation;

* Refer to 2014 Guidelines on the approved method process (resolution MEPC.243(66)).

† Refer to 2014 Guidelines in respect of the information to be submitted by an Administration to the Organization covering the certification of an approved method as required under regulation 13.7.1 of MARPOL Annex VI (resolution MEPC.242(66)).

.2 the engine has been certified pursuant to paragraph 7.1.2 of this regulation;

.3 an approved method is not yet commercially available as described in paragraph 7.2 of this regulation; or

.4 an approved method is not applicable.

7.4 Subject to regulation 3 of this Annex, the operation of a marine diesel engine described in paragraph 7.1 of this regulation is prohibited, except when the emission of nitrogen oxides (calculated as the total weighted emission of NO_2) from the engine is within the following limits, where n = rated engine speed (crankshaft revolutions per minute):

.1 17.0 g/kWh when n is less than 130 rpm;

.2 $45 \cdot n^{(-0.2)}$ g/kWh when n is 130 or more but less than 2,000 rpm; and

.3 9.8 g/kWh when n is 2,000 rpm or more.

7.5 Certification of an approved method shall be in accordance with chapter 7 of the revised NO_x Technical Code 2008 and shall include verification:

.1 by the designer of the base marine diesel engine to which the approved method applies that the calculated effect of the approved method will not decrease engine rating by more than 1.0%, increase fuel consumption by more than 2.0% as measured according to the appropriate test cycle set forth in the revised NO_x Technical Code 2008, or adversely affect engine durability or reliability; and

.2 that the cost of the approved method is not excessive, which is determined by a comparison of the amount of NO_x reduced by the approved method to achieve the standard set forth in paragraph 7.4 of this regulation and the cost of purchasing and installing such approved method.*

Certification

8 The revised NO_x Technical Code 2008 shall be applied in the certification, testing and measurement procedures for the standards set forth in this regulation.

9 The procedures for determining NO_x emissions set out in the revised NO_x Technical Code 2008 are intended to be representative of the normal operation of the engine. Defeat devices and irrational emission control strategies undermine this intention and shall not be allowed. This regulation shall not prevent the use of auxiliary control devices that are used to protect the engine and/or its ancillary equipment against operating conditions that could result in damage or failure or that are used to facilitate the starting of the engine.

Regulation 14
Sulphur oxides (SO_x) and particulate matter

General requirements

1 The sulphur content of any fuel oil used on board ships shall not exceed the following limits:

.1 4.50% m/m prior to 1 January 2012;

.2 3.50% m/m on and after 1 January 2012; and

.3 0.50% m/m on and after 1 January 2020.

* The cost of an approved method shall not exceed 375 Special Drawing Rights/metric tonne NO_x calculated in accordance with the cost-effectiveness (Ce) formula below:

$$Ce = \frac{\text{Cost of approved method} \cdot 10^6}{\text{Power (kW)} \cdot 0.768 \cdot 6{,}000 \text{ (hours/year)} \cdot 5 \text{ (years)} \cdot \Delta NO_x \text{ (g/kWh)}}$$

Refer to Definitions for the cost-effectiveness formula in regulation 13.7.5 of the revised MARPOL Annex VI (MEPC.1/Circ.678).

2 The worldwide average sulphur content of residual fuel oil supplied for use on board ships shall be monitored taking into account guidelines developed by the Organization.*

Requirements within emission control areas

3 For the purpose of this regulation, emission control areas shall include:

.1 the Baltic Sea area as defined in regulation 1.11.2 of Annex I and the North Sea as defined in regulation 1.14.6 of Annex V;

.2 the North American area as described by the coordinates provided in appendix VII to this Annex;

.3 the United States Caribbean Sea area as described by the coordinates provided in appendix VII to this Annex; and

.4 any other sea area, including any port area, designated by the Organization in accordance with the criteria and procedures set forth in appendix III to this Annex.

4 While ships are operating within an emission control area, the sulphur content of fuel oil used on board ships shall not exceed the following limits:

.1 1.50% m/m prior to 1 July 2010;

.2 1.00% m/m on and after 1 July 2010; and

.3 0.10% m/m on and after 1 January 2015.

.4 Prior to 1 January 2020, the sulphur content of fuel oil referred to in paragraph 4 of this regulation shall not apply to ships operating in the North American area or the United States Caribbean Sea area defined in paragraph 3, built on or before 1 August 2011 that are powered by propulsion boilers that were not originally designed for continued operation on marine distillate fuel or natural gas.

5 The sulphur content of fuel oil referred to in paragraph 1 and paragraph 4 of this regulation shall be documented by its supplier as required by regulation 18 of this Annex.

6 Those ships using separate fuel oils to comply with paragraph 4 of this regulation and entering or leaving an emission control area set forth in paragraph 3 of this regulation shall carry a written procedure showing how the fuel oil changeover is to be done, allowing sufficient time for the fuel oil service system to be fully flushed of all fuel oils exceeding the applicable sulphur content specified in paragraph 4 of this regulation prior to entry into an emission control area. The volume of low sulphur fuel oils in each tank as well as the date, time and position of the ship when any fuel oil changeover operation is completed prior to the entry into an emission control area or commenced after exit from such an area shall be recorded in such logbook as prescribed by the Administration.

7 During the first 12 months immediately following entry into force of an amendment designating a specific emission control area under paragraph 3 of this regulation, ships operating in that emission control area are exempt from the requirements in paragraphs 4 and 6 of this regulation and from the requirements of paragraph 5 of this regulation insofar as they relate to paragraph 4 of this regulation.

Review provision†

8 A review of the standard set forth in paragraph 1.3 of this regulation shall be completed by 2018 to determine the availability of fuel oil to comply with the fuel oil standard set forth in that paragraph and shall take into account the following elements:

.1 the global market supply and demand for fuel oil to comply with paragraph 1.3 of this regulation that exist at the time that the review is conducted;

* Refer to 2010 Guidelines for monitoring the worldwide average sulphur content of fuel oils supplied for use on board ships (resolution MEPC.192(61), as amended by resolution MEPC.273(69)).

† Refer to Effective date of implementation of the fuel oil standard in regulation 14.1.3 of MARPOL Annex VI (resolution MEPC.280(70)).

.2 an analysis of the trends in fuel oil markets; and

.3 any other relevant issue.

9 The Organization shall establish a group of experts, comprising representatives with the appropriate expertise in the fuel oil market and appropriate maritime, environmental, scientific and legal expertise, to conduct the review referred to in paragraph 8 of this regulation. The group of experts shall develop the appropriate information to inform the decision to be taken by the Parties.

10 The Parties, based on the information developed by the group of experts, may decide whether it is possible for ships to comply with the date in paragraph 1.3 of this regulation. If a decision is taken that it is not possible for ships to comply, then the standard in that paragraph shall become effective on 1 January 2025.

Regulation 15
Volatile organic compounds (VOCs)

1 If the emissions of VOCs from a tanker are to be regulated in a port or ports or a terminal or terminals under the jurisdiction of a Party, they shall be regulated in accordance with the provisions of this regulation.

2 A Party regulating tankers for VOC emissions shall submit a notification to the Organization.[*] This notification shall include information on the size of tankers to be controlled, the cargoes requiring vapour emission control systems and the effective date of such control. The notification shall be submitted at least six months before the effective date.

3 A Party that designates ports or terminals at which VOC emissions from tankers are to be regulated shall ensure that vapour emission control systems, approved by that Party taking into account the safety standards for such systems developed by the Organization,[†] are provided in any designated port and terminal and are operated safely and in a manner so as to avoid undue delay to a ship.

4 The Organization shall circulate a list of the ports and terminals designated by Parties to other Parties and Member States of the Organization for their information.

5 A tanker to which paragraph 1 of this regulation applies shall be provided with a vapour emission collection system approved by the Administration taking into account the safety standards for such systems developed by the Organization,[†] and shall use this system during the loading of relevant cargoes. A port or terminal that has installed vapour emission control systems in accordance with this regulation may accept tankers that are not fitted with vapour collection systems for a period of three years after the effective date identified in paragraph 2 of this regulation.

6 A tanker carrying crude oil shall have on board and implement a VOC management plan approved by the Administration.[‡] Such a plan shall be prepared taking into account the guidelines developed by the Organization. The plan shall be specific to each ship and shall at least:

.1 provide written procedures for minimizing VOC emissions during the loading, sea passage and discharge of cargo;

.2 give consideration to the additional VOC generated by crude oil washing;

.3 identify a person responsible for implementing the plan; and

.4 for ships on international voyages, be written in the working language of the master and officers and, if the working language of the master and officers is not English, French or Spanish, include a translation into one of these languages.

[*] Refer to Notification to the Organization on ports or terminals where volatile organic compounds (VOCs) emissions are to be regulated (MEPC.1/Circ.509).

[†] Refer to Standards for vapour emission control systems (MSC/Circ.585).

[‡] Refer to Guidelines for the development of a VOC management plan (resolution MEPC.185(59)). Refer also to Technical information on systems and operation to assist development of VOC management plans (MEPC.1/Circ.680), and Technical information on a vapour pressure control system to facilitate the development and update of VOC management plans (MEPC.1/Circ.719).

ANNEX I ANNEX II ANNEX III ANNEX IV ANNEX V ANNEX VI UNIFIED INTERPRETATIONS

7 This regulation shall also apply to gas carriers only if the types of loading and containment systems allow safe retention of non-methane VOCs on board or their safe return ashore.*

SEE INTERPRETATION 8

Regulation 16
Shipboard incineration

1 Except as provided in paragraph 4 of this regulation, shipboard incineration shall be allowed only in a shipboard incinerator.

2 Shipboard incineration of the following substances shall be prohibited:

.1 residues of cargoes subject to Annex I, II or III or related contaminated packing materials;

.2 polychlorinated biphenyls (PCBs);

.3 garbage, as defined by Annex V, containing more than traces of heavy metals;

.4 refined petroleum products containing halogen compounds;

.5 sewage sludge and sludge oil either of which is not generated on board the ship; and

.6 exhaust gas cleaning system residues.

3 Shipboard incineration of polyvinyl chlorides (PVCs) shall be prohibited, except in shipboard incinerators for which IMO Type Approval Certificates† have been issued.

4 Shipboard incineration of sewage sludge and sludge oil generated during normal operation of a ship may also take place in the main or auxiliary power plant or boilers, but in those cases, shall not take place inside ports, harbours and estuaries.

5 Nothing in this regulation neither:

.1 affects the prohibition in, or other requirements of, the Convention on the Prevention of Marine Pollution by Dumping of Wastes and Other Matter, 1972, as amended, and the 1996 Protocol thereto, nor

.2 precludes the development, installation and operation of alternative design shipboard thermal waste treatment devices that meet or exceed the requirements of this regulation.

6.1 Except as provided in paragraph 6.2 of this regulation, each incinerator on a ship constructed on or after 1 January 2000 or incinerator that is installed on board a ship on or after 1 January 2000 shall meet the requirements contained in appendix IV to this Annex. Each incinerator subject to this paragraph shall be approved by the Administration taking into account the standard specification for shipboard incinerators developed by the Organization;‡ or

6.2 The Administration may allow exclusion from the application of paragraph 6.1 of this regulation to any incinerator installed on board a ship before 19 May 2005, provided that the ship is solely engaged in voyages within waters subject to the sovereignty or jurisdiction of the State the flag of which the ship is entitled to fly.

7 Incinerators installed in accordance with the requirements of paragraph 6.1 of this regulation shall be provided with a manufacturer's operating manual, which is to be retained with the unit and which shall specify how to operate the incinerator within the limits described in paragraph 2 of appendix IV of this Annex.

---

* Refer to the International Code for the construction and equipment of ships carrying liquefied gases in bulk (resolution MSC.370(93)).

† Type Approval Certificates issued in accordance with Revised guidelines for the implementation of Annex V of MARPOL 73/78 (resolution MEPC.59(33), as amended by resolution MEPC.92(45)), or Standard specification for shipboard incinerators (resolution MEPC.76(40), as amended by resolution MEPC.93(45)), or 2014 Standard specification for shipboard incinerators (resolution MEPC 244(66)).

‡ Refer to 2014 Standard specification for shipboard incinerators (resolution MEPC.244(66)), or Standard specification for shipboard incinerators (resolution MEPC.76(40), as amended by resolution MEPC.93(45)), and Type approval of shipboard incinerators (MEPC.1/Circ.793).

8 Personnel responsible for the operation of an incinerator installed in accordance with the requirements of paragraph 6.1 of this regulation shall be trained to implement the guidance provided in the manufacturer's operating manual as required by paragraph 7 of this regulation.

9 For incinerators installed in accordance with the requirements of paragraph 6.1 of this regulation the combustion chamber gas outlet temperature shall be monitored at all times the unit is in operation. Where that incinerator is of the continuous-feed type, waste shall not be fed into the unit when the combustion chamber gas outlet temperature is below 850°C. Where that incinerator is of the batch-loaded type, the unit shall be designed so that the combustion chamber gas outlet temperature shall reach 600°C within five minutes after start-up and will thereafter stabilize at a temperature not less than 850°C.

SEE INTERPRETATION 9

Regulation 17
Reception facilities

1 Each Party undertakes to ensure the provision of facilities adequate to meet the:

.1 needs of ships using its repair ports for the reception of ozone-depleting substances and equipment containing such substances when removed from ships;

.2 needs of ships using its ports, terminals or repair ports for the reception of exhaust gas cleaning residues from an exhaust gas cleaning system;

without causing undue delay to ships, and

.3 needs in ship-breaking facilities for the reception of ozone-depleting substances and equipment containing such substances when removed from ships.

2 Small Island Developing States[*] may satisfy the requirements in paragraph 1 of this regulation through regional arrangements when, because of those States' unique circumstances, such arrangements are the only practical means to satisfy these requirements. Parties participating in a regional arrangement shall develop a Regional Reception Facilities Plan, taking into account the guidelines developed by the Organization.[†]

The Government of each Party participating in the arrangement shall consult with the Organization for circulation to the Parties of the present Convention:

.1 how the Regional Reception Facilities Plan takes into account the Guidelines;

.2 particulars of the identified Regional Ships Waste Reception Centres; and

.3 particulars of those ports with only limited facilities.

3 If a particular port or terminal of a Party is, taking into account the guidelines to be developed by the Organization, remotely located from, or lacking in, the industrial infrastructure necessary to manage and process those substances referred to in paragraph 1 of this regulation and therefore cannot accept such substances, then the Party shall inform the Organization of any such port or terminal so that this information may be circulated to all Parties and Member States of the Organization for their information and any appropriate action. Each Party that has provided the Organization with such information shall also notify the Organization of its ports and terminals where reception facilities are available to manage and process such substances.

4 Each Party shall notify the Organization for transmission to the Members of the Organization of all cases where the facilities provided under this regulation are unavailable or alleged to be inadequate.

[*] Refer to 2012 Guidelines for the development of a regional reception facilities plan (resolution MEPC.221(63)).

[†] Refer to 2011 Guidelines for reception facilities under MARPOL Annex VI (resolution MEPC.199(62)).

Regulation 18
Fuel oil availability and quality

Fuel oil availability

1 Each Party shall take all reasonable steps to promote the availability of fuel oils that comply with this Annex and inform the Organization of the availability of compliant fuel oils in its ports and terminals.

2.1 If a ship is found by a Party not to be in compliance with the standards for compliant fuel oils set forth in this Annex, the competent authority of the Party is entitled to require the ship to:

.1 present a record of the actions taken to attempt to achieve compliance; and

.2 provide evidence that it attempted to purchase compliant fuel oil in accordance with its voyage plan and, if it was not made available where planned, that attempts were made to locate alternative sources for such fuel oil and that despite best efforts to obtain compliant fuel oil, no such fuel oil was made available for purchase.

2.2 The ship should not be required to deviate from its intended voyage or to delay unduly the voyage in order to achieve compliance.

2.3 If a ship provides the information set forth in paragraph 2.1 of this regulation, a Party shall take into account all relevant circumstances and the evidence presented to determine the appropriate action to take, including not taking control measures.

2.4 A ship shall notify its Administration and the competent authority of the relevant port of destination when it cannot purchase compliant fuel oil.

2.5 A Party shall notify the Organization when a ship has presented evidence of the non-availability of compliant fuel oil.

Fuel oil quality

3 Fuel oil for combustion purposes delivered to and used on board ships to which this Annex applies shall meet the following requirements:

.1 except as provided in paragraph 3.2 of this regulation:

.1.1 the fuel oil shall be blends of hydrocarbons derived from petroleum refining. This shall not preclude the incorporation of small amounts of additives intended to improve some aspects of performance;

.1.2 the fuel oil shall be free from inorganic acid; and

.1.3 the fuel oil shall not include any added substance or chemical waste that:

.1.3.1 jeopardizes the safety of ships or adversely affects the performance of the machinery, or

.1.3.2 is harmful to personnel, or

.1.3.3 contributes overall to additional air pollution.

.2 fuel oil for combustion purposes derived by methods other than petroleum refining shall not:

.2.1 exceed the applicable sulphur content set forth in regulation 14 of this Annex;

.2.2 cause an engine to exceed the applicable NO_x emission limit set forth in paragraphs 3, 4, 5.1.1 and 7.4 of regulation 13;

.2.3 contain inorganic acid; or

.2.4.1 jeopardize the safety of ships or adversely affect the performance of the machinery, or

.2.4.2 be harmful to personnel, or

.2.4.3 contribute overall to additional air pollution.

4 This regulation does not apply to coal in its solid form or nuclear fuels. Paragraphs 5, 6, 7.1, 7.2, 8.1, 8.2, 9.2, 9.3, and 9.4 of this regulation do not apply to gas fuels such as liquefied natural gas, compressed natural gas or liquefied petroleum gas. The sulphur content of gas fuels delivered to a ship specifically for combustion purposes on board that ship shall be documented by the supplier.

5 For each ship subject to regulations 5 and 6 of this Annex, details of fuel oil for combustion purposes delivered to and used on board shall be recorded by means of a bunker delivery note that shall contain at least the information specified in appendix V to this Annex.

SEE INTERPRETATION 10

6 The bunker delivery note shall be kept on board the ship in such a place as to be readily available for inspection at all reasonable times. It shall be retained for a period of three years after the fuel oil has been delivered on board.

SEE INTERPRETATION 10

7.1 The competent authority of a Party may inspect the bunker delivery notes on board any ship to which this Annex applies while the ship is in its port or offshore terminal, may make a copy of each delivery note, and may require the master or person in charge of the ship to certify that each copy is a true copy of such bunker delivery note. The competent authority may also verify the contents of each note through consultations with the port where the note was issued.

7.2 The inspection of the bunker delivery notes and the taking of certified copies by the competent authority under paragraph 7.1 shall be performed as expeditiously as possible without causing the ship to be unduly delayed.

8.1 The bunker delivery note shall be accompanied by a representative sample of the fuel oil delivered taking into account guidelines developed by the Organization.* The sample is to be sealed and signed by the supplier's representative and the master or officer in charge of the bunker operation on completion of bunkering operations and retained under the ship's control until the fuel oil is substantially consumed, but in any case for a period of not less than 12 months from the time of delivery.

8.2 If an Administration requires the representative sample to be analysed, it shall be done in accordance with the verification procedure set forth in appendix VI to determine whether the fuel oil meets the requirements of this Annex.

9 Parties undertake to ensure that appropriate authorities designated by them:

.1 maintain a register of local suppliers of fuel oil;

.2 require local suppliers to provide the bunker delivery note and sample as required by this regulation, certified by the fuel oil supplier that the fuel oil meets the requirements of regulations 14 and 18 of this Annex;

.3 require local suppliers to retain a copy of the bunker delivery note for at least three years for inspection and verification by the port State as necessary;

.4 take action as appropriate against fuel oil suppliers that have been found to deliver fuel oil that does not comply with that stated on the bunker delivery note;

.5 inform the Administration of any ship receiving fuel oil found to be non-compliant with the requirements of regulation 14 or 18 of this Annex; and

.6 inform the Organization for transmission to Parties and Member States of the Organization of all cases where fuel oil suppliers have failed to meet the requirements specified in regulations 14 or 18 of this Annex.

* Refer to 2009 Guidelines for the sampling of fuel oil for determination of compliance with the revised MARPOL Annex VI (resolution MEPC.182(59)).

10 In connection with port State inspections carried out by Parties, the Parties further undertake to:

.1 inform the Party or non-Party under whose jurisdiction a bunker delivery note was issued of cases of delivery of non-compliant fuel oil, giving all relevant information; and

.2 ensure that remedial action as appropriate is taken to bring non-compliant fuel oil discovered into compliance.

11 For every ship of 400 gross tonnage and above on scheduled services with frequent and regular port calls, an Administration may decide after application and consultation with affected States that compliance with paragraph 6 of this regulation may be documented in an alternative manner that gives similar certainty of compliance with regulations 14 and 18 of this Annex.

Chapter 4 – Regulations on energy efficiency for ships

Regulation 19

Application

1 This chapter shall apply to all ships of 400 gross tonnage and above.

2 The provisions of this chapter shall not apply to:

.1 ships solely engaged in voyages within waters subject to the sovereignty or jurisdiction of the State the flag of which the ship is entitled to fly. However, each Party should ensure, by the adoption of appropriate measures, that such ships are constructed and act in a manner consistent with the requirements of chapter 4 of this Annex, so far as is reasonable and practicable.

.2 ships not propelled by mechanical means, and platforms including FPSOs and FSUs and drilling rigs, regardless of their propulsion.

3 Regulations 20 and 21 of this Annex shall not apply to ships which have non-conventional propulsion, except that regulations 20 and 21 shall apply to cruise passenger ships having non-conventional propulsion and LNG carriers having conventional or non-conventional propulsion, delivered on or after 1 September 2019, as defined in paragraph 43 of regulation 2. Regulations 20 and 21 shall not apply to cargo ships having ice-breaking capability.

4 Notwithstanding the provisions of paragraph 1 of this regulation, the Administration may waive the requirement for a ship of 400 gross tonnage and above from complying with regulations 20 and 21 of this Annex.

5 The provision of paragraph 4 of this regulation shall not apply to ships of 400 gross tonnage and above:

.1 for which the building contract is placed on or after 1 January 2017; or

.2 in the absence of a building contract, the keel of which is laid or which is at a similar stage of construction on or after 1 July 2017; or

.3 the delivery of which is on or after 1 July 2019; or

.4 in cases of a major conversion of a new or existing ship, as defined in regulation 2.24 of this Annex, on or after 1 January 2017, and in which regulations 5.4.2 and 5.4.3 of this Annex apply.

6 The Administration of a Party to the present Convention which allows application of paragraph 4, or suspends, withdraws or declines the application of that paragraph, to a ship entitled to fly its flag shall forthwith communicate to the Organization for circulation to the Parties to the present Protocol particulars thereof, for their information.

Regulation 20

Attained Energy Efficiency Design Index (attained EEDI)

1 The attained EEDI shall be calculated for:

.1 each new ship;

.2 each new ship which has undergone a major conversion; and

.3 each new or existing ship which has undergone a major conversion, that is so extensive that the ship is regarded by the Administration as a newly-constructed ship, which falls into one or more of the categories in regulations 2.25 to 2.35, 2.38 and 2.39 of this Annex. The attained EEDI shall be specific to each ship and shall indicate the estimated performance of the ship in terms of energy efficiency, and be accompanied by the EEDI technical file that contains the information necessary for the calculation of the attained EEDI and that shows the process of calculation. The attained EEDI shall be verified, based on the EEDI technical file, either by the Administration or by any organization duly authorized by it.*

2 The attained EEDI shall be calculated taking into account guidelines† developed by the Organization.

Regulation 21
Required EEDI

1 For each:

.1 new ship,

.2 new ship which has undergone a major conversion, and

.3 new or existing ship which has undergone a major conversion that is so extensive that the ship is regarded by the Administration as a newly-constructed ship, which falls into one of the categories in regulations 2.25 to 2.31, 2.33 to 2.35, 2.38 and 2.39 and to which this chapter is applicable, the attained EEDI shall be as follows:

$$\text{Attained EEDI} \leq \text{Required EEDI} = \left(1 - \frac{X}{100}\right) \cdot \text{Reference line value}$$

where X is the reduction factor specified in table 1 for the required EEDI compared to the EEDI reference line.

2 For each new and existing ship that has undergone a major conversion which is so extensive that the ship is regarded by the Administration as a newly constructed ship, the attained EEDI shall be calculated and meet the requirement of paragraph 21.1 with the reduction factor applicable corresponding to the ship type and size of the converted ship at the date of the contract of the conversion, or in the absence of a contract, the commencement date of the conversion.

* Refer to Guidelines for the authorization of organizations acting on behalf of the Administration (resolution A.739(18), as amended by resolution MSC.208(81)), and Specifications on the survey and certification functions of recognized organizations acting on behalf of the Administration (resolution A.789(19), as may be amended).

† Refer to 2014 Guidelines on the method of calculation of the Energy Efficiency Design Index for new ships (resolution MEPC.245(66), as amended by resolutions MEPC.263(68) and MEPC.281(70)).

Table 1 – *Reduction factors (in percentage) for the EEDI relative to the EEDI reference line*

Ship type	Size	Phase 0 1 Jan 2013 – 31 Dec 2014	Phase 1 1 Jan 2015 – 31 Dec 2019	Phase 2 1 Jan 2020 – 31 Dec 2024	Phase 3 1 Jan 2025 and onwards
Bulk carrier	20,000 DWT and above	0	10	20	30
	10,000 – 20,000 DWT	n/a	0–10*	0–20*	0–30*
Gas carrier	10,000 DWT and above	0	10	20	30
	2,000 – 10,000 DWT	n/a	0–10*	0–20*	0–30*
Tanker	20,000 DWT and above	0	10	20	30
	4,000 – 20,000 DWT	n/a	0–10*	0–20*	0–30*
Containership	15,000 DWT and above	0	10	20	30
	10,000 – 15,000 DWT	n/a	0–10*	0–20*	0–30*
General cargo ships	15,000 DWT and above	0	10	15	30
	3,000 – 15,000 DWT	n/a	0–10*	0–15*	0–30*
Refrigerated cargo carrier	5,000 DWT and above	0	10	15	30
	3,000 – 5,000 DWT	n/a	0–10*	0–15*	0–30*
Combination carrier	20,000 DWT and above	0	10	20	30
	4,000 – 20,000 DWT	n/a	0–10*	0–20*	0–30*
LNG carrier***	10,000 DWT and above	n/a	10**	20	30
Ro-ro cargo ship (vehicle carrier)***	10,000 DWT and above	n/a	5**	15	30
Ro-ro cargo ship***	2,000 DWT and above	n/a	5**	20	30
	1,000 – 2,000 DWT	n/a	0–5* **	0–20*	0–30*
Ro-ro passenger ship***	1,000 DWT and above	n/a	5**	20	30
	250 – 1,000 DWT	n/a	0–5* **	0–20*	0–30*
Cruise passenger ship*** having non-conventional propulsion	85,000 GT and above	n/a	5**	20	30
	25,000 – 85,000 GT	n/a	0–5* **	0–20*	0–30*

* Reduction factor to be linearly interpolated between the two values dependent upon vessel size. The lower value of the reduction factor is to be applied to the smaller ship size.

** Phase 1 commences for those ships on 1 September 2015.

*** Reduction factor applies to those ships delivered on or after 1 September 2019, as defined in paragraph 43 of regulation 2.

Note: n/a means that no required EEDI applies.

3 The reference line values shall be calculated as follows:

$$\text{Reference line value} = a \cdot b^{-c}$$

where a, b and c are the parameters given in table 2.

4 If the design of a ship allows it to fall into more than one of the above ship type definitions specified in table 2, the required EEDI for the ship shall be the most stringent (the lowest) required EEDI.

5 For each ship to which this regulation applies, the installed propulsion power shall not be less than the propulsion power needed to maintain the manoeuvrability of the ship under adverse conditions as defined in the guidelines to be developed by the Organization.*

6 At the beginning of phase 1 and at the midpoint of phase 2, the Organization shall review the status of technological developments and, if proven necessary, amend the time periods, the EEDI reference line parameters for relevant ship types and reduction rates set out in this regulation.

* Refer to 2013 Interim Guidelines for determining minimum propulsion power to maintain the manoeuvrability of ships in adverse conditions (resolution MEPC.232(65), as amended by resolutions MEPC.255(67) and MEPC.262(68)).

Table 2 – *Parameters for determination of reference values for the different ship types*

Ship type defined in regulation 2	a	b	c
2.25 Bulk carrier	961.79	DWT of the ship	0.477
2.26 Gas carrier	1,120.00	DWT of the ship	0.456
2.27 Tanker	1,218.80	DWT of the ship	0.488
2.28 Containership	174.22	DWT of the ship	0.201
2.29 General cargo ship	107.48	DWT of the ship	0.216
2.30 Refrigerated cargo carrier	227.01	DWT of the ship	0.244
2.31 Combination carrier	1,219.00	DWT of the ship	0.488
2.33 Ro-ro cargo ship (vehicle carrier)	$(DWT/GT)^{-0.7} \cdot 780.36$ where DWT/GT < 0.3 1,812.63 where DWT/GT ≥ 0.3	DWT of the ship	0.471
2.34 Ro-ro cargo ship	1,405.15	DWT of the ship	0.498
2.35 Ro-ro passenger ship	752.16	DWT of the ship	0.381
2.38 LNG carrier	2,253.7	DWT of the ship	0.474
2.39 Cruise passenger ship having non-conventional propulsion	170.84	GT of the ship	0.214

Regulation 22
Ship Energy Efficiency Management Plan (SEEMP)

1 Each ship shall keep on board a ship specific Ship Energy Efficiency Management Plan (SEEMP). This may form part of the ship's Safety Management System (SMS).

SEE INTERPRETATION 4

2 The SEEMP shall be developed taking into account guidelines adopted by the Organization.*

Regulation 23
Promotion of technical cooperation and transfer of technology relating to the improvement of energy efficiency of ships†

1 Administrations shall, in cooperation with the Organization and other international bodies, promote and provide, as appropriate, support directly or through the Organization to States, especially developing States, that request technical assistance.

2 The Administration of a Party shall cooperate actively with other Parties, subject to its national laws, regulations and policies, to promote the development and transfer of technology and exchange of information to States which request technical assistance, particularly developing States, in respect of the implementation of measures to fulfil the requirements of chapter 4 of this Annex, in particular regulations 19.4 to 19.6.

* Refer to 2016 Guidelines for the development of a Ship Energy Efficiency Management Plan (SEEMP) (resolution MEPC.282(70)).

† Refer to Promotion of technical cooperation and transfer of technology relating to the improvement of energy efficiency of ships (resolution MEPC.229(65)), and Model agreement between governments on technological cooperation for the implementation of the regulations in chapter 4 of MARPOL Annex VI (MEPC.1/Circ.861).

Chapter 5 – Verification of compliance with the provisions of this Annex

Regulation 24
Application

Parties shall use the provisions of the Code for Implementation in the execution of their obligations and responsibilities contained in this Annex.

Regulation 25
Verification of compliance

1 Every Party shall be subject to periodic audits by the Organization in accordance with the audit standard to verify compliance with and implementation of this Annex.

2 The Secretary-General of the Organization shall have responsibility for administering the Audit Scheme, based on the guidelines developed by the Organization.*

3 Every Party shall have responsibility for facilitating the conduct of the audit and implementation of a programme of actions to address the findings, based on the guidelines developed by the Organization.*

4 Audit of all Parties shall be:

.1 based on an overall schedule developed by the Secretary-General of the Organization, taking into account the guidelines developed by the Organization;* and

.2 conducted at periodic intervals, taking into account the guidelines developed by the Organization.*

* Refer to Framework and Procedures for the IMO Member State Audit Scheme (resolution A.1067(28)).

ANNEX I ANNEX II ANNEX III ANNEX IV ANNEX V ANNEX VI UNIFIED INTERPRETATIONS

Appendices to Annex VI

Appendix I
Form of International Air Pollution Prevention (IAPP) Certificate (regulation 8)

INTERNATIONAL AIR POLLUTION PREVENTION CERTIFICATE

Issued under the provisions of the Protocol of 1997, as amended, to amend the International Convention for the Prevention of Pollution from Ships, 1973, as modified by the Protocol of 1978 related thereto (hereinafter referred to as "the Convention") under the authority of the Government of:

. . . .

(full designation of the country)

by. . . . .

(full designation of the competent person or organization authorized under the provisions of the Convention)

Particulars of ship[*]

Name of ship. . . . .

Distinctive number or letters. . . . .

IMO Number[†] . . . .

Port of registry . . . .

Gross tonnage. . . . .

THIS IS TO CERTIFY:

1 That the ship has been surveyed in accordance with regulation 5 of Annex VI of the Convention; and

2 That the survey shows that the equipment, systems, fittings, arrangements and materials fully comply with the applicable requirements of Annex VI of the Convention.

This Certificate is valid until (dd/mm/yyyy)[‡] . . . .
subject to surveys in accordance with regulation 5 of Annex VI of the Convention.

Completion date of the survey on which this Certificate is based (dd/mm/yyyy). . . . .

Issued at . . . .

(place of issue of Certificate)

Date (dd/mm/yyyy) . . . .

(date of issue) *(signature of duly authorized official issuing the Certificate)*

(seal or stamp of the authority, as appropriate)

[*] Alternatively, the particulars of the ship may be placed horizontally in boxes.

[†] In accordance with the IMO ship identification number scheme (resolution A.1078(28)).

[‡] Insert the date of expiry as specified by the Administration in accordance with regulation 9.1 of Annex VI of the Convention. The day and the month of this date correspond to the anniversary date as defined in regulation 2.3 of Annex VI of the Convention, unless amended in accordance with regulation 9.8 of Annex VI of the Convention.

ENDORSEMENT FOR ANNUAL AND INTERMEDIATE SURVEYS

THIS IS TO CERTIFY that, at a survey required by regulation 5 of Annex VI of the Convention, the ship was found to comply with the relevant provisions of that Annex:

Annual survey

Signed. . . . .
(signature of duly authorized official)

Place . . . .

Date (dd/mm/yyyy) . . . .

(seal or stamp of the authority, as appropriate)

Annual/Intermediate* survey

Signed. . . . .
(signature of duly authorized official)

Place . . . .

Date (dd/mm/yyyy) . . . .

(seal or stamp of the authority, as appropriate)

Annual/Intermediate* survey

Signed. . . . .
(signature of duly authorized official)

Place . . . .

Date (dd/mm/yyyy) . . . .

(seal or stamp of the authority, as appropriate)

Annual survey

Signed. . . . .
(signature of duly authorized official)

Place . . . .

Date (dd/mm/yyyy) . . . .

(seal or stamp of the authority, as appropriate)

ANNUAL/INTERMEDIATE SURVEY IN ACCORDANCE WITH REGULATION 9.8.3

THIS IS TO CERTIFY that, at an annual/intermediate* survey in accordance with regulation 9.8.3 of Annex VI of the Convention, the ship was found to comply with the relevant provisions of that Annex:

Signed. . . . .
(signature of duly authorized official)

Place . . . .

Date (dd/mm/yyyy) . . . .

(seal or stamp of the authority, as appropriate)

ENDORSEMENT TO EXTEND THE CERTIFICATE IF VALID FOR LESS THAN 5 YEARS WHERE REGULATION 9.3 APPLIES

The ship complies with the relevant provisions of the Annex, and this Certificate shall, in accordance with regulation 9.3 of Annex VI of the Convention, be accepted as valid until (dd/mm/yyyy) . . . .

Signed. . . . .
(signature of duly authorized official)

Place . . . .

Date (dd/mm/yyyy) . . . .

(seal or stamp of the authority, as appropriate)

* Delete as appropriate.

ENDORSEMENT WHERE THE RENEWAL SURVEY HAS BEEN COMPLETED AND REGULATION 9.4 APPLIES

The ship complies with the relevant provisions of the Annex, and this Certificate shall, in accordance with regulation 9.4 of Annex VI of the Convention, be accepted as valid until (dd/mm/yyyy) . . . .

Signed. . . . .

(signature of duly authorized official)

Place . . . .

Date (dd/mm/yyyy) . . . .

(seal or stamp of the authority, as appropriate)

ENDORSEMENT TO EXTEND THE VALIDITY OF THE CERTIFICATE UNTIL REACHING THE PORT OF SURVEY OR FOR A PERIOD OF GRACE WHERE REGULATION 9.5 OR 9.6 APPLIES

This Certificate shall, in accordance with regulation 9.5 or 9.6* of Annex VI of the Convention, be accepted as valid until (dd/mm/yyyy) . . . .

Signed. . . . .

(signature of duly authorized official)

Place . . . .

Date (dd/mm/yyyy) . . . .

(seal or stamp of the authority, as appropriate)

ENDORSEMENT FOR ADVANCEMENT OF ANNIVERSARY DATE WHERE REGULATION 9.8 APPLIES

In accordance with regulation 9.8 of Annex VI of the Convention, the new anniversary date is (dd/mm/yyyy) . . . .

Signed. . . . .

(signature of duly authorized official)

Place . . . .

Date (dd/mm/yyyy) . . . .

(seal or stamp of the authority, as appropriate)

In accordance with regulation 9.8 of Annex VI of the Convention, the new anniversary date is (dd/mm/yyyy) . . . .

Signed. . . . .

(signature of duly authorized official)

Place . . . .

Date (dd/mm/yyyy) . . . .

(seal or stamp of the authority, as appropriate)

* Delete as appropriate.

SUPPLEMENT TO INTERNATIONAL AIR POLLUTION PREVENTION CERTIFICATE (IAPP CERTIFICATE)

RECORD OF CONSTRUCTION AND EQUIPMENT

Notes:

1 This Record shall be permanently attached to the IAPP Certificate. The IAPP Certificate shall be available on board the ship at all times.

2 The Record shall be at least in English, French or Spanish. If an official language of the issuing country is also used, this shall prevail in case of a dispute or discrepancy.

3 Entries in boxes shall be made by inserting either: a cross (x) for the answers "yes" and "applicable"; or a dash (–) for the answers "no" and "not applicable", as appropriate.

4 Unless otherwise stated, regulations mentioned in this Record refer to regulations of Annex VI of the Convention and resolutions or circulars refer to those adopted by the International Maritime Organization.

1 Particulars of ship

1.1 Name of ship . . . .

1.2 IMO Number. . . . .

1.3 Date on which keel was laid or ship was at a similar stage of construction (dd/mm/yyyy). . . . .

1.4 Length (*L*)* metres . . . .

2 Control of emissions from ships

2.1 *Ozone-depleting substances* (regulation 12)

2.1.1 The following fire-extinguishing systems, other systems and equipment containing ozone-depleting substances, other than hydrochlorofluorocarbons (HCFCs), installed before 19 May 2005 may continue in service:

System or equipment	Location on board	Substance

2.1.2 The following systems containing HCFCs installed before 1 January 2020 may continue in service:

System or equipment	Location on board	Substance

* Completed only in respect of ships constructed on or after 1 January 2016 that are specially designed, and used solely for recreational purposes and to which, in accordance with regulation 13.5.2.1 or regulation 13.5.2.3, the NOx emission limit as given by regulation 13.5.1.1 will not apply.

2.2 *Nitrogen oxides (NO_x)* (regulation 13)

2.2.1 The following marine diesel engines installed on this ship are in accordance with the requirements of regulation 13, as indicated:

	Applicable regulation of MARPOL Annex VI (NTC = NO_x Technical Code 2008) (AM = Approved Method)		Engine #1	Engine #2	Engine #3	Engine #4	Engine #5	Engine #6
1	Manufacturer and model							
2	Serial number							
3	Use (applicable application cycle(s) – NTC 3.2)							
4	Rated power (kW) (NTC 1.3.11)							
5	Rated speed (rpm) (NTC 1.3.12)							
6	Identical engine installed ≥ 1/1/2000 exempted by 13.1.1.2		☐	☐	☐	☐	☐	☐
7	Identical engine installation date (dd/mm/yyyy) as per 13.1.1.2							
8a	Major conversion (dd/mm/yyyy)	13.2.1.1 & 13.2.2						
8b	Major conversion (dd/mm/yyyy)	13.2.1.2 & 13.2.3						
8c	Major conversion (dd/mm/yyyy)	13.2.1.3 & 13.2.3						
9a	Tier I	13.3	☐	☐	☐	☐	☐	☐
9b	Tier I	13.2.2	☐	☐	☐	☐	☐	☐
9c	Tier I	13.2.3.1	☐	☐	☐	☐	☐	☐
9d	Tier I	13.2.3.2	☐	☐	☐	☐	☐	☐
9e	Tier I	13.7.1.2	☐	☐	☐	☐	☐	☐
10a	Tier II	13.4	☐	☐	☐	☐	☐	☐
10b	Tier II	13.2.2	☐	☐	☐	☐	☐	☐
10c	Tier II	13.2.2 (Tier III not possible)	☐	☐	☐	☐	☐	☐
10d	Tier II	13.2.3.2	☐	☐	☐	☐	☐	☐
10e	Tier II	13.5.2 (Exemptions)	☐	☐	☐	☐	☐	☐
10f	Tier II	13.7.1.2	☐	☐	☐	☐	☐	☐
11a	Tier III (ECA-NO_x only)	13.5.1.1	☐	☐	☐	☐	☐	☐
11b	Tier III (ECA-NO_x only)	13.2.2	☐	☐	☐	☐	☐	☐
11c	Tier III (ECA-NO_x only)	13.2.3.2	☐	☐	☐	☐	☐	☐
11d	Tier III (ECA-NO_x only)	13.7.1.2	☐	☐	☐	☐	☐	☐
12	AM*	installed	☐	☐	☐	☐	☐	☐
13	AM*	not commercially available at this survey	☐	☐	☐	☐	☐	☐
14	AM*	not applicable	☐	☐	☐	☐	☐	☐

* Refer to 2014 Guidelines on the approved method process (resolution MEPC.243(66)).

2.3 *Sulphur oxides (SO_x) and particulate matter* (regulation 14)

2.3.1 When the ship operates outside of an emission control area specified in regulation 14.3, the ship uses:

.1 fuel oil with a sulphur content as documented by bunker delivery notes that does not exceed the limit value of:

- 4.50% m/m (not applicable on or after 1 January 2012); or. . . . . ☐
- 3.50% m/m (not applicable on or after 1 January 2020); or . . . . ☐
- 0.50% m/m, and/or. . . . . ☐

.2 an equivalent arrangement approved in accordance with regulation 4.1 as listed in 2.6 that is at least as effective in terms of SO_x emission reductions as compared to using a fuel oil with a sulphur content limit value of:

- 4.50% m/m (not applicable on or after 1 January 2012); or. . . . . ☐
- 3.50% m/m (not applicable on or after 1 January 2020); or . . . . ☐
- 0.50% m/m . . . . ☐

2.3.2 When the ship operates inside an emission control area specified in regulation 14.3, the ship uses:

.1 fuel oil with a sulphur content as documented by bunker delivery notes that does not exceed the limit value of:

- 1.00% m/m (not applicable on or after 1 January 2015); or. . . . . ☐
- 0.10% m/m, and/or . . . . ☐

.2 an equivalent arrangement approved in accordance with regulation 4.1 as listed in 2.6 that is at least as effective in terms of SO_x emission reductions as compared to using a fuel oil with a sulphur content limit value of:

- 1.00% m/m (not applicable on or after 1 January 2015); or. . . . . ☐
- 0.10% m/m . . . . ☐

SEE INTERPRETATION 5

2.4 *Volatile organic compounds (VOCs)* (regulation 15)

2.4.1 The tanker has a vapour collection system installed and approved in accordance with MSC/Circ.585.. . . . . ☐

2.4.2.1 For a tanker carrying crude oil, there is an approved VOC management plan. . . . . ☐

2.4.2.2 VOC management plan approval reference . . . .

2.5 *Shipboard incineration* (regulation 16)

The ship has an incinerator:

.1 installed on or after 1 January 2000 that complies with:

.1 resolution MEPC.76(40), as amended* . . . . ☐

.2 resolution MEPC.244(66) . . . . ☐

.2 installed before 1 January 2000 that complies with:

.1 resolution MEPC.59(33), as amended† . . . . ☐

.2 resolution MEPC.76(40), as amended* . . . . ☐

* As amended by resolution MEPC.93(45).

† As amended by resolution MEPC.92(45).

2.6 *Equivalents* (regulation 4)

The ship has been allowed to use the following fitting, material, appliance or apparatus to be fitted in a ship or other procedures, alternative fuel oils, or compliance methods used as an alternative to that required by this Annex:

System or equipment	Equivalent used	Approval reference

THIS IS TO CERTIFY that this Record is correct in all respects.

Issued at . . . .
(place of issue of the Record)

Date (dd/mm/yyyy) . . . . *(date of issue)*

. . . .
(signature of duly authorized official issuing the Record)

(seal or stamp of the authority, as appropriate)

Appendix II

Test cycles and weighting factors (regulation 13)

The following test cycles and weighting factors shall be applied for verification of compliance of marine diesel engines with the applicable NO_x limit in accordance with regulation 13 of this Annex using the test procedure and calculation method as specified in the revised NO_x Technical Code 2008.

.1 For constant-speed marine engines for ship main propulsion, including diesel-electric drive, test cycle E2 shall be applied;

.2 For controllable-pitch propeller sets test cycle E2 shall be applied;

.3 For propeller-law-operated main and propeller-law-operated auxiliary engines the test cycle E3 shall be applied;

.4 For constant-speed auxiliary engines test cycle D2 shall be applied; and

.5 For variable-speed, variable-load auxiliary engines, not included above, test cycle C1 shall be applied.

Test cycle for *constant-speed main propulsion* application
(including diesel-electric drive and all controllable-pitch propeller installations)

Test cycle type E2	Speed	100%	100%	100%	100%
	Power	100%	75%	50%	25%
	Weighting factor	0.2	0.5	0.15	0.15

Test cycle for *propeller-law-operated main* and *propeller-law-operated auxiliary engine* application

Test cycle type E3	Speed	100%	91%	80%	63%
	Power	100%	75%	50%	25%
	Weighting factor	0.2	0.5	0.15	0.15

Test cycle for *constant-speed auxiliary engine* application

Test cycle type D2	Speed	100%	100%	100%	100%	100%
	Power	100%	75%	50%	25%	10%
	Weighting factor	0.05	0.25	0.3	0.3	0.1

Test cycle for *variable-speed and variable-load auxiliary engine* application

Test cycle type C1	Speed	Rated				Intermediate			Idle
	Torque	100%	75%	50%	10%	100%	75%	50%	0%
	Weighting factor	0.15	0.15	0.15	0.1	0.1	0.1	0.1	0.15

In the case of an engine to be certified in accordance with paragraph 5.1.1 of regulation 13, the specific emission at each individual mode point shall not exceed the applicable NO_x emission limit value by more than 50% except as follows:

.1 The 10% mode point in the D2 test cycle.

.2 The 10% mode point in the C1 test cycle.

.3 The idle mode point in the C1 test cycle.

Appendix III

Criteria and procedures for designation of emission control areas (regulations 13.6 and 14.3)

1 Objectives

1.1 The purpose of this appendix is to provide the criteria and procedures to Parties for the formulation and submission of proposals for the designation of emission control areas and to set forth the factors to be considered in the assessment of such proposals by the Organization.

1.2 Emissions of NO_x, SO_x and particulate matter from ocean-going ships contribute to ambient concentrations of air pollution in cities and coastal areas around the world. Adverse public health and environmental effects associated with air pollution include premature mortality, cardiopulmonary disease, lung cancer, chronic respiratory ailments, acidification and eutrophication.

1.3 An emission control area should be considered for adoption by the Organization if supported by a demonstrated need to prevent, reduce and control emissions of NO_x or SO_x and particulate matter or all three types of emissions (hereinafter emissions) from ships.

2 Process for the designation of emission control areas

2.1 A proposal to the Organization for designation of an emission control area for NO_x or SO_x and particulate matter or all three types of emissions may be submitted only by Parties. Where two or more Parties have a common interest in a particular area, they should formulate a coordinated proposal.

2.2 A proposal to designate a given area as an emission control area should be submitted to the Organization in accordance with the rules and procedures established by the Organization.

3 Criteria for designation of an emission control area

3.1 The proposal shall include:

.1 a clear delineation of the proposed area of application, along with a reference chart on which the area is marked;

.2 the type or types of emission(s) that is or are being proposed for control (i.e. NO_x or SO_x and particulate matter or all three types of emissions);

.3 a description of the human populations and environmental areas at risk from the impacts of ship emissions;

.4 an assessment that emissions from ships operating in the proposed area of application are contributing to ambient concentrations of air pollution or to adverse environmental impacts. Such assessment shall include a description of the impacts of the relevant emissions on human health and the environment, such as adverse impacts to terrestrial and aquatic ecosystems, areas of natural productivity, critical habitats, water quality, human health, and areas of cultural and scientific significance, if applicable. The sources of relevant data including methodologies used shall be identified;

.5 relevant information, pertaining to the meteorological conditions in the proposed area of application, to the human populations and environmental areas at risk, in particular prevailing wind patterns, or to topographical, geological, oceanographic, morphological or other conditions that contribute to ambient concentrations of air pollution or adverse environmental impacts;

.6 the nature of the ship traffic in the proposed emission control area, including the patterns and density of such traffic;

.7 a description of the control measures taken by the proposing Party or Parties addressing land-based sources of NO_x, SO_x and particulate matter emissions affecting the human populations and environmental areas at risk that are in place and operating concurrent with the consideration of measures to be adopted in relation to provisions of regulations 13 and 14 of Annex VI; and

.8 the relative costs of reducing emissions from ships when compared with land-based controls, and the economic impacts on shipping engaged in international trade.

3.2 The geographical limits of an emission control area will be based on the relevant criteria outlined above, including emissions and deposition from ships navigating in the proposed area, traffic patterns and density, and wind conditions.

4 Procedures for the assessment and adoption of emission control areas by the Organization

4.1 The Organization shall consider each proposal submitted to it by a Party or Parties.

4.2 In assessing the proposal, the Organization shall take into account the criteria that are to be included in each proposal for adoption as set forth in section 3 above.

4.3 An emission control area shall be designated by means of an amendment to this Annex, considered, adopted and brought into force in accordance with article 16 of the present Convention.

5 Operation of emission control areas

5.1 Parties that have ships navigating in the area are encouraged to bring to the Organization any concerns regarding the operation of the area.

Appendix IV

Type approval and operating limits for shipboard incinerators (regulation 16)

1 Shipboard incinerators described in regulation 16.6.1 shall possess an IMO Type Approval Certificate for each incinerator. In order to obtain such certificate, the incinerator shall be designed and built to an approved standard as described in regulation 16.6.1. Each model shall be subject to a specified type approval test operation at the factory or an approved test facility, and under the responsibility of the Administration, using the following standard fuel/waste specification for the type approval test for determining whether the incinerator operates within the limits specified in paragraph 2 of this appendix:

Sludge oil consisting of:	75% sludge oil from heavy fuel oil (HFO);
	5% waste lubricating oil; and
	20% emulsified water.
Solid waste consisting of:	50% food waste;
	50% rubbish containing:
	approx. 30% paper,
	" 40% cardboard,
	" 10% rags,
	" 20% plastic.
	The mixture will have up to 50% moisture and 7% incombustible solids.

2 Incinerators described in regulation 16.6.1 shall operate within the following limits:

O_2 in combustion chamber:	6–12%
CO in flue gas maximum average:	200 mg/MJ
Soot number maximum average:	Bacharach 3 or Ringelman 1 (20% opacity) (a higher soot number is acceptable only during very short periods such as starting up)
Unburned components in ash residues:	Maximum 10% by weight
Combustion chamber flue gas outlet temperature range:	850–1,200°C

Appendix V
Information to be included in the bunker delivery note (regulation 18.5)

Name and IMO Number of receiving ship

Port

Date of commencement of delivery

Name, address and telephone number of marine fuel oil supplier

Product name(s)

Quantity in metric tonnes

Density at 15°C, kg/m^3*

Sulphur content (% m/m)†

A declaration signed and certified by the fuel oil supplier's representative that the fuel oil supplied is in conformity with the applicable paragraph of regulation 14.1 or 14.4 and regulation 18.3 of this Annex.

* Fuel oil shall be tested in accordance with ISO 3675:1998 or ISO 12185:1996.

† Fuel oil shall be tested in accordance with ISO 8754:2003.

Appendix VI

Fuel verification procedure for MARPOL Annex VI fuel oil samples (regulation 18.8.2)

The following procedure shall be used to determine whether the fuel oil delivered to and used on board ships is compliant with the sulphur limits required by regulation 14 of Annex VI.

1 General requirements

1.1 The representative fuel oil sample, which is required by paragraph 8.1 of regulation 18 (the "MARPOL sample") shall be used to verify the sulphur content of the fuel oil supplied to a ship.

1.2 An Administration, through its competent authority, shall manage the verification procedure.

1.3 The laboratories responsible for the verification procedure set forth in this appendix shall be fully accredited* for the purpose of conducting the tests.

2 Verification procedure stage 1

2.1 The MARPOL sample shall be delivered by the competent authority to the laboratory.

2.2 The laboratory shall:

- **.1** record the details of the seal number and the sample label on the test record;
- **.2** confirm that the condition of the seal on the MARPOL sample is that it has not been broken; and
- **.3** reject any MARPOL sample where the seal has been broken.

2.3 If the seal of the MARPOL sample has not been broken, the laboratory shall proceed with the verification procedure and shall:

- **.1** ensure that the MARPOL sample is thoroughly homogenized;
- **.2** draw two subsamples from the MARPOL sample; and
- **.3** reseal the MARPOL sample and record the new reseal details on the test record.

2.4 The two subsamples shall be tested in succession, in accordance with the specified test method referred to in appendix V (second footnote). For the purposes of this verification procedure, the results of the test analysis shall be referred to as "A" and "B":

- **.1** If the results of "A" and "B" are within the repeatability (*r*) of the test method, the results shall be considered valid.
- **.2** If the results of "A" and "B" are not within the repeatability (*r*) of the test method, both results shall be rejected and two new subsamples should be taken by the laboratory and analysed. The sample bottle should be resealed in accordance with paragraph 2.3.3 above after the new subsamples have been taken.

2.5 If the test results of "A" and "B" are valid, an average of these two results should be calculated thus giving the result referred to as "X":

- **.1** If the result of "X" is equal to or falls below the applicable limit required by Annex VI, the fuel oil shall be deemed to meet the requirements.
- **.2** If the result of "X" is greater than the applicable limit required by Annex VI, verification procedure stage 2 should be conducted; however, if the result of "X" is greater than the specification limit by $0.59R$ (where R is the reproducibility of the test method), the fuel oil shall be considered non-compliant and no further testing is necessary.

---

* Accreditation is in accordance with ISO 17025 or an equivalent standard.

3 Verification procedure stage 2

3.1 If stage 2 of the verification procedure is necessary in accordance with paragraph 2.5.2 above, the competent authority shall send the MARPOL sample to a second accredited laboratory.

3.2 Upon receiving the MARPOL sample, the laboratory shall:

.1 record the details of the reseal number applied in accordance with 2.3.3 above and the sample label on the test record;

.2 draw two subsamples from the MARPOL sample; and

.3 reseal the MARPOL sample and record the new reseal details on the test record.

3.3 The two subsamples shall be tested in succession, in accordance with the test method specified in appendix V (second footnote). For the purposes of this verification procedure, the results of the test analysis shall be referred to as "C" and "D":

.1 If the results of "C" and "D" are within the repeatability (*r*) of the test method, the results shall be considered valid.

.2 If the results of "C" and "D" are not within the repeatability (*r*) of the test method, both results shall be rejected and two new subsamples shall be taken by the laboratory and analysed. The sample bottle should be resealed in accordance with paragraph 3.2.3 above after the new subsamples have been taken.

3.4 If the test results of "C" and "D" are valid, and the results of "A", "B", "C", and "D" are within the reproducibility (*R*) of the test method then the laboratory shall average the results, which is referred to as "Y":

.1 If the result of "Y" is equal to or falls below the applicable limit required by Annex VI, the fuel oil shall be deemed to meet the requirements.

.2 If the result of "Y" is greater than the applicable limit required by Annex VI, then the fuel oil fails to meet the standards required by Annex VI.

3.5 If the results of "A", "B", "C" and "D" are not within the reproducibility (*R*) of the test method then the Administration may discard all of the test results and, at its discretion, repeat the entire testing process.

3.6 The results obtained from the verification procedure are final.

Appendix VII
Emission control areas (regulations 13.6 and 14.3)

1 The boundaries of emission control areas designated under regulations 13.6 and 14.3, other than the Baltic Sea and the North Sea areas, are set forth in this appendix.

2 The North American area comprises:

.1 the sea area located off the Pacific coasts of the United States and Canada, enclosed by geodesic lines connecting the following coordinates:

Point	Latitude	Longitude
1	32°32′.10 N	117°06′.11 W
2	32°32′.04 N	117°07′.29 W
3	32°31′.39 N	117°14′.20 W
4	32°33′.13 N	117°15′.50 W
5	32°34′.21 N	117°22′.01 W
6	32°35′.23 N	117°27′.53 W
7	32°37′.38 N	117°49′.34 W
8	31°07′.59 N	118°36′.21 W
9	30°33′.25 N	121°47′.29 W
10	31°46′.11 N	123°17′.22 W
11	32°21′.58 N	123°50′.44 W
12	32°56′.39 N	124°11′.47 W
13	33°40′.12 N	124°27′.15 W
14	34°31′.28 N	125°16′.52 W
15	35°14′.38 N	125°43′.23 W
16	35°44′.00 N	126°18′.53 W
17	36°16′.25 N	126°45′.30 W
18	37°01′.35 N	127°07′.18 W
19	37°45′.39 N	127°38′.02 W
20	38°25′.08 N	127°53′.00 W
21	39°25′.05 N	128°31′.23 W
22	40°18′.47 N	128°45′.46 W
23	41°13′.39 N	128°40′.22 W
24	42°12′.49 N	129°00′.38 W
25	42°47′.34 N	129°05′.42 W
26	43°26′.22 N	129°01′.26 W
27	44°24′.43 N	128°41′.23 W
28	45°30′.43 N	128°40′.02 W
29	46°11′.01 N	128°49′.01 W
30	46°33′.55 N	129°04′.29 W
31	47°39′.55 N	131°15′.41 W
32	48°32′.32 N	132°41′.00 W

Point	Latitude	Longitude
33	48°57′.47 N	133°14′.47 W
34	49°22′.39 N	134°15′.51 W
35	50°01′.52 N	135°19′.01 W
36	51°03′.18 N	136°45′.45 W
37	51°54′.04 N	137°41′.54 W
38	52°45′.12 N	138°20′.14 W
39	53°29′.20 N	138°40′.36 W
40	53°40′.39 N	138°48′.53 W
41	54°13′.45 N	139°32′.38 W
42	54°39′.25 N	139°56′.19 W
43	55°20′.18 N	140°55′.45 W
44	56°07′.12 N	141°36′.18 W
45	56°28′.32 N	142°17′.19 W
46	56°37′.19 N	142°48′.57 W
47	58°51′.04 N	153°15′.03 W

.2 the sea areas located off the Atlantic coasts of the United States, Canada and France (Saint-Pierre-et-Miquelon), and the Gulf of Mexico coast of the United States enclosed by geodesic lines connecting the following coordinates:

Point	Latitude	Longitude
1	60°00′.00 N	64°09′.36 W
2	60°00′.00 N	56°43′.00 W
3	58°54′.01 N	55°38′.05 W
4	57°50′.52 N	55°03′.47 W
5	57°35′.13 N	54°00′.59 W
6	57°14′.20 N	53°07′.58 W
7	56°48′.09 N	52°23′.29 W
8	56°18′.13 N	51°49′.42 W
9	54°23′.21 N	50°17′.44 W
10	53°44′.54 N	50°07′.17 W
11	53°04′.59 N	50°10′.05 W
12	52°20′.06 N	49°57′.09 W
13	51°34′.20 N	48°52′.45 W
14	50°40′.15 N	48°16′.04 W
15	50°02′.28 N	48°07′.03 W
16	49°24′.03 N	48°09′.35 W
17	48°39′.22 N	47°55′.17 W
18	47°24′.25 N	47°46′.56 W
19	46°35′.12 N	48°00′.54 W
20	45°19′.45 N	48°43′.28 W
21	44°43′.38 N	49°16′.50 W
22	44°16′.38 N	49°51′.23 W

Point	Latitude	Longitude
23	43°53′.15 N	50°34′.01 W
24	43°36′.06 N	51°20′.41 W
25	43°23′.59 N	52°17′.22 W
26	43°19′.50 N	53°20′.13 W
27	43°21′.14 N	54°09′.20 W
28	43°29′.41 N	55°07′.41 W
29	42°40′.12 N	55°31′.44 W
30	41°58′.19 N	56°09′.34 W
31	41°20′.21 N	57°05′.13 W
32	40°55′.34 N	58°02′.55 W
33	40°41′.38 N	59°05′.18 W
34	40°38′.33 N	60°12′.20 W
35	40°45′.46 N	61°14′.03 W
36	41°04′.52 N	62°17′.49 W
37	40°36′.55 N	63°10′.49 W
38	40°17′.32 N	64°08′.37 W
39	40°07′.46 N	64°59′.31 W
40	40°05′.44 N	65°53′.07 W
41	39°58′.05 N	65°59′.51 W
42	39°28′.24 N	66°21′.14 W
43	39°01′.54 N	66°48′.33 W
44	38°39′.16 N	67°20′.59 W
45	38°19′.20 N	68°02′.01 W
46	38°05′.29 N	68°46′.55 W
47	37°58′.14 N	69°34′.07 W
48	37°57′.47 N	70°24′.09 W
49	37°52′.46 N	70°37′.50 W
50	37°18′.37 N	71°08′.33 W
51	36°32′.25 N	71°33′.59 W
52	35°34′.58 N	71°26′.02 W
53	34°33′.10 N	71°37′.04 W
54	33°54′.49 N	71°52′.35 W
55	33°19′.23 N	72°17′.12 W
56	32°45′.31 N	72°54′.05 W
57	31°55′.13 N	74°12′.02 W
58	31°27′.14 N	75°15′.20 W
59	31°03′.16 N	75°51′.18 W
60	30°45′.42 N	76°31′.38 W
61	30°12′.48 N	77°18′.29 W
62	29°25′.17 N	76°56′.42 W
63	28°36′.59 N	76°48′.00 W
64	28°17′.13 N	76°40′.10 W

Point	Latitude	Longitude
65	28°17′.12 N	79°11′.23 W
66	27°52′.56 N	79°28′.35 W
67	27°26′.01 N	79°31′.38 W
68	27°16′.13 N	79°34′.18 W
69	27°11′.54 N	79°34′.56 W
70	27°05′.59 N	79°35′.19 W
71	27°00′.28 N	79°35′.17 W
72	26°55′.16 N	79°34′.39 W
73	26°53′.58 N	79°34′.27 W
74	26°45′.46 N	79°32′.41 W
75	26°44′.30 N	79°32′.23 W
76	26°43′.40 N	79°32′.20 W
77	26°41′.12 N	79°32′.01 W
78	26°38′.13 N	79°31′.32 W
79	26°36′.30 N	79°31′.06 W
80	26°35′.21 N	79°30′.50 W
81	26°34′.51 N	79°30′.46 W
82	26°34′.11 N	79°30′.38 W
83	26°31′.12 N	79°30′.15 W
84	26°29′.05 N	79°29′.53 W
85	26°25′.31 N	79°29′.58 W
86	26°23′.29 N	79°29′.55 W
87	26°23′.21 N	79°29′.54 W
88	26°18′.57 N	79°31′.55 W
89	26°15′.26 N	79°33′.17 W
90	26°15′.13 N	79°33′.23 W
91	26°08′.09 N	79°35′.53 W
92	26°07′.47 N	79°36′.09 W
93	26°06′.59 N	79°36′.35 W
94	26°02′.52 N	79°38′.22 W
95	25°59′.30 N	79°40′.03 W
96	25°59′.16 N	79°40′.08 W
97	25°57′.48 N	79°40′.38 W
98	25°56′.18 N	79°41′.06 W
99	25°54′.04 N	79°41′.38 W
100	25°53′.24 N	79°41′.46 W
101	25°51′.54 N	79°41′.59 W
102	25°49′.33 N	79°42′.16 W
103	25°48′.24 N	79°42′.23 W
104	25°48′.20 N	79°42′.24 W
105	25°46′.26 N	79°42′.44 W
106	25°46′.16 N	79°42′.45 W

Point	Latitude	Longitude
107	25°43′.40 N	79°42′.59 W
108	25°42′.31 N	79°42′.48 W
109	25°40′.37 N	79°42′.27 W
110	25°37′.24 N	79°42′.27 W
111	25°37′.08 N	79°42′.27 W
112	25°31′.03 N	79°42′.12 W
113	25°27′.59 N	79°42′.11 W
114	25°24′.04 N	79°42′.12 W
115	25°22′.21 N	79°42′.20 W
116	25°21′.29 N	79°42′.08 W
117	25°16′.52 N	79°41′.24 W
118	25°15′.57 N	79°41′.31 W
119	25°10′.39 N	79°41′.31 W
120	25°09′.51 N	79°41′.36 W
121	25°09′.03 N	79°41′.45 W
122	25°03′.55 N	79°42′.29 W
123	25°03′.00 N	79°42′.56 W
124	25°00′.30 N	79°44′.05 W
125	24°59′.03 N	79°44′.48 W
126	24°55′.28 N	79°45′.57 W
127	24°44′.18 N	79°49′.24 W
128	24°43′.04 N	79°49′.38 W
129	24°42′.36 N	79°50′.50 W
130	24°41′.47 N	79°52′.57 W
131	24°38′.32 N	79°59′.58 W
132	24°36′.27 N	80°03′.51 W
133	24°33′.18 N	80°12′.43 W
134	24°33′.05 N	80°13′.21 W
135	24°32′.13 N	80°15′.16 W
136	24°31′.27 N	80°16′.55 W
137	24°30′.57 N	80°17′.47 W
138	24°30′.14 N	80°19′.21 W
139	24°30′.06 N	80°19′.44 W
140	24°29′.38 N	80°21′.05 W
141	24°28′.18 N	80°24′.35 W
142	24°28′.06 N	80°25′.10 W
143	24°27′.23 N	80°27′.20 W
144	24°26′.30 N	80°29′.30 W
145	24°25′.07 N	80°32′.22 W
146	24°23′.30 N	80°36′.09 W
147	24°22′.33 N	80°38′.56 W
148	24°22′.07 N	80°39′.51 W

Point	Latitude	Longitude
149	24°19′.31 N	80°45′.21 W
150	24°19′.16 N	80°45′.47 W
151	24°18′.38 N	80°46′.49 W
152	24°18′.35 N	80°46′.54 W
153	24°09′.51 N	80°59′.47 W
154	24°09′.48 N	80°59′.51 W
155	24°08′.58 N	81°01′.07 W
156	24°08′.30 N	81°01′.51 W
157	24°08′.26 N	81°01′.57 W
158	24°07′.28 N	81°03′.06 W
159	24°02′.20 N	81°09′.05 W
160	24°00′.00 N	81°11′.16 W
161	23°55′.32 N	81°12′.55 W
162	23°53′.52 N	81°19′.43 W
163	23°50′.52 N	81°29′.59 W
164	23°50′.02 N	81°39′.59 W
165	23°49′.05 N	81°49′.59 W
166	23°49′.05 N	82°00′.11 W
167	23°49′.42 N	82°09′.59 W
168	23°51′.14 N	82°24′.59 W
169	23°51′.14 N	82°39′.59 W
170	23°49′.42 N	82°48′.53 W
171	23°49′.32 N	82°51′.11 W
172	23°49′.24 N	82°59′.59 W
173	23°49′.52 N	83°14′.59 W
174	23°51′.22 N	83°25′.49 W
175	23°52′.27 N	83°33′.01 W
176	23°54′.04 N	83°41′.35 W
177	23°55′.47 N	83°48′.11 W
178	23°58′.38 N	83°59′.59 W
179	24°09′.37 N	84°29′.27 W
180	24°13′.20 N	84°38′.39 W
181	24°16′.41 N	84°46′.07 W
182	24°23′.30 N	84°59′.59 W
183	24°26′.37 N	85°06′.19 W
184	24°38′.57 N	85°31′.54 W
185	24°44′.17 N	85°43′.11 W
186	24°53′.57 N	85°59′.59 W
187	25°10′.44 N	86°30′.07 W
188	25°43′.15 N	86°21′.14 W
189	26°13′.13 N	86°06′.45 W
190	26°27′.22 N	86°13′.15 W

Point	Latitude	Longitude
191	26°33′.46 N	86°37′.07 W
192	26°01′.24 N	87°29′.35 W
193	25°42′.25 N	88°33′.00 W
194	25°46′.54 N	90°29′.41 W
195	25°44′.39 N	90°47′.05 W
196	25°51′.43 N	91°52′.50 W
197	26°17′.44 N	93°03′.59 W
198	25°59′.55 N	93°33′.52 W
199	26°00′.32 N	95°39′.27 W
200	26°00′.33 N	96°48′.30 W
201	25°58′.32 N	96°55′.28 W
202	25°58′.15 N	96°58′.41 W
203	25°57′.58 N	97°01′.54 W
204	25°57′.41 N	97°05′.08 W
205	25°57′.24 N	97°08′.21 W
206	25°57′.24 N	97°08′.47 W

.3 the sea area located off the coasts of the Hawaiian Islands of Hawai'i, Maui, Oahu, Moloka'i, Ni'ihau, Kaua'i, Lāna'i and Kaho'olawe, enclosed by geodesic lines connecting the following coordinates:

Point	Latitude	Longitude
1	22°32′.54 N	153°00′.33 W
2	23°06′.05 N	153°28′.36 W
3	23°32′.11 N	154°02′.12 W
4	23°51′.47 N	154°36′.48 W
5	24°21′.49 N	155°51′.13 W
6	24°41′.47 N	156°27′.27 W
7	24°57′.33 N	157°22′.17 W
8	25°13′.41 N	157°54′.13 W
9	25°25′.31 N	158°30′.36 W
10	25°31′.19 N	159°09′.47 W
11	25°30′.31 N	159°54′.21 W
12	25°21′.53 N	160°39′.53 W
13	25°00′.06 N	161°38′.33 W
14	24°40′.49 N	162°13′.13 W
15	24°15′.53 N	162°43′.08 W
16	23°40′.50 N	163°13′.00 W
17	23°03′.20 N	163°32′.58 W
18	22°20′.09 N	163°44′.41 W
19	21°36′.45 N	163°46′.03 W
20	20°55′.26 N	163°37′.44 W
21	20°13′.34 N	163°19′.13 W

Point	Latitude	Longitude
22	19°39′.03 N	162°53′.48 W
23	19°09′.43 N	162°20′.35 W
24	18°39′.16 N	161°19′.14 W
25	18°30′.31 N	160°38′.30 W
26	18°29′.31 N	159°56′.17 W
27	18°10′.41 N	159°14′.08 W
28	17°31′.17 N	158°56′.55 W
29	16°54′.06 N	158°30′.29 W
30	16°25′.49 N	157°59′.25 W
31	15°59′.57 N	157°17′.35 W
32	15°40′.37 N	156°21′.06 W
33	15°37′.36 N	155°22′.16 W
34	15°43′.46 N	154°46′.37 W
35	15°55′.32 N	154°13′.05 W
36	16°46′.27 N	152°49′.11 W
37	17°33′.42 N	152°00′.32 W
38	18°30′.16 N	151°30′.24 W
39	19°02′.47 N	151°22′.17 W
40	19°34′.46 N	151°19′.47 W
41	20°07′.42 N	151°22′.58 W
42	20°38′.43 N	151°31′.36 W
43	21°29′.09 N	151°59′.50 W
44	22°06′.58 N	152°31′.25 W
45	22°32′.54 N	153°00′.33 W

3 The United States Caribbean Sea area includes:

.1 the sea area located off the Atlantic and Caribbean coasts of the Commonwealth of Puerto Rico and the United States Virgin Islands, enclosed by geodesic lines connecting the following coordinates:

Point	Latitude	Longitude
1	17°18′.37 N	67°32′.14 W
2	19°11′.14 N	67°26′.45 W
3	19°30′.28 N	65°16′.48 W
4	19°12′.25 N	65°06′.08 W
5	18°45′.13 N	65°00′.22 W
6	18°41′.14 N	64°59′.33 W
7	18°29′.22 N	64°53′.51 W
8	18°27′.35 N	64°53′.22 W
9	18°25′.21 N	64°52′.39 W
10	18°24′.30 N	64°52′.19 W
11	18°23′.51 N	64°51′.50 W
12	18°23′.42 N	64°51′.23 W

Point	Latitude	Longitude
13	18°23′.36 N	64°50′.17 W
14	18°23′.48 N	64°49′.41 W
15	18°24′.11 N	64°49′.00 W
16	18°24′.28 N	64°47′.57 W
17	18°24′.18 N	64°47′.01 W
18	18°23′.13 N	64°46′.37 W
19	18°22′.37 N	64°45′.20 W
20	18°22′.39 N	64°44′.42 W
21	18°22′.42 N	64°44′.36 W
22	18°22′.37 N	64°44′.24 W
23	18°22′.39 N	64°43′.42 W
24	18°22′.30 N	64°43′.36 W
25	18°22′.25 N	64°42′.58 W
26	18°22′.26 N	64°42′.28 W
27	18°22′.15 N	64°42′.03 W
28	18°22′.22 N	64°40′.60 W
29	18°21′.57 N	64°40′.15 W
30	18°21′.51 N	64°38′.23 W
31	18°21′.22 N	64°38′.16 W
32	18°20′.39 N	64°38′.33 W
33	18°19′.15 N	64°38′.14 W
34	18°19′.07 N	64°38′.16 W
35	18°17′.23 N	64°39′.38 W
36	18°16′.43 N	64°39′.41 W
37	18°11′.33 N	64°38′.58 W
38	18°03′.02 N	64°38′.03 W
39	18°02′.56 N	64°29′.35 W
40	18°02′.51 N	64°27′.02 W
41	18°02′.30 N	64°21′.08 W
42	18°02′.31 N	64°20′.08 W
43	18°02′.03 N	64°15′.57 W
44	18°00′.12 N	64°02′.29 W
45	17°59′.58 N	64°01′.04 W
46	17°58′.47 N	63°57′.01 W
47	17°57′.51 N	63°53′.54 W
48	17°56′.38 N	63°53′.21 W
49	17°39′.40 N	63°54′.53 W
50	17°37′.08 N	63°55′.10 W
51	17°30′.21 N	63°55′.56 W
52	17°11′.36 N	63°57′.57 W
53	17°05′.00 N	63°58′.41 W
54	16°59′.49 N	63°59′.18 W
55	17°18′.37 N	67°32′.14 W

Appendix VIII
Form of International Energy Efficiency (IEE) Certificate

INTERNATIONAL ENERGY EFFICIENCY CERTIFICATE

Issued under the provisions of the Protocol of 1997, as amended, to amend the International Convention for the Prevention of Pollution by Ships, 1973, as modified by the Protocol of 1978 related thereto (hereinafter referred to as "the Convention") under the authority of the Government of:

. . . .

(full designation of the Party)

by. . . . .

(full designation of the competent person or organization authorized under the provisions of the Convention)

Particulars of ship[*]

Name of ship. . . . .

Distinctive number or letters. . . . .

IMO Number[†] . . . .

Port of registry . . . .

Gross tonnage. . . . .

THIS IS TO CERTIFY:

1 That the ship has been surveyed in accordance with regulation 5.4 of Annex VI of the Convention; and

2 That the survey shows that the ship complies with the applicable requirements in regulation 20, regulation 21 and regulation 22.

Completion date of survey on which this Certificate is based. . . . . (dd/mm/yyyy)

Issued at . . . .

(place of issue of Certificate)

Date (dd/mm/yyyy) . . . .

(date of issue) *(signature of duly authorized official issuing the Certificate)*

(seal or stamp of the authority, as appropriate)

[*] Alternatively, the particulars of the ship may be placed horizontally in boxes.

[†] In accordance with the IMO ship identification number scheme (resolution A.1078(28)).

SUPPLEMENT TO INTERNATIONAL ENERGY EFFICIENCY CERTIFICATE (IEE CERTIFICATE)

RECORD OF CONSTRUCTION RELATING TO ENERGY EFFICIENCY

Notes:

1 This Record shall be permanently attached to the IEE Certificate. The IEE Certificate shall be available on board the ship at all times.

2 The Record shall be at least in English, French or Spanish. If an official language of the issuing Party is also used, this shall prevail in case of a dispute or discrepancy.

3 Entries in boxes shall be made by inserting either: a cross (x) for the answers "yes" and "applicable"; or a dash (–) for the answers "no" and "not applicable", as appropriate.

4 Unless otherwise stated, regulations mentioned in this Record refer to regulations in Annex VI of the Convention, and resolutions or circulars refer to those adopted by the International Maritime Organization.

1 Particulars of ship

1.1 Name of ship . . . .

1.2 IMO number. . . . .

1.3 Date of building contract. . . . .

1.4 Gross tonnage. . . . .

1.5 Deadweight . . . .

1.6 Type of ship* . . . .

2 Propulsion system

2.1 Diesel propulsion. . . . . ☐

2.2 Diesel-electric propulsion . . . . ☐

2.3 Turbine propulsion. . . . . ☐

2.4 Hybrid propulsion . . . . ☐

2.5 Propulsion system other than any of the above . . . . ☐

3 Attained Energy Efficiency Design Index (EEDI)

3.1 The Attained EEDI in accordance with regulation 20.1 is calculated based on the information contained in the EEDI Technical File which also shows the process of calculating the Attained EEDI . . . . ☐

The Attained EEDI is: . . . . grams-CO_2/tonne-mile

3.2 The Attained EEDI is not calculated as:

3.2.1 the ship is exempt under regulation 20.1 as it is not a new ship as defined in regulation 2.23 . . . . ☐

3.2.2 the type of propulsion system is exempt in accordance with regulation 19.3 . . . . ☐

3.2.3 the requirement of regulation 20 is waived by the ship's Administration in accordance with regulation 19.4 . . . . ☐

3.2.4 the type of ship is exempt in accordance with regulation 20.1 . . . . ☐

* Insert ship type in accordance with definitions specified in regulation 2. Ships falling into more than one of the ship types defined in regulation 2 should be considered as being the ship type with the most stringent (the lowest) required EEDI. If ship does not fall into the ship types defined in regulation 2, insert "Ship other than any of the ship type defined in regulation 2".

4 Required EEDI

4.1 Required EEDI is: . . . . grams-CO_2/tonne-mile

4.2 The required EEDI is not applicable as:

4.2.1 the ship is exempt under regulation 21.1 as it is not a new ship as defined in regulation 2.23. . . . . ☐

4.2.2 the type of propulsion system is exempt in accordance with regulation 19.3 . . . . ☐

4.2.3 the requirement of regulation 21 is waived by the ship's Administration in accordance with regulation 19.4 . . . . ☐

4.2.4 the type of ship is exempt in accordance with regulation 21.1 . . . . ☐

4.2.5 the ship's capacity is below the minimum capacity threshold in Table 1 of regulation 21.2 . . . . . . ☐

5 Ship Energy Efficiency Management Plan

5.1 The ship is provided with a Ship Energy Efficiency Management Plan (SEEMP) in compliance with regulation 22 . . . . ☐

6 EEDI Technical File

6.1 The IEE Certificate is accompanied by the EEDI Technical File in compliance with regulation 20.1. . . ☐

6.2 The EEDI Technical File identification/verification number. . . . .

6.3 The EEDI Technical File verification date . . . .

THIS IS TO CERTIFY that this Record is correct in all respects.

Issued at . . . .
(place of issue of the Record)

Date (dd/mm/yyyy) . . . .
(date of issue) *(signature of duly authorized official issuing the Record)*

(seal or stamp of the issuing authority, as appropriate)

Unified Interpretations of Annex VI

1 Definition of "new ship"

Reg. 2.23

1.1 For the application of the definition "new ship" as specified in regulation 2.23 to each Phase specified in table 1 of regulation 21, it should be interpreted as follows:

.1 the date specified in regulation 2.23.1 should be replaced with the start date of each Phase;

.2 the date specified in regulation 2.23.2 should be replaced with the date six months after the start date of each Phase; and

.3 the date specified in regulation 2.23.3 should, for Phase 1, 2 and 3, be replaced with the date 48 months after the start date of each Phase.

1.2 With the above interpretations, the required EEDI of each Phase is applied to the following new ship which falls into one of the categories defined in regulations 2.25 to 2.31 and to which chapter 4 is applicable:

.1 the required EEDI of Phase 0 is applied to the following new ship:

.1 the building contract of which is placed in Phase 0, and the delivery is before 1 January 2019; or

.2 the building contract of which is placed before Phase 0, and the delivery is on or after 1 July 2015 and before 1 January 2019; or

in the absence of a building contract,

.3 the keel of which is laid or which is at a similar stage of construction on or after 1 July 2013 and before 1 July 2015, and the delivery is before 1 January 2019; or

.4 the keel of which is laid or which is at a similar stage of construction before 1 July 2013, and the delivery is on or after 1 July 2015 and before 1 January 2019.

.2 the required EEDI of Phase 1 is applied to the following new ship:

.1 the building contract of which is placed in Phase 1, and the delivery is before 1 January 2024; or

.2 the building contract of which is placed before Phase 1, and the delivery is on or after 1 January 2019 and before 1 January 2024; or

in the absence of a building contract,

.3 the keel of which is laid or which is at a similar stage of construction on or after 1 July 2015 and before 1 July 2020, and the delivery is before 1 January 2024; or

.4 the keel of which is laid or which is at a similar stage of construction before 1 July 2015, and the delivery is on or after 1 January 2019 and before 1 January 2024.

.3 the required EEDI of Phase 2 is applied to the following new ship:

.1 the building of which contract is placed in Phase 2, and the delivery is before 1 January 2029; or

.2 the building contract of which is placed before Phase 2, and the delivery is on or after 1 January 2024 and before 1 January 2029; or

in the absence of a building contract,

.3 the keel of which is laid or which is at a similar stage of construction on or after 1 July 2020 and before 1 July 2025, and the delivery is before 1 January 2029; or

.4 the keel of which is laid or which is at a similar stage of construction before 1 July 2020, and the delivery is on or after 1 January 2024 and before 1 January 2029.

.4 the required EEDI of Phase 3 is applied to the following new ship:

.1 the building of which contract is placed in Phase 3; or

.2 in the absence of a building contract, the keel of which is laid or which is at a similar stage of construction on or after 1 July 2025; or

.3 the delivery of which is on or after 1 January 2029.

2 Major conversion

Reg. 2.24

2.1 For regulation 2.24.1, any substantial change in hull dimensions and/or capacity (e.g. change of length between perpendiculars (LPP) or change of assigned freeboard) should be considered a major conversion. Any substantial increase of total engine power for propulsion (e.g. 5% or more) should be considered a major conversion. In any case, it is the Administration's authority to evaluate and decide whether an alteration should be considered as major conversion, consistent with chapter 4.

Note: Notwithstanding paragraph 2.1, assuming no alteration to the ship structure, both decrease of assigned freeboard and temporary increase of assigned freeboard due to the limitation of deadweight or draft at calling port should not be construed as a major conversion. However, an increase of assigned freeboard, except a temporary increase, should be construed as a major conversion.

2.2 Notwithstanding paragraph 2.1, for regulation 2.24.5, the effect on Attained EEDI as a result of any change of ships' parameters, particularly any increase in total engine power for propulsion, should be investigated. In any case, it is the Administration's authority to evaluate and decide whether an alteration should be considered as major conversion, consistent with chapter 4.

2.3 A company may, at any time, voluntarily request re-certification of the EEDI, with IEE Certificate reissuance, on the basis of any new improvements to the ships' efficiency that are not considered to be major conversions.

2.4 In regulation 2.24.4, the terms "new ship" and "existing ship" should be understood as they are used in MARPOL Annex I, regulation 1.9.1.4, rather than as the defined terms in regulations 2.22 and 2.23.

2.5 The term "a ship" referred to in regulation 5.4.2 is interpreted as "new ship".

3 Ships dedicated to the carriage of fruit juice in refrigerated cargo tanks

Reg. 2.30

Ships dedicated to the carriage of fruit juice in refrigerated cargo tanks should be categorized as refrigerated cargo carrier.

4 Timing for existing ships to have on board a SEEMP

Regs. 5.4.4, 6.4, 22.1

4.1 The International Energy Efficiency Certificate (IEEC) should be issued for both new and existing ships to which chapter 4 applies. Ships which are not required to keep an SEEMP on board are not required to be issued with an IECC.

4.2 The SEEMP required by regulation 22.1 is not required to be placed on board an existing ship to which this regulation applies until the verification survey specified in regulation 5.4.4 is carried out.

4.3 For existing ships, a SEEMP required in accordance with regulation 22 should be verified on board according to regulation 5.4.4, and an IEEC should be issued, not later than the first intermediate or renewal survey, in accordance with chapter 2, whichever is earlier, on or after 1 January 2013, i.e. a survey connected to an intermediate/renewal survey of the IAPP Certificate.

4.4 The intermediate or renewal survey referenced in paragraph 4.3 relates solely to the timing of the verification of the SEEMP on board, i.e. these IAPP Certificate survey windows will also become the IEEC initial survey date for existing ships. The SEEMP is, however, a survey item solely under chapter 4 and is not a survey item relating to IAPP Certificate surveys.

4.5 In the event that the SEEMP is not available on board during the first intermediate/renewal survey of the IAPP Certificate on or after 1 January 2013, the RO should seek the advice of the Administration concerning the issuance of an IEEC and be guided accordingly. However, the validity of the IAPP Certificate is not impacted by the lack of a SEEMP as the SEEMP is a survey item solely under chapter 4 and not under the IAPP Certificate surveys.

4.6 With respect to ships required to keep on board a SEEMP, such ships exclude platforms (including FPSOs and FSUs) and drilling rigs, regardless of their propulsion, and any other ship without means of propulsion.

4.7 The SEEMP should be written in a working language or languages understood by ships' personnel.

5 Section 2.3 of the supplement to IAPP Certificate

Reg. 8.1, appendix 1

Section 2.3 of the supplement ("as documented by bunker delivery notes") allows for an "x" to be entered in advance of the dates indicated in all of the relevant check boxes recognizing that the bunker delivery notes, required to be retained on board for a minimum period of three years, provide the subsequent means to check that a ship is actually operating in a manner consistent with the intent as given in section 2.3.

6 Identical replacement engines

Regs. 13.1.1.2, 13.2.2

6.1 In regulation 13.1.1.2, the term "identical" (and hence, by application of the converse, in regulation 13.2.2 the term "non-identical") as applied to engines under regulation 13 should be taken as:

6.2 An "identical engine" is, as compared to the engine being replaced,* an engine which is of the same:

.1 design and model;

.2 rated power;

.3 rated speed;

.4 use;

.5 number of cylinders; and

.6 fuel system type (including, if applicable, injection control software):

  .1 for engines without EIAPP certification, have the same NO_x critical components and settings;† or

  .2 for engines with EIAPP certification, belonging to the same Engine Group/Engine Family.

7 Time of replacement of an engine

Reg. 13.2.2

7.1 The term "time of the replacement or addition" of the engine in regulation 13.2.2 should be taken as the date of:

.1 the contractual delivery date of the engine to the ship;‡ or

.2 in the absence of a contractual delivery date, the actual delivery date of the engine to the ship,‡ provided that the date is confirmed by a delivery receipt; or

.3 in the event the engine is fitted on board and tested for its intended purpose on or after 1 July 2016, the actual date that the engine is tested on board for its intended purpose applies in determining the standards in this regulation in force at the time of the replacement or addition of the engine.

7.2 The date in paragraph 7.1 above, provided the conditions associated with those dates apply, is the "Date of major conversion – According to regulation 13.2.2" to be entered in the Supplement of IAPP Certificate. In this case, the "Date of installation", which applies only for identical replacement engines, should be filled in with "N.A.".

7.3 If the engine is delivered in accordance with either paragraphs 7.1.1 or 7.1.2 above before 1 January 2016, but not tested before 1 July 2016 due to unforeseen circumstances beyond the control of the shipowner, then the provisions of "unforeseen delay in delivery" may be considered by the Administration in a manner similar to UI4 of MARPOL Annex I.

---

* In those instances where the replaced engine will not be available to be directly compared with the replacing engine at the time of updating the Supplement to the IAPP Certificate reflecting that engine change it is to be ensured that the necessary records in respect of the replaced engine are available in order that it can be confirmed that the replacing engine represents "an identical engine".

† For engines without EIAPP Certification there will not be the defining NO_x critical component markings or setting values as usually given in the approved Technical File. Consequently, in these instances, the assessment of "... same NO_x critical components and settings ..." shall be established on the basis that the following components and settings are the same:

Fuel system:

.1 fuel pump model and injection timing; and

.2 injection nozzle model;

Charge air:

.1 configuration and, if applicable, turbocharger model and auxiliary blower specification; and

.2 cooling medium (seawater/freshwater).

‡ The engine is to be fitted on board and tested for its intended purpose before 1 July 2016.

8 VOC management plan

Regs. 15.6, 15.7

The requirement for a VOC management plan applies only to a tanker carrying crude oil.

9 Continuous-feed type shipboard incinerators

Reg. 16.9

For the application of this regulation, the term "waste shall not be fed into the unit" should be interpreted as follows:

The introduction of sludge oil, generated during normal operation of a ship, into a continuous-feed type incinerator during the warm-up process at combustion chamber temperatures above 500°C* in order to achieve the normal operation combustion chamber temperature of 850°C is allowed. The combustion chamber flue gas outlet temperature should reach 850°C within the period of time specified in the manufacturer's operations manual but should not be more than five minutes.

10 Applicability of the requirements for a bunker delivery note

Regs. 18.5, 18.6

For the application of these regulations, they should be interpreted as being applicable to all ships of 400 gross tonnage or above and, at the Administration's discretion, to ships of less than 400 gross tonnage.

---

* For the introduction of sludge oil into the incinerator, two conditions need to be fulfilled to secure smokeless and complete combustion:

.1 the combustion chamber flue gas outlet temperature has to be above 850°C as required by regulation 16.9 to ensure smokeless combustion; and

.2 the combustion chamber temperature (material temperature of the fire brickwork) has to be above 500°C to ensure a sufficient evaporation of the burnable components of the sludge oil.

Additional information

1

List of MEPC resolutions

MEPC.1(II)*	Resolution on establishment of the list of substances to be annexed to the Protocol relating to Intervention on the High Seas in Cases of Marine Pollution by Substances other than Oil
MEPC.2(VI)	Recommendation on international effluent standards and guidelines for performance tests for sewage treatment plants
MEPC.3(XII)	Recommendation on the standard format for the crude oil washing operations and equipment manual
MEPC.4(XIII)	Recommendation regarding acceptance of oil content meters in oil tankers
MEPC.5(XIII)	Specification for oil/water interface detectors
MEPC.6(XIV)	Application of the provisions of Annex I of the International Convention for the Prevention of Pollution from Ships, 1973, as modified by the Protocol of 1978 relating thereto on the discharge of oil in the Baltic Sea area
MEPC.7(XV)	Entries in oil record books on methods of disposal of residue
MEPC.8(XVI)	Discharge of oils not specified by the International Convention for the Prevention of Pollution of the Sea by Oil, 1954, as amended in 1962 and 1969
MEPC.9(17)	Application of the provisions of Annex V of MARPOL 73/78 on the discharge of garbage in the Baltic Sea area
MEPC.10(18)	Application scheme for oil discharge monitoring and control systems
MEPC.11(18)	Guidelines for surveys under Annex I of the International Convention for the Prevention of Pollution from Ships, 1973, as modified by the Protocol of 1978 relating thereto
MEPC.12(18)	Regional arrangements for combating major incidents of marine pollution
MEPC.13(19)	Guidelines for plan approval and installation survey of oil discharge monitoring and control systems for oil tankers and environmental testing of control sections thereof
MEPC.14(20)	Adoption of amendments to Annex I of MARPOL 73/78
MEPC.15(21)	Installation of oil discharge monitoring and control systems in existing oil tankers
MEPC.16(22)	Adoption of amendments to Annex II of MARPOL 73/78
MEPC.17(22)	Implementation of Annex II of MARPOL 73/78
MEPC.18(22)	Adoption of the standards for procedures and arrangements for the discharge of noxious liquid substances
MEPC.19(22)	Adoption of the International code for the construction and equipment of ships carrying dangerous chemicals in bulk (IBC Code)
MEPC.20(22)	Adoption of the Code for the construction and equipment of ships carrying dangerous chemicals in bulk (BCH Code)

* Roman or Arabic figures in brackets show the session number, and the texts of these resolutions are annexed to the MEPC report of that session.

MEPC.21(22)	Adoption of amendments to Protocol I to MARPOL 73/78 and the text of the Protocol, as amended, annexed thereto
MEPC.22(22)	Adoption of guidelines for reporting incidents involving harmful substances and the text of guidelines annexed thereto
MEPC.23(22)	The application of Annex II of MARPOL 73/78 on the discharge of noxious liquid substances in the Baltic Sea area
MEPC.24(22)	Adoption of amendments to the Revised guidelines and specifications for oil discharge monitoring and control systems for oil tankers as adopted by the Organization by resolution A.586(14) and to the Recommendation on international performance and test specifications for oily-water separating equipment and oil content meters adopted by the Organization by resolution A.393(X)
MEPC.25(23)	Guidelines for surveys under Annex II of the International Convention for the Prevention of Pollution from Ships, 1973, as modified by the Protocol of 1978 relating thereto (MARPOL 73/78)
MEPC.26(23)	Procedures for the control of ships and discharges under Annex II of the International Convention for the Prevention of Pollution from Ships, 1973, as modified by the Protocol of 1978 relating thereto (MARPOL 73/78)
MEPC.27(23)	Categorization of liquid substances
MEPC.28(24)	Compliance with Annex II of MARPOL 73/78
MEPC.29(25)	Adoption of amendments to the Annex of the Protocol of 1978 relating to the International Convention for the Prevention of Pollution from Ships, 1973 (designation of the Gulf of Aden as a special area)
MEPC.30(25)	Guidelines for reporting incidents involving harmful substances
MEPC.31(26)	Establishment of the date of application of the provisions of regulation 5 of Annex V of the International Convention for the Prevention of Pollution from Ships, 1973 as modified by the Protocol of 1978 relating thereto on the discharge of garbage in the Baltic Sea area
MEPC.32(27)	Adoption of amendments to the International code for the construction and equipment of ships carrying dangerous chemicals in bulk (IBC Code)
MEPC.33(27)	Adoption of amendments to the Code for the construction and equipment of ships carrying dangerous chemicals in bulk (BCH Code)
MEPC.34(27)	Adoption of amendments to the Annex of the Protocol of 1978 relating to the International Convention for the Prevention of Pollution from Ships, 1973 (Appendices II and III of Annex II of MARPOL 73/78)
MEPC.35(27)	Implementation of Annex III of MARPOL 73/78
MEPC.36(28)	Adoption of amendments to the Annex of the Protocol of 1978 relating to the International Convention for the Prevention of Pollution from Ships, 1973 (Amendments to Annex V of MARPOL 73/78)
MEPC.37(28)	Establishment of the date of application of the provisions of regulation 5 of Annex V of the International Convention for the Prevention of Pollution from Ships, 1973 as modified by the Protocol of 1978 relating thereto on the discharge of garbage in the North Sea area
MEPC.38(29)	Application of the provisions of Annex IV of the International Convention for the Prevention of Pollution from Ships, 1973 as modified by the Protocol of 1978 relating thereto on the discharge of sewage in the Baltic Sea area

MEPC.39(29)	Adoption of amendments to the Annex of the Protocol of 1978 relating to the International Convention for the Prevention of Pollution from Ships, 1973 (Introduction of the Harmonized System of Survey and Certification to Annexes I and II of MARPOL 73/78)
MEPC.40(29)	Adoption of amendments to the International Code for the Construction and Equipment of Ships Carrying Dangerous Chemicals in Bulk (IBC Code) (Harmonized System of Survey and Certification)
MEPC.41(29)	Adoption of amendments to the Code for the Construction and Equipment of Ships Carrying Dangerous Chemicals in Bulk (BCH Code) (Harmonized System of Survey and Certification)
MEPC.42(30)	Adoption of amendments to the Annex of the Protocol of 1978 relating to the International Convention for the Prevention of Pollution from Ships, 1973 (Designation of Antarctic area as a special area under Annexes I and V of MARPOL 73/78)
MEPC.43(30)	Prevention of pollution by garbage in the Mediterranean
MEPC.44(30)	Identification of the Great Barrier Reef region as a particularly sensitive area
MEPC.45(30)	Protection of the Great Barrier Reef region
MEPC.46(30)	Measures to control potential adverse impacts associated with use of tributyl tin compounds in anti-fouling paints
MEPC.47(31)	Adoption of amendments to the Annex of the Protocol of 1978 relating to the International Convention for the Prevention of Pollution from Ships, 1973 (new regulation 26 and other amendments to Annex I of MARPOL 73/78)
MEPC.48(31)	Adoption of amendments to the Annex of the Protocol of 1978 relating to the International Convention for the Prevention of Pollution from Ships, 1973 (designation of the Wider Caribbean area as a special area under Annex V of MARPOL 73/78)
MEPC.49(31)	Revision of the list of substances to be annexed to the Protocol relating to intervention on the high seas in cases of marine pollution by substances other than oil, 1973
MEPC.50(31)	Guidelines for preventing the introduction of unwanted aquatic organisms and pathogens from ships' ballast water and sediment discharges
MEPC.51(32)	Adoption of amendments to the Annex of the Protocol of 1978 relating to the International Convention for the Prevention of Pollution from Ships, 1973 (Discharge criteria of Annex I of MARPOL 73/78)
MEPC.52(32)	Adoption of amendments to the Annex of the Protocol of 1978 relating to the International Convention for the Prevention of Pollution from Ships, 1973 (New regulations 13F and 13G and related amendments to Annex I of MARPOL 73/78)
MEPC.53(32)	Development of the capacity of ship scrapping for the smooth implementation of the amendments to Annex I of MARPOL 73/78
MEPC.54(32)	Guidelines for the development of shipboard oil pollution emergency plans
MEPC.55(33)	Adoption of amendments to the International code for the construction and equipment of ships carrying dangerous chemicals in bulk (IBC Code)
MEPC.56(33)	Adoption of amendments to the Code for the construction and equipment of ships carrying dangerous chemicals in bulk (BCH Code)
MEPC.57(33)	Adoption of amendments to the Annex of the Protocol of 1978 relating to the International Convention for the Prevention of Pollution from Ships, 1973 (Designation of the Antarctic area as a special area and lists of liquid substances in Annex II)

MEPC.58(33)	Adoption of amendments to the Annex of the Protocol of 1978 relating to the International Convention for the Prevention of Pollution from Ships, 1973 (Revised Annex III)
MEPC.59(33)	Revised guidelines for the implementation of Annex V of MARPOL 73/78
MEPC.60(33)	Guidelines and specifications for pollution prevention equipment for machinery space bilges of ships
MEPC.61(34)	Visibility limits of oil discharges of Annex I of MARPOL 73/78
MEPC.62(35)	Amendments to the standards for procedures and arrangements for the discharge of noxious liquid substances
MEPC.63(36)	Oil tanker stability, operational safety and protection of the marine environment
MEPC.64(36)	Guidelines for approval of alternative structural or operational arrangements as called for in regulation 13G(7) of Annex I of MARPOL 73/78
MEPC.65(37)	Amendments to the Annex of the Protocol of 1978 relating to the International Convention for the Prevention of Pollution from Ships, 1973 (Amendments to regulation 2 and new regulation 9 of Annex V)
MEPC.66(37)	Interim Guidelines for the approval of alternative methods of design and construction of oil tankers under regulation 13F(5) of Annex I of MARPOL 73/78
MEPC.67(37)	Guidelines on application of the precautionary approach
MEPC.68(38)	Amendments to the Annex of the Protocol of 1978 relating to the International Convention for the Prevention of Pollution from Ships, 1973 (Amendments to Protocol I)
MEPC.69(38)	Amendments to the International code for the construction and equipment of ships carrying dangerous chemicals in bulk (IBC Code)
MEPC.70(38)	Amendments to the Code for the construction and equipment of ships carrying dangerous chemicals in bulk (BCH Code)
MEPC.71(38)	Guidelines for the development of garbage management plans
MEPC.72(38)	Revision of the list of substances to be annexed to the Protocol relating to the Intervention on the High Seas in Cases of Marine Pollution by Substances other than Oil
MEPC.73(39)	Amendments to the International code for the construction and equipment of ships carrying dangerous chemicals in bulk (IBC Code) (Vague expressions)
MEPC.74(40)	Identification of the Archipelago of Sabana-Camagüey as a Particularly Sensitive Sea Area
MEPC.75(40)	Amendments to the Annex of the Protocol of 1978 relating to the International Convention for the Prevention of Pollution from Ships, 1973
MEPC.76(40)	Standard specification for shipboard incinerators
MEPC.77(41)	Establishment of the date on which the amendments to regulation 10 of Annex I of MARPOL 73/78 in respect of the North-West European Waters special area shall take effect
MEPC.78(43)	Amendments to the Annex of the Protocol of 1978 relating to the International Convention for the Prevention of Pollution from Ships, 1973
MEPC.79(43)	Amendments to the International code for the construction and equipment of ships carrying dangerous chemicals in bulk (IBC Code)
MEPC.80(43)	Amendments to the Code for the construction and equipment of ships carrying dangerous chemicals in bulk (BCH Code)

MEPC.81(43)	Amendments to section 9 of the standard format for the COW manual (resolution MEPC.3(XII))
MEPC.82(43)	Guidelines for monitoring the world-wide average sulphur content of residual fuel oils supplied for use on board ships
MEPC.83(44)	Guidelines for ensuring the adequacy of port waste reception facilities
MEPC.84(44)	Amendments to the Annex of the Protocol of 1978 relating to the International Convention for the Prevention of Pollution from Ships, 1973
MEPC.85(44)	Guidelines for the development of shipboard marine pollution emergency plans for oil and/or noxious liquid substances
MEPC.86(44)	Amendments to the Guidelines for the development of shipboard oil pollution emergency plans
MEPC.87(44)	Use of Spanish under IMO conventions relating to pollution prevention
MEPC.88(44)	Implementation of Annex IV of MARPOL 73/78
MEPC.89(45)	Amendments to the Annex of the Protocol of 1978 relating to the International Convention for the Prevention of Pollution from Ships, 1973
MEPC.90(45)	Amendments to the International code for the construction and equipment of ships carrying dangerous chemicals in bulk (IBC Code)
MEPC.91(45)	Amendments to the Code for the construction and equipment of ships carrying dangerous chemicals in bulk (BCH Code)
MEPC.92(45)	Amendments to the Revised Guidelines for the implementation of Annex V of MARPOL 73/78 (resolution MEPC.59(33))
MEPC.93(45)	Amendments to the Standard Specification for Shipboard Incinerators
MEPC.94(46)	Condition Assessment Scheme
MEPC.95(46)	Amendments to the Annex of the Protocol of 1978 relating to the International Convention for the Prevention of Pollution from Ships, 1973
MEPC.96(47)	Guidelines for the sampling of fuel oil for determination of compliance with Annex VI of MARPOL 73/78
MEPC.97(47)	Identification of the sea area around Malpelo Island as a Particularly Sensitive Sea Area
MEPC.98(47)	Identification of the sea area around the Florida Keys as a Particularly Sensitive Sea Area
MEPC.99(48)	Amendments to the Condition Assessment Scheme (CAS)
MEPC.100(48)	Revision of the list of substances annexed to the Protocol relating to intervention on the high seas in cases of pollution by substances other than oil, 1973
MEPC.101(48)	Identification of the Wadden Sea as a Particularly Sensitive Sea Area
MEPC.102(48)	Guidelines for survey and certification of anti-fouling systems on ships
MEPC.103(49)	Guidelines for on-board NO_x verification procedure – direct measurement and monitoring method
MEPC.104(49)	Guidelines for brief sampling of anti-fouling systems on ships
MEPC.105(49)	Guidelines for inspection of anti-fouling systems in ships
MEPC.106(49)	Designation of the Paracas National Reserve as a Particularly Sensitive Sea Area
MEPC.107(49)	Revised Guidelines and specifications for pollution prevention equipment for machinery space bilges of ships
MEPC.108(49)	Revised Guidelines and specifications for oil discharge monitoring and control systems for oil tankers

MEPC.109(49)	Tripartite agreements
MEPC.110(49)	Revised interim Guidelines for the approval of alternative methods of design and construction of oil tankers under regulation 13F(5) of Annex I of MARPOL 73/78
MEPC.111(50)	Amendments to the Annex of the Protocol of 1978 relating to the International Convention for the Prevention of Pollution From Ships, 1973 (Amendments to regulation 13G, addition of new regulation 13H and consequential amendments to the IOPP Certificate of Annex I of MARPOL 73/78)
MEPC.112(50)	Amendments to the Condition Assessment Scheme (CAS)
MEPC.113(50)	Ship recycling for the smooth implementation of the amendments to Annex I of MARPOL 73/78
MEPC.114(50)	Early and effective application of the amendments to Annex I of MARPOL 73/78 (Revised regulation 13G and new regulation 13H)
MEPC.115(51)	Amendments to the Annex of the Protocol of 1978 relating to the International Convention for the Prevention of Pollution from Ships, 1973 (Revised Annex IV of MARPOL 73/78)
MEPC.116(51)	Amendments to the Annex of the Protocol of 1978 relating to the International Convention for the Prevention of Pollution from Ships, 1973 (Amendments to the appendix to Annex V of MARPOL 73/78)
MEPC.117(52)	Amendments to the Annex of the Protocol of 1978 relating to the International Convention for the Prevention of Pollution from Ships, 1973 (Revised Annex I of MARPOL 73/78)
MEPC.118(52)	Amendments to the Annex of the Protocol of 1978 relating to the International Convention for the Prevention of Pollution from Ships, 1973 (Revised Annex II of MARPOL 73/78)
MEPC.119(52)	2004 Amendments to the International code for the construction and equipment of ships carrying dangerous chemicals in bulk (IBC Code)
MEPC.120(52)	Guidelines for the transport of vegetable oils in deeptanks or in independent tanks specially designed for the carriage of such vegetable oils in general dry cargo ships
MEPC.121(52)	Designation of the Western European Waters as a Particularly Sensitive Sea Area
MEPC.122(52)	Explanatory notes on matters related to the accidental oil outflow performance under regulation 23 of the revised MARPOL Annex I
MEPC.123(53)	Guidelines for ballast water management equivalent compliance (G3)
MEPC.124(53)	Guidelines for ballast water exchange (G6)
MEPC.125(53)	Guidelines for approval of ballast water management systems (G8)
MEPC.126(53)	Procedure for approval of ballast water management systems that make use of active substances (G9)
MEPC.127(53)	Guidelines for ballast water management and development of ballast water management plans (G4)
MEPC.128(53)	Amendments to the revised survey guidelines under the harmonized system of survey and certification (resolution A.948(23)) for the purpose of MARPOL Annex VI
MEPC.129(53)	Guidelines for port State control under MARPOL Annex VI
MEPC.130(53)	Guidelines for on-board exhaust gas-SO_x cleaning systems
MEPC.131(53)	Amendments to the Condition Assessment Scheme (CAS)

MEPC.132(53)	Amendments to the Annex of the Protocol of 1997 to amend the International Convention for the Prevention of Pollution from Ships, 1973, as modified by the Protocol of 1978 relating thereto (Amendments to MARPOL Annex VI and the NO_x Technical Code)
MEPC.133(53)	Designation of the Torres Strait as an extension of the Great Barrier Reef Particularly Sensitive Sea Area
MEPC.134(53)	Designation of the Canary Islands as a Particularly Sensitive Sea Area
MEPC.135(53)	Designation of the Galapagos Archipelago as a Particularly Sensitive Sea Area
MEPC.136(53)	Designation of the Baltic Sea area as a Particularly Sensitive Sea Area
MEPC.137(53)	Amendments to the Guidelines for the development of shipboard marine pollution emergency plans for oil and/or noxious liquid substances (resolution MEPC.85(44))
MEPC.138(53)	Amendments to the General principles for ship reporting systems and ship reporting requirements, including guidelines for reporting incidents involving dangerous goods, harmful substances and/or marine pollutants (resolution A.851(20))
MEPC.139(53)	Guidelines for the application of the revised MARPOL Annex I requirements to floating production, storage and offloading facilities (FPSOs) and floating storage units (FSUs)
MEPC.140(54)	Guidelines for approval and oversight of prototype ballast water treatment technology programmes (G10)
MEPC.141(54)	Amendments to the Annex of the Protocol of 1978 relating to the International Convention for the Prevention of Pollution from Ships, 1973 (amendments to regulation 1, addition to regulation 12A, consequential amendments to the IOPP Certificate and amendments to regulation 21 of the revised Annex I of MARPOL 73/78)
MEPC.142(54)	Amendments to the Guidelines for the application of the revised MARPOL Annex I requirements to floating production, storage and offloading facilities (FPSOs) and floating storage units (FSUs) (resolution MEPC.139(53))
MEPC.143(54)	Amendments to the Annex of the Protocol of 1978 relating to the International Convention for the Prevention of Pollution from Ships, 1973 (addition of regulation 13 to Annex IV of MARPOL 73/78)
MEPC.144(54)	Amendments to the Code for the Construction and Equipment of Ships carrying Dangerous Chemicals in Bulk (BCH Code)
MEPC.145(54)	Early and effective application of the 2006 amendments to the Code for the construction and equipment of ships carrying dangerous chemicals in bulk (BCH Code)
MEPC.146(54)	Amendments to the explanatory notes on matters related to the accidental oil outflow performance under regulation 23 of the revised MARPOL Annex I
MEPC.147(54)	Guidelines on the assessment of residual fillet weld between deck plating and longitudinals
MEPC.148(54)	Revised Guidelines for the transport of vegetable oils in deeptanks or in independent tanks specially designed for the carriage of such vegetable oils in general dry cargo ships
MEPC.149(55)	Guidelines for ballast water exchange design and construction standards
MEPC.150(55)	Guidelines on design and construction to facilitate sediment control on ships
MEPC.151(55)	Guidelines on designation of areas for ballast water exchange
MEPC.152(55)	Guidelines for sediment reception facilities
MEPC.153(55)	Guidelines for ballast water reception facilities

MEPC.154(55)	Amendments to the Annex of the Protocol of 1978 relating to the International Convention for the Prevention of Pollution from Ships, 1973 (Designation of the southern South African waters as a special area)
MEPC.155(55)	Amendments to the Condition Assessment Scheme (CAS)
MEPC.156(55)	Amendments to the Annex of the Protocol of 1978 relating to the International Convention for the Prevention of Pollution from Ships, 1973 (Revised Annex III of MARPOL 73/78)
MEPC.157(55)	Recommendation on standards for the rate of discharge of untreated sewage from ships
MEPC.158(55)	Amendments to the Guidelines for the transport and handling of limited amounts of hazardous and noxious liquid substances in bulk on off-shore support vessels (Resolution A.673(16))
MEPC.159(55)	Revised Guidelines on implementation of effluent standards and performance tests for sewage treatment plants
MEPC.160(55)	Implications of the revised Annex II to MARPOL 73/78 for the reference in article 1.5(a)(ii) of the HNS convention to "Noxious Liquid Substances carried in bulk"
MEPC.161(56)	Guidelines for additional measures regarding ballast water management including emergency situations
MEPC.162(56)	Guidelines for risk assessment under regulation A-4 of the BWM convention
MEPC.163(56)	Guidelines for ballast water exchange in the Antarctic treaty area
MEPC.164(56)	Amendments to the Annex of the Protocol of 1978 relating to the International Convention for the Prevention of Pollution from Ships, 1973 (Reception facilities outside special areas and discharge of sewage)
MEPC.165(56)	Amendments to the list of substances annexed to the Protocol relating to intervention on the high seas in cases of pollution by substances other than oil, 1973
MEPC.166(56)	2007 Amendments to the International code for the construction and equipment of ships carrying dangerous chemicals in bulk (IBC Code)
MEPC.167(56)	Establishment of the date on which the amendments to Regulation 1.11 of MARPOL Annex I, in respect of the southern South African waters special area, shall take effect
MEPC.168(56)	Establishment of the date on which regulation 1.11.5 of MARPOL Annex I and regulation 5(1)(e) of MARPOL Annex V, in respect of the Gulfs area special area, shall take effect
MEPC.169(57)	Procedure for approval of ballast water management systems that make use of active substances
MEPC.170(57)	Guidelines for exhaust gas cleaning systems
MEPC.171(57)	Designation of the Papahānaumokuākea Marine National Monument as a Particularly Sensitive Sea Area
MEPC.172(57)	Establishment of the date on which regulation 5(1)(A) of MARPOL Annex V in respect of the Mediterranean sea area special area shall take effect
MEPC.173(58)	Guidelines for ballast water sampling
MEPC.174(58)	Guidelines for approval of ballast water management systems
MEPC.175(58)	Information reporting on type approved ballast water management systems
MEPC.176(58)	Amendments to the Annex of the Protocol of 1997 to amend the International Convention for the Prevention of Pollution from Ships, 1973, as modified by the Protocol of 1978 relating thereto (Revised MARPOL Annex VI)

MEPC.177(58)	Amendments to the Technical code on control of emission of nitrogen oxides from marine diesel engines (NO_x Technical Code 2008)
MEPC.178(59)	Calculation of recycling capacity for meeting the entry-into-force conditions of the Hong Kong international convention for the safe and environmentally sound recycling of ships, 2009
MEPC.179(59)	Guidelines for the development of the inventory of hazardous materials
MEPC.180(59)	Amendments to the Survey guidelines under the harmonized system of survey and certification for the revised MARPOL Annex VI
MEPC.181(59)	2009 Guidelines for port State control under the revised MARPOL Annex VI
MEPC.182(59)	2009 Guidelines for the sampling of fuel oil for determination of compliance with the revised MARPOL Annex VI
MEPC.183(59)	2009 Guidelines for monitoring the worldwide average sulphur content of residual fuel oils supplied for use on board ships
MEPC.184(59)	2009 Guidelines for exhaust gas cleaning systems
MEPC.185(59)	Guidelines for the development of a VOC management plan
MEPC.186(59)	Amendments to the Annex of the Protocol of 1978 relating to the International Convention for the Prevention of Pollution from Ships, 1973
MEPC.187(59)	Amendments to the Annex of the Protocol of 1978 relating to the International Convention for the Prevention of Pollution from Ships, 1973
MEPC.188(60)	Installation of ballast water management systems on new ships in accordance with the application dates contained in the ballast water management convention (BWM Convention)
MEPC.189(60)	Amendments to the Annex of the Protocol of 1978 relating to the International Convention for the Prevention of Pollution from Ships, 1973 (addition of a new chapter 9 to MARPOL Annex I)
MEPC.190(60)	Amendments to the Annex of the Protocol of 1997 to amend the International Convention for the Prevention of Pollution from Ships, 1973, as modified by the Protocol of 1978 relating thereto (North American emission control area)
MEPC.191(60)	Establishment of the date on which regulation 5(1)(h) of MARPOL Annex V in respect of the Wider Caribbean region special area shall take effect
MEPC.192(61)	2010 Guidelines for monitoring the worldwide average sulphur content of fuel oils supplied for use on board ships
MEPC.193(61)	Amendments to the Annex of the Protocol of 1978 relating to the International Convention for the Prevention of Pollution from Ships, 1973 (Revised MARPOL Annex III)
MEPC.194(61)	Amendments to the Annex of the Protocol of 1997 to amend the International Convention for the Prevention of Pollution from Ships, 1973, as modified by the Protocol of 1978 relating thereto (Revised form of Supplement to the IAPP Certificate)
MEPC.195(61)	2010 Guidelines for survey and certification of anti-fouling systems on ships
MEPC.196(62)	2011 Guidelines for the development of the Ship Recycling Plan
MEPC.197(62)	2011 Guidelines for the development of the Inventory of Hazardous Materials
MEPC.198(62)	2011 Guidelines addressing additional aspects to the NO_x Technical Code 2008 with regard to particular requirements related to marine diesel engines fitted with Selective Catalytic Reduction (SCR) systems
MEPC.199(62)	2011 Guidelines for reception facilities under MARPOL Annex VI

MEPC.200(62)	Amendments to the Annex of the Protocol of 1978 relating to the International Convention for the Prevention of Pollution from Ships, 1973 (Special area provisions and the designation of the Baltic Sea as a special area under MARPOL Annex IV)
MEPC.201(62)	Amendments to the Annex of the Protocol of 1978 relating to the International Convention for the Prevention of Pollution from Ships, 1973 (Revised MARPOL Annex V)
MEPC.202(62)	Amendments to the Annex of the Protocol of 1997 relating to the International Convention for the Prevention of Pollution from Ships, 1973, as modified by the Protocol of 1978 relating thereto (designation of the United States Caribbean Sea emission control area and exemption of certain ships operating in the North American emission control area and the United States Caribbean Sea emission control area under regulations 13 and 14 and appendix VII of MARPOL Annex VI)
MEPC.203(62)	Amendments to the Annex of the Protocol of 1997 relating to the International Convention for the Prevention of Pollution from Ships, 1973, as modified by the Protocol of 1978 relating thereto (inclusion of regulations on energy efficiency for ships in MARPOL Annex VI)
MEPC.204(62)	Designation of the Strait of Bonifacio as a Particularly Sensitive Sea Area
MEPC.205(62)	2011 Guidelines and specifications for add-on equipment for upgrading resolution MEPC.60(33)-compliant oil filtering equipment
MEPC.206(62)	Procedure for approving other methods of ballast water management in accordance with regulation B-3.7 of the BWM Convention
MEPC.207(62)	2011 Guidelines for the control and management of ships' biofouling to minimize the transfer of invasive aquatic species
MEPC.208(62)	2011 Guidelines for inspection of anti-fouling systems on ships
MEPC.209(63)	2012 Guidelines on design and construction to facilitate sediment control on ships (G12)
MEPC.210(63)	2012 Guidelines for safe and environmentally sound ship recycling
MEPC.211(63)	2012 Guidelines for the authorization of ship recycling facilities
MEPC.212(63)	2012 Guidelines on the method of calculation of the attained energy efficiency design index (EEDI) for new ships
MEPC.213(63)	2012 Guidelines for the development of a ship energy efficiency management plan (SEEMP)
MEPC.214(63)	2012 Guidelines on survey and certification of the energy efficiency design index (EEDI)
MEPC.215(63)	Guidelines for calculation of reference lines for use with the energy efficiency design index (EEDI)
MEPC.216(63)	Amendments to the Annex of the Protocol of 1978 relating to the International Convention for the Prevention of Pollution from Ships, 1973 (Regional arrangements for port reception facilities under MARPOL Annexes I, II, IV and V)
MEPC.217(63)	Amendments to the Annex of the Protocol of 1997 to amend the International Convention for the Prevention of Pollution from Ships, 1973, as modified by the Protocol of 1978 relating thereto (Regional arrangements for port reception facilities under MARPOL Annex VI and certification of marine diesel engines fitted with selective catalytic reduction systems under the NO_x Technical Code 2008)
MEPC.218(63)	Development of technical onboard equipment in relation to designation of the Baltic Sea as a special area under MARPOL Annex IV
MEPC.219(63)	2012 Guidelines for implementation of MARPOL Annex V

MEPC.220(63)	2012 Guidelines for the development of garbage management plans
MEPC.221(63)	2012 Guidelines for the development of a regional reception facilities plan
MEPC.222(64)	2012 Guidelines for the survey and certification of ships under the Hong Kong Convention
MEPC.223(64)	2012 Guidelines for the inspection of ships under the Hong Kong Convention
MEPC.224(64)	Amendments to the 2012 Guidelines on the method of calculation of the attained energy efficiency design index (EEDI) for new ships
MEPC.225(64)	2012 Amendments to the International Code for the Construction and Equipment of Ships Carrying Dangerous Chemicals in Bulk (IBC Code)
MEPC.226(64)	Designation of the Saba Bank as a Particularly Sensitive Sea Area
MEPC.227(64)	2012 Guidelines on implementation of effluent standards and performance tests for sewage treatment plants
MEPC.228(65)	Information reporting on type approved ballast water management systems
MEPC.229(65)	Promotion of technical cooperation and transfer of technology relating to the improvement of energy efficiency of ships
MEPC.230(65)	2013 Guidelines as required by regulation 13.2.2 of MARPOL Annex VI in respect of non-identical replacement engines not required to meet the Tier III limit
MEPC.231(65)	2013 Guidelines for calculation of reference lines for use with the energy efficiency design index (EEDI)
MEPC.232(65)	2013 Interim Guidelines for determining minimum propulsion power to maintain the manoeuvrability of ships in adverse conditions
MEPC.233(65)	2013 Guidelines for calculation of reference lines for use with the energy efficiency design index (EEDI) for cruise passenger ships having non-conventional propulsion
MEPC.234(65)	Amendments to 2012 Guidelines on survey and certification of the energy efficiency design index (EEDI) (resolution MEPC.214(63)), as amended
MEPC.235(65)	Amendments to the Annex of the Protocol of 1978 relating to the International Convention for the Prevention of Pollution from Ships, 1973 (Amendments to Form A and Form B of Supplements to the IOPP Certificate under MARPOL Annex I)
MEPC.236(65)	Amendments to the Condition Assessment Scheme under MARPOL Annex I
MEPC.237(65)	Adoption of the Code for Recognized Organizations (RO Code)
MEPC.238(65)	Amendments to the Annex of the Protocol of 1978 relating to the International Convention for the Prevention of Pollution from Ships, 1973 (Amendments to MARPOL Annexes I and II to make the RO Code mandatory)
MEPC.239(65)	Amendments to the 2012 Guidelines for the implementation of MARPOL Annex V
MEPC.240(65)	2013 Amendments to the Revised Guidelines and specifications for oil discharge monitoring and control systems for oil tankers (resolution MEPC.108(49))
MEPC.241(65)	Appreciation of the services to the Marine Environment Protection Committee by Mr. Andreas Chrysostomou
MEPC.242(66)	2014 Guidelines in respect of the information to be submitted by an Administration to the Organization covering the certification of an approved method as required under regulation 13.7.1 of MARPOL Annex VI
MEPC.243(66)	2014 Guidelines on the approved method process
MEPC.244(66)	2014 Standard specification for shipboard incinerators
MEPC.245(66)	2014 Guidelines on the method of calculation of the attained energy efficiency design index (EEDI) for new ships

MEPC.246(66)	Amendments to MARPOL Annexes I, II, III, IV and V to make the use of the III Code mandatory
MEPC.247(66)	Amendments to MARPOL Annex VI to make the use of the III Code mandatory
MEPC.248(66)	Amendments to MARPOL Annex I on mandatory carriage requirements for a stability instrument
MEPC.249(66)	Amendments to the BCH Code on cargo containment and Form of Certificate of Fitness
MEPC.250(66)	Amendments to the IBC Code on General, Ship survival capability and location of cargo tanks, Cargo tank venting and gas-freeing arrangements, Environmental control, Fire protection and fire extinction, Special requirements, Summary of minimum requirements, and Form of Certificate of Fitness
MEPC.251(66)	Amendments to MARPOL Annex VI and the NO_x Technical Code 2008
MEPC.252(67)	Guidelines for port State control under the BMW Convention
MEPC.253(67)	Measures to be taken to facilitate entry into force of the International Convention for the Control and Management of Ships' Ballast Water and Sediments, 2004
MEPC.254(67)	2014 Guidelines on survey and certification of the energy efficiency design index (EEDI)
MEPC.255(67)	Amendments to the 2013 Interim Guidelines for determining minimum propulsion power to maintain the manoeuvrability of ships in adverse conditions (resolution MEPC.232(65))
MEPC.256(67)	Amendments to MARPOL Annex I (Amendments to regulation 43)
MEPC.257(67)	Amendments to MARPOL Annex III (Amendments to the Appendix on criteria for the identification of harmful substances in packaged form)
MEPC.258(67)	Amendments to MARPOL Annex VI (Amendments to regulations 2 and 13 and the Supplement to the IAPP Certificate)
MEPC.259(68)	2015 Guidelines for exhaust gas cleaning systems
MEPC.260(68)	Amendments to the 2011 Guidelines addressing additional aspects to the NO_x Technical Code 2008 with regard to particular requirements related to marine diesel engines fitted with selective catalytic reduction (SCR) systems (resolution MEPC.198(62))
MEPC.261(68)	Amendments to the 2014 Guidelines on survey and certification of the energy efficiency design index (EEDI) (resolution MEPC.254(67))
MEPC.262(68)	Amendments to the 2013 Interim Guidelines for determining minimum propulsion power to maintain manoeuvrability of ships in adverse conditions (resolution MEPC.232(65)), as amended by resolution MEPC.255(67))
MEPC.263(68)	Amendments to the 2014 Guidelines on the method of calculation of the attained energy efficiency design index (EEDI) for new ships (resolution MEPC.245(66))
MEPC.264(68)	International Code for Ships Operating in Polar Waters (Polar Code)
MEPC.265(68)	Amendments to MARPOL Annexes I, II, IV and V (Making the use of the environment-related provisions of the Polar Code mandatory)
MEPC.266(68)	Amendments to regulation 12 of MARPOL Annex I
MEPC.267(68)	Amendments to the Revised Guidelines for the identification and designation of Particularly Sensitive Sea Areas (resolution A.982(24))
MEPC.268(68)	Designating the south-west part of the Coral Sea as an extension to the Great Barrier Reef and Torres Strait PSSA
MEPC.269(68)	2015 Guidelines for the development of the inventory of hazardous materials

MEPC.270(69)	Amendments to MARPOL Annex II (Revised GESAMP Hazard Evaluation Procedure)
MEPC.271(69)	Amendments to regulation 13 of MARPOL Annex VI (Record requirements for operational compliance with NO_x Tier III emission control areas)
MEPC.272(69)	Amendments to the NO_x Technical Code 2008 (Testing of gas-fuelled and dual fuel engines)
MEPC.273(69)	Amendments to the 2010 Guidelines for monitoring the worldwide average sulphur content of fuel oils supplied for use on board ships (resolution MEPC.192(61))
MEPC.274(69)	Amendments to MARPOL Annex IV (Baltic Sea special area and form of ISPP Certificate)
MEPC.275(69)	Establishment of the date on which regulation 11.3 of MARPOL Annex IV in respect of the Baltic Sea special area shall take effect
MEPC.276(70)	Amendments to MARPOL Annex I (Form B of the supplement to the International Oil Pollution Prevention Certificate)
MEPC.277(70)	Amendments to MARPOL Annex V (HME substances and form of Garbage Record Book)
MEPC.278(70)	Amendments to MARPOL Annex VI (Data collection system for fuel oil consumption of ships)
MEPC.279(70)	2016 Guidelines for approval of ballast water management systems (G8)
MEPC.280(70)	Effective date of implementation of the fuel oil standard in regulation 14.1.3 of MARPOL Annex VI
MEPC.281(70)	Amendments to the 2014 Guidelines on the method of calculation of the attained energy efficiency design index (EEDI) for new ships (resolution MEPC.245(66), as amended by resolution MEPC.263(68))
MEPC.282(70)	2016 Guidelines for the development of a ship energy efficiency management plan (SEEMP)
MEPC.283(70)	Designation of the Jomard Entrance as a particularly sensitive sea area
MEPC.284(70)	Amendments to the 2012 Guidelines on implementation of effluent standards and performance tests for sewage treatment plants
MEPC.285(70)	Amendments to the revised guidelines and specifications for pollution prevention equipment for machinery space bilges of ships (resolution MEPC.107(49))

2

Certificates and documents required to be carried on board ships[*]

(Note: All certificates to be carried on board must be valid and drawn up in the form corresponding to the model where required by the relevant international convention or instrument)

1 All ships to which the referenced Convention applies	Reference
International Tonnage Certificate (1969)	
An International Tonnage Certificate (1969) shall be issued to every ship, the gross and net tonnage of which have been determined in accordance with the Convention.	Tonnage Convention, article 7
International Load Line Certificate	
An International Load Line Certificate shall be issued under the provisions of the International Convention on Load Lines, 1966, to every ship which has been surveyed and marked in accordance with the Convention or the Convention as modified by the 1988 LL Protocol, as appropriate.	LL Convention, article 16; 1988 LL Protocol, article 16
International Load Line Exemption Certificate	
An International Load Line Exemption Certificate shall be issued to any ship to which an exemption has been granted under and in accordance with article 6 of the Load Line Convention or the Convention as modified by the 1988 LL Protocol, as appropriate.	LL Convention, article 16; 1988 LL Protocol, article 16
Exemption Certificate[†]	
When an exemption is granted to a ship under and in accordance with the provisions of SOLAS 1974, a certificate called an Exemption Certificate shall be issued in addition to the certificates listed above.	SOLAS 1974, regulation I/12; 1988 SOLAS Protocol, regulation I/12

[*] Refer to List of certificates required to be carried on board ships, 2017 (FAL.2/Circ.131 - MEPC.1/Circ.873 - MSC.1/Circ.1586 - LEG.2/Circ.3).

[†] SLS.14/Circ.115, Add.1, Add.2 and Add.3 refer to the issue of exemption certificate.

Coating Technical File

A Coating Technical File, containing specifications of the coating system applied, where applicable, to dedicated seawater ballast tanks in all types of ships and double-side skin spaces of bulk carriers of 150 m in length and upwards and cargo oil tanks of crude oil tankers, record of the shipyard's and shipowner's coating work, detailed criteria for coating sections, job specifications, inspection, maintenance and repair, shall be kept on board and maintained throughout the life of the ship.

SOLAS 1974, regulations II-1/3-2 and II-1/3-11; Performance standard for protective coatings for dedicated seawater ballast tanks in all types of ships and double-side skin spaces of bulk carriers, and cargo oil tanks of crude oil tankers (resolution MSC.215(82) as amended by resolution MSC.341(91) and MSC.1/Circ.1381; and resolution MSC.288(87) as modified by circular MSC.1/Circ.1381 and amended by resolution MSC.342(91))

Emergency Towing Procedure

All ships shall be provided with a ship-specific emergency towing procedure. Such a procedure shall be carried on board the ship for use in emergency situations and shall be developed based on the guidelines developed by the Organization.

SOLAS regulation II-1/3-4; MSC.1/Circ.1255

Construction drawings

A set of as-built construction drawings and other plans showing any subsequent structural alterations shall be kept on board a ship constructed on or after 1 January 2007.

SOLAS 1974, regulation II-1/3-7; MSC/Circ.1135 on As-built construction drawings to be maintained on board the ship and ashore

Ship Construction File

A Ship Construction File with specific information should be kept on board oil tankers of 150 m in length and above and bulk carriers of 150 m in length and above, constructed with single deck, top-side tanks and hopper side tanks in cargo spaces, excluding ore carriers and combination carriers:

.1 for which the building contract is placed on or after 1 July 2016;

.2 in the absence of a building contract, the keels of which are laid or which are at a similar stage of construction on or after 1 July 2017; or

.3 the delivery of which is on or after 1 July 2020;

shall carry a Ship Construction File containing information in accordance with regulations and guidelines, and updated as appropriate throughout the ship's life in order to facilitate safe operation, maintenance, survey, repair and emergency measures.

SOLAS 1974, regulation II-1/3-10; MSC.1/Circ.1343 on Guidelines for the information to be included in a Ship Construction File

Noise Survey Report	
Applicable to new ships of 1,600 gross tonnage and above, excluding dynamically supported crafts, high-speed crafts, fishing vessels, pipe-laying barges, crane barges, mobile offshore drilling units, pleasure yachts not engaged in trade, ships of war and troopships, ships not propelled by mechanical means, pile driving vessels and dredgers.	SOLAS 1974, regulation II-1/3-12; Code on noise levels on board ships, section 4.3
A noise survey report shall always be carried on board and be accessible for the crew.	
For existing ships, refer to section "Other certificates and documents which are not mandatory – Noise Survey Report" (resolution A.468(XII).	
Stability information	
Every passenger ship regardless of size and every cargo ship of 24 m and over shall be inclined on completion and the elements of their stability determined. The master shall be supplied with stability information containing such information as is necessary to enable him, by rapid and simple procedures, to obtain accurate guidance as to the stability of the ship under varying conditions of service to maintain the required intact stability and stability after damage. For bulk carriers, the information required in a bulk carrier booklet may be contained in the stability information.	SOLAS 1974, regulations II-1/5 and II 1/5-1; LL Convention regulation 10; 1988 LL Protocol, regulation 10
Damage control plans and booklets	
On passenger and cargo ships, there shall be permanently exhibited plans showing clearly for each deck and hold the boundaries of the watertight compartments, the openings therein with the means of closure and position of any controls thereof, and the arrangements for the correction of any list due to flooding. Booklets containing the aforementioned information shall be made available to the officers of the ship.	SOLAS 1974, regulation II-1/19; MSC.1/Circ.1245
Manoeuvring booklet	
The stopping times, ship headings and distances recorded on trials, together with the results of trials to determine the ability of ships having multiple propellers to navigate and manoeuvre with one or more propellers inoperative, shall be available on board for the use of the master or designated personnel.	SOLAS 1974, regulation II-1/28
Evaluation of the alternative design and arrangements	
Where applicable, a copy of the documentation, as approved by the Administration, indicating that the alternative design and arrangements comply with this regulation shall be carried onboard the ship.	SOLAS regulations II-1/55.4.2, II-2/17.4.2, and chapter III/38.4.2
Maintenance plans	
The maintenance plan shall include the necessary information about fire protection systems and fire-fighting systems and appliances as required by regulation II-2/14.2.2. For tankers, additional requirements are referred to in regulation II-2/14.4.	SOLAS 1974, regulations II-2/14.2.2, II-2/14.3 and II-2/14.4
For passenger ships carrying more than 36 passengers, the maintenance plan should include low-location lighting and public address system as required by SOLAS regulation II-2/14.3	
Onboard training and drills record	
Fire drills shall be conducted and recorded in accordance with the provisions of regulations III/19.3 and III/19.5.	SOLAS 1974, regulation II-2/15.2.2.5

Fire safety training manual

A training manual shall be written in the working language of the ship and shall be provided in each crew mess room and recreation room or in each crew cabin. The manual shall contain the instructions and information required in regulation II-2/15.2.3.4. Part of such information may be provided in the form of audio-visual aids in lieu of the manual.	SOLAS 1974, regulation II-2/15.2.3

Fire control plan/booklet

General arrangement plans shall be permanently exhibited for the guidance of the ship's officers, showing clearly for each deck the control stations, the various fire sections together with particulars of the fire detection and fire alarm systems and the fire-extinguishing appliances, etc. Alternatively, at the discretion of the Administration, the aforementioned details may be set out in a booklet, a copy of which shall be supplied to each officer, and one copy shall at all times be available on board in an accessible position. Plans and booklets shall be kept up to date; any alterations shall be recorded as soon as practicable. A duplicate set of fire control plans or a booklet containing such plans shall be permanently stored in a prominently marked weathertight enclosure outside the deckhouse for the assistance of shoreside fire-fighting personnel.	SOLAS 1974, regulations II-2/15.2.4 and II-2/15.3.2

Fire safety operational booklet

The fire safety operational booklet shall contain the necessary information and instructions for the safe operation of the ship and cargo handling operations in relation to fire safety. The booklet shall be written in the working language of the ship and be provided in each crew mess room and recreation room or in each crew cabin. The booklet may be combined with the fire safety training manuals required in regulation II-2/15.2.3.	SOLAS 1974, regulation II-2/16.2

Operations manual for helicopter facility

Each helicopter facility, if fitted, shall have an operations manual, including a description and a checklist of safety precautions, procedures and equipment requirements. This manual may be part of the ship's emergency response procedures.	SOLAS 1974, regulation II-2/18.8.1

Statement of acceptance of the installation of replacement release and retrieval system to an existing lifeboat

For all ships, no later than the first scheduled dry-docking after 1 July 2014, but no later than 1 July 2019, lifeboat on-load release mechanisms not complying with paragraphs 4.4.7.6.4 to 4.4.7.6.6 of the LSA Code shall be replaced with equipment that complies with the Code.	SOLAS regulation III/1.5; LSA Code, paragraph 4.4.7.6; MSC.1/Circ.1392/Corr.1

Muster list and emergency instructions

All ships shall be provided with muster list and emergency instructions, which shall comply with the requirements of regulation 37 and be exhibited in conspicuous places throughout the ship including the navigation bridge, engine room and crew accommodation spaces. In the case of passenger ships, these instructions shall be drawn up in the language(s) required by its flag State and in the English language.	SOLAS regulation III/8 and III/37

Ship-specific Plans and Procedures for Recovery of Persons from the Water	
All ships shall have ship-specific plans and procedures for recovery of persons from the water. Ships constructed before 1 July 2014 shall comply with this requirement by the first periodical or renewal safety equipment survey of the ship to be carried out after 1 July 2014, whichever comes first. Ro-ro passenger ships which comply with regulation III/26.4 shall be deemed to comply with this regulation. The Plans and Procedures should be considered as a part of the emergency preparedness plan required by paragraph 8 of the ISM Code.	SOLAS 1974 regulation, III/17-1; resolution MSC.346(91); MSC.1/Circ.1447
Training manual	
The training manual, which may comprise several volumes, shall contain instructions and information, in easily understood terms illustrated wherever possible, on the life-saving appliances provided in the ship and on the best methods of survival. Any part of such information may be provided in the form of audio-visual aids in lieu of the manual.	SOLAS 1974, regulation III/35
Radio record	
A record shall be kept, to the satisfaction of the Administration and as required by the Radio Regulations, of all incidents connected with the radiocommunication service which appear to be of importance to safety of life at sea.	SOLAS 1974, regulation IV/17
Minimum safe manning document	
Every ship to which chapter I of the Convention applies shall be provided with an appropriate safe manning document or equivalent issued by the Administration as evidence of the minimum safe manning.	SOLAS 1974, regulation V/14.2
Voyage data recorder system – certificate of compliance	
The voyage data recorder system, including all sensors, shall be subjected to an annual performance test. The test shall be conducted by an approved testing or servicing facility to verify the accuracy, duration and recoverability of the recorded data. In addition, tests and inspections shall be conducted to determine the serviceability of all protective enclosures and devices fitted to aid location. A copy of the certificate of compliance issued by the testing facility, stating the date of compliance and the applicable performance standards, shall be retained on board the ship.	SOLAS 1974, regulation V/18.8
AIS test report	
The Automatic Identification System (AIS) shall be subjected to an annual test by an approved surveyor or an approved testing or servicing facility. A copy of the test report shall be retained on board and should be in accordance with a model form set out in the annex to MSC.1/Circ.1252	SOLAS 1974, regulation V/18.9; MSC.1/Circ.1252
Nautical charts and nautical publications	
Nautical charts and nautical publications for the intended voyage shall be adequate and up to date. An electronic chart display and information system (ECDIS) is also accepted as meeting the chart carriage requirements of this subparagraph.	SOLAS 1974, regulations V/19.2.1.4 and V/27

LRIT conformance test report

A conformance test report should be issued, on satisfactory completion of a conformance test, by the Administration or the ASP who conducted the test acting on behalf of the Administration and should be in accordance with the model set out in appendix 2 of MSC.1/Circ.1307.	SOLAS 1974, regulation V/19-1; MSC.1/Circ.1307

International Code of Signals and a copy of Volume III of IAMSAR Manual

All ships required to carry a radio installation shall carry the International Code of Signals; all ships shall carry an up-to-date copy of Volume III of the International Aeronautical and Maritime Search and Rescue (IAMSAR) Manual.	SOLAS 1974, regulation V/21

Records for pilot ladders used for pilot transfer

All pilot ladders used for pilot transfer shall be clearly identified with tags or other permanent marking so as to enable identification of each appliance for the purposes of survey, inspection and record keeping. A record shall be kept on the ship as to the date the identified ladder is placed into service and any repairs effected.	SOLAS regulation V/23.2.4

Records of navigational activities

All ships engaged on international voyages shall keep on board a record of navigational activities and incidents including drills and pre-departure tests. When such information is not maintained in the ship's logbook, it shall be maintained in another form approved by the Administration.	SOLAS 1974, regulations V/26 and V/28.1

Cargo Securing Manual

All cargoes other than solid and liquid bulk cargoes, cargo units and cargo transport units, shall be loaded, stowed and secured throughout the voyage in accordance with the Cargo Securing Manual approved by the Administration. In ships with ro-ro spaces, as defined in regulation II-2/3.41, all securing of such cargoes, cargo units and cargo transport units, in accordance with the Cargo Securing Manual, shall be completed before the ship leaves the berth. The Cargo Securing Manual is required on all types of ships engaged in the carriage of all cargoes other than solid and liquid bulk cargoes, which shall be drawn up to a standard at least equivalent to the guidelines developed by the Organization.	SOLAS 1974, regulations VI/5.6 and VII/5; MSC.1/Circ.1353/Rev.1

Material Safety Data Sheets (MSDS)

Ships carrying oil or oil fuel, as defined in regulation 1 of annex 1 of the International Convention for the Prevention of Pollution from Ships, 1973, as modified by the Protocol of 1978 relating thereto, shall be provided with material safety data sheets, based on the recommendations developed by the Organization, prior to the loading of such oil as cargo in bulk or bunkering of oil fuel.	SOLAS 1974, regulation VI/5-1; resolution MSC.286(86)

Safety Management Certificate

A Safety Management Certificate shall be issued to every ship by the Administration or an organization recognized by the Administration. The Administration or an organization recognized by it shall, before issuing the Safety Management Certificate, verify that the company and its shipboard management operate in accordance with the approved safety management system.	SOLAS 1974, regulation IX/4; ISM Code, paragraph 13

Document of Compliance

A document of compliance shall be issued to every company which complies with the requirements of the ISM Code. A copy of the document shall be kept on board.

SOLAS 1974, regulation IX/4; ISM Code, paragraph 13

Continuous Synopsis Record (CSR)

Every ship to which chapter I of the Convention applies shall be issued with a Continuous Synopsis Record. The Continuous Synopsis Record provides an onboard record of the history of the ship with respect to the information recorded therein.

SOLAS 1974, regulation XI-1/5

Ship Security Plan and associated records

Each ship shall carry on board a ship security plan approved by the Administration. The plan shall make provisions for the three security levels as defined in part A of the ISPS Code. Records of the following activities addressed in the ship security plan shall be kept on board for at least the minimum period specified by the Administration:

SOLAS 1974, regulation XI-2/9; ISPS Code, part A, sections 9 and 10

.1 training, drills and exercises;

.2 security threats and security incidents;

.3 breaches of security;

.4 changes in security level;

.5 communications relating to the direct security of the ship such as specific threats to the ship or to port facilities the ship is, or has been, in;

.6 internal audits and reviews of security activities;

.7 periodic review of the ship security assessment;

.8 periodic review of the ship security plan;

.9 implementation of any amendments to the plan; and

.10 maintenance, calibration and testing of any security equipment provided on board, including testing of the ship security alert system.

International Ship Security Certificate (ISSC) or Interim International Ship Security Certificate

An International Ship Security Certificate (ISSC) shall be issued to every ship by the Administration or an organization recognized by it to verify that the ship complies with the maritime security provisions of SOLAS chapter XI-2 and part A of the ISPS Code. An interim ISSC may be issued under the ISPS Code, part A, section 19.4.

SOLAS 1974, regulation XI-2/9.1.1; ISPS Code, part A, section 19 and appendices

International Oil Pollution Prevention Certificate

An International Oil Pollution Prevention Certificate shall be issued, after survey in accordance with regulation 6 of Annex I of MARPOL, to any oil tanker of 150 gross tonnage and above and any other ship of 400 gross tonnage and above which is engaged in voyages to ports or offshore terminals under the jurisdiction of other Parties to MARPOL. The certificate is supplemented with a Record of Construction and Equipment for Ships other than Oil Tankers (Form A) or a Record of Construction and Equipment for Oil Tankers (Form B), as appropriate.

MARPOL Annex I, regulation 7

Oil Record Book

Every oil tanker of 150 gross tonnage and above and every ship of 400 gross tonnage and above other than an oil tanker shall be provided with an Oil Record Book, Part I (Machinery space operations). Every oil tanker of 150 gross tonnage and above shall also be provided with an Oil Record Book, Part II (Cargo/ballast operations).

MARPOL Annex I, regulations 17 and 36

Shipboard Oil Pollution Emergency Plan

Every oil tanker of 150 gross tonnage and above and every ship other than an oil tanker of 400 gross tonnage and above shall carry on board a Shipboard Oil Pollution Emergency Plan approved by the Administration.

MARPOL Annex I, regulation 37; resolution MEPC.54(32), as amended by resolution MEPC.86(44)

International Sewage Pollution Prevention Certificate

An International Sewage Pollution Prevention Certificate shall be issued, after an initial or renewal survey in accordance with the provisions of regulation 4 of Annex IV of MARPOL, to any ship which is required to comply with the provisions of that Annex and is engaged in voyages to ports or offshore terminals under the jurisdiction of other Parties to the Convention.

MARPOL Annex IV, regulation 5; MEPC/Circ.408

Document of approval for the rate of sewage discharge

Untreated sewage from ships other than passenger ships in all areas and from passenger ships outside special areas that has been stored in holding tanks shall be discharged at a moderate rate approved by the Administration based upon the standards developed by the Organization.

MARPOL Annex IV, regulation 11.1.1; resolution MEPC.157(55)

Garbage Management Plan

Every ship of 100 gross tonnage and above and every ship which is certified to carry 15 persons or more shall carry a garbage management plan which the crew shall follow.

MARPOL Annex V, regulation 10; resolution MEPC.220(63)

Garbage Record Book

Every ship of 400 gross tonnage and above and every ship which is certified to carry 15 persons or more engaged in voyages to ports or offshore terminals under the jurisdiction of other Parties to the Convention and every fixed and floating platform engaged in exploration and exploitation of the seabed shall be provided with a Garbage Record Book.

MARPOL Annex V, regulation 10

International Air Pollution Prevention Certificate

Ships constructed before the date of entry into force of the Protocol of 1997 shall be issued with an International Air Pollution Prevention Certificate. Any ship of 400 gross tonnage and above engaged in voyages to ports or offshore terminals under the jurisdiction of other Parties and platforms and drilling rigs engaged in voyages to waters under the sovereignty or jurisdiction of other Parties to the Protocol of 1997 shall be issued with an International Air Pollution Prevention Certificate.

MARPOL Annex VI, regulation 6

International Energy Efficiency Certificate

An International Energy Efficiency Certificate for the ship shall be issued after a survey in accordance with the provisions of regulation 5.4 to any ships of 400 gross tonnage and above before that ship may engage in voyages to ports or offshore terminals under the jurisdiction of other Parties.

MARPOL Annex VI, regulation 6

Ozone-depleting Substances Record Book

Each ship subject to MARPOL Annex VI, regulation 6.1 that has rechargeable systems that contain ozone-depleting substances shall maintain an ozone-depleting substances record book.	MARPOL Annex VI, regulation 12.6

Fuel Oil Changeover Procedure and Logbook (record of fuel changeover)

Those ships using separate fuel oils to comply with MARPOL Annex VI, regulation 14.4 and entering or leaving an emission control area shall carry a written procedure showing how the fuel oil changeover is to be done. The volume of low-sulphur fuel oils in each tank as well as the date, time and position of the ship when any fuel oil changeover operation is completed prior to the entry into an emission control area or commenced after exit from such an area shall be recorded in such logbook as prescribed by the Administration.	MARPOL Annex VI, regulation 14.6

Manufacturer's Operating Manual for Incinerators

Incinerators installed in accordance with the requirements of MARPOL Annex VI, regulation 16.6.1 shall be provided with a Manufacturer's Operating Manual, which is to be retained with the unit.	MARPOL Annex VI, regulation 16.7

Bunker Delivery Note and Representative Sample

Bunker Delivery Note and representative sample of the fuel oil delivered shall be kept on board in accordance with requirements of MARPOL Annex VI, regulations 18.6 and 18.8.1.	MARPOL Annex VI, regulations 18.6 and 18.8.1

EEDI Technical File

Applicable to ships falling into one or more of categories in MARPOL Annex VI, regulations 2.25 to 2.35.	MARPOL Annex VI, regulation 20

Ship Energy Efficiency Management Plan (SEEMP)

All ships of 400 gross tonnage and above, excluding platforms (including FPSOs and FSUs) and drilling rigs, regardless of their propulsion, shall keep on board a ship specific Ship Energy Efficiency Management Plan (SEEMP). This may form part of the ship's Safety management System (SMS).	MARPOL Annex VI, regulation 22; MEPC.1/Circ.795

Technical File

Every marine diesel engine installed on board a ship shall be provided with a Technical File. The Technical File shall be prepared by the applicant for engine certification and approved by the Administration, and is required to accompany an engine throughout its life on board ships. The Technical File shall contain the information as specified in paragraph 2.4.1 of the NO_x Technical Code, 2008.	NO_x Technical Code, 2008, paragraph 2.3.4

Record Book of Engine Parameters

Where the Engine Parameter Check method in accordance with paragraph 6.2 of the NO_x Technical Code, 2008 is used to verify compliance, if any adjustments or modifications are made to an engine after its pre-certification, a full record of such adjustments or modifications shall be recorded in the engine's Record Book of Engine Parameters.	NO_x Technical Code, 2008, paragraph 2.3.7

Certificates for masters, officers or ratings	
Certificates for masters, officers or ratings shall be issued to those candidates who, to the satisfaction of the Administration, meet the requirements for service, age, medical fitness, training, qualifications and examinations in accordance with the appropriate provisions of the 1978 STCW Convention and STCW Code. Formats of certificates are given in section A I/2 of the STCW Code. Certificates must be kept available in their original form on board the ships on which the holder is serving.	STCW 1978, article VI, regulation I/2; STCW Code, section A-I/2
Fishing vessel personnel serving on board seagoing fishing vessels shall be certificated in accordance with the provisions of STCW-F Convention 1995. Formats of certificates are given in appendices 1, 2 and 3 of the Convention.	STCW-F 1995, article 6, regulation 3
Records of daily hours of rest	
Records of daily hours of rest of seafarers shall be maintained on board.	STCW Code, section A-VIII/1; Maritime Labour Convention, 2006; IMO/ILO Guidelines for the development of tables of seafarers' shipboard working arrangements and formats of records of seafarers' hours of work or hours of rest
International Anti-fouling System Certificate	
Ships of 400 gross tonnage and above engaged in international voyages, excluding fixed or floating platforms, FSUs, and FPSOs, shall be issued after inspection and survey an international Anti-fouling System Certificate together with a Record of Anti-fouling Systems.	AFS Convention, regulation 2(1) of annex 4
Declaration on Anti-fouling System	
Ships of 24 m or more in length, but less than 400 gross tonnage engaged in international voyages, excluding fixed or floating platforms, FSUs, and FPSOs, shall carry a declaration signed by the owner or owner's authorized agents. Such a declaration shall be accompanied by appropriate documentation (such as a paint receipt or a contractor invoice) or contain appropriate endorsement.	AFS Convention, regulation 5(1) of annex 4
International Ballast Water Management Certificate	
Ships of 400 gross tonnage and above to which the BWM 2004 applies, excluding floating platforms, FSUs and FPSOs, shall be issued the certificate after successful completion of a survey conducted in accordance with regulation E-1.	BWM 2004, regulation E-2
Note: The item was added by the Secretariat as per the relevant requirements of the International Convention for the Control and Management of Ships' Ballast Water and Sediments, 2004 (BWM 2004), which will enter into force on 8 September 2017.	

Ballast Water Management Plan	
Each ship shall have on board and implement a Ballast Water Management plan. Such a plan shall be approved by the Administration taking into account guidelines developed by the Organization.	BWM 2004, regulation B-1; resolution MEPC.127(53)
Note: The item was added by the Secretariat as per the relevant requirements of the International Convention for the Control and Management of Ships' Ballast Water and Sediments, 2004 (BWM 2004), which will enter into force on 8 September 2017.	
Ballast Water Record Book	
Each ship shall have on board a Ballast Water record book that may be an electronic record system, or that may be integrated into another record book or system and which shall at least contain the information specified in appendix II of the Convention. The Ballast water record book entries shall be maintained on board the ship for a minimum period of two years after the last entry has been made and thereafter in the Company's control for a minimum period of three years.	BWM 2004, Regulation B-2
Note: The item was added by the Secretariat as per the relevant requirements of the International Convention for the Control and Management of Ships' Ballast Water and Sediments, 2004 (BWM 2004), which will enter into force on 8 September 2017.	
Certificate of insurance or other financial security in respect of civil liability for bunker oil pollution damage	
Certificate attesting that insurance or other financial security is in force in accordance with the provisions of this Convention shall be issued to each ship having a gross tonnage greater than 1,000 after the appropriate authority of a State Party has determined that the requirements of article 7, paragraph 1 have been complied with. With respect to a ship registered in a State Party such certificate shall be issued or certified by the appropriate authority of the State of the ship's registry; with respect to a ship not registered in a State Party it may be issued or certified by the appropriate authority of any State Party. A State Party may authorize either an institution or an organization recognized by it to issue the certificate referred to in article 7, paragraph 2. This compulsory insurance certificate shall be in the form of the model set out in the annex to the Convention.	Bunkers Convention 2001, article 7
Certificate of insurance or other financial security in respect of liability for the removal of wrecks	
Certificate attesting that insurance or other financial security is in force in accordance with the provisions of the Convention shall be issued to each ship of 300 gross tonnage and above by the appropriate authority of the State of the ship's registry after determining that the requirements of article 12.1 have been complied with. With respect to a ship registered in a State Party, such certificate shall be issued or certified by the appropriate authority of the State of the ship's registry; with respect to a ship not registered in a State Party it may be issued or certified by the appropriate authority of any State Party. This compulsory insurance certificate shall be in the form of the model set out in the annex to the Convention.	Nairobi WRC 2007, article 12

2 In addition to the certificates listed in section 1 above, passenger ships shall carry:

Passenger Ship Safety Certificate

A certificate called a Passenger Ship Safety Certificate shall be issued after inspection and survey to a passenger ship which complies with the requirements of chapters II 1, II 2, III, IV and V and any other relevant requirements of SOLAS 1974. A Record of Equipment for the Passenger Ship Safety Certificate (Form P) shall be permanently attached.	SOLAS 1974, regulation I/12; 1988 SOLAS Protocol, regulation I/12

Decision support system for masters

In all passenger ships, a decision support system for emergency management shall be provided on the navigation bridge.	SOLAS 1974, regulation III/29

Search and rescue cooperation plan

Passenger ships to which chapter I of the Convention applies shall have on board a plan for cooperation with appropriate search and rescue services in event of an emergency.	SOLAS 1974, regulation V/7.3

List of operational limitations

Passenger ships to which chapter I of the Convention applies shall keep on board a list of all limitations on the operation of the ship, including exemptions from any of the SOLAS regulations, restrictions in operating areas, weather restrictions, sea state restrictions, restrictions in permissible loads, trim, speed and any other limitations, whether imposed by the Administration or established during the design or the building stages.	SOLAS 1974, regulation V/30

Special Trade Passenger Ship Safety Certificate,
Special Trade Passenger Ship Space Certificate

A Special Trade Passenger Ship Safety Certificate issued under the provisions of the Special Trade Passenger Ships Agreement, 1971.	STP 71, rule 5
A certificate called a Special Trade Passenger Ship Space Certificate shall be issued under the provisions of the Protocol on Space Requirements for Special Trade Passenger Ships, 1973.	SSTP 73, rule 5

Certificate of insurance or other financial security in respect of liability for the death of and personal injury to passengers

A certificate attesting that insurance or other financial security is in force in accordance with the provisions of this Convention shall be issued to each ship that is licensed to carry more than twelve passengers, after the appropriate authority of a State Party has determined that the requirements of Article 4bis paragraph 1 have been complied with. With respect to a ship registered in a State Party, such certificate shall be issued or certified by the appropriate authority of the State of the ship's registry; with respect to a ship not registered in a State Party it may be issued or certified by the appropriate authority of any State Party. A State Party may authorize an institution or an organization recognized by it to issue the certificate. The certificate shall be in the form of the model set out in the annex to the Convention.	PAL 1974 as modified by PAL Protocol 2002, article 4bis; resolution A.988(24); Circular Letter No. 2758

Pursuant to resolution A.988(24), States are recommended to ratify the Athens Protocol as soon as possible with the reservation that they reserve the right to issue and accept insurance certificates with such special exceptions and limitations as the insurance market conditions at the time of issue of the certificate may necessitate, examples being the biochemical clause and terrorism-related clauses (Circular Letter No.2758 refers).

3 In addition to the certificates listed in section 1 above, cargo ships shall carry:

Cargo Ship Safety Construction Certificate

A certificate called a Cargo Ship Safety Construction Certificate shall be issued after survey to a cargo ship of 500 gross tonnage and over which satisfies the requirements for cargo ships on survey, set out in regulation I/10 of SOLAS 1974, and complies with the applicable requirements of chapters II-1 and II-2, other than those relating to fire-extinguishing appliances and fire control plans.

SOLAS 1974, regulation I/12; 1988 SOLAS Protocol, regulation I/12

Cargo Ship Safety Equipment Certificate

A certificate called a Cargo Ship Safety Equipment Certificate shall be issued after survey to a cargo ship of 500 gross tonnage and over which complies with the relevant requirements of chapters II-1 and II-2, III and V and any other relevant requirements of SOLAS 1974. A Record of Equipment for the Cargo Ship Safety Equipment Certificate (Form E) shall be permanently attached.

SOLAS 1974, regulation I/12; 1988 SOLAS Protocol, regulation I/12

Cargo Ship Safety Radio Certificate

A certificate called a Cargo Ship Safety Radio Certificate shall be issued after survey to a cargo ship of 300 gross tonnage and over, fitted with a radio installation, including those used in life-saving appliances, which complies with the requirements of chapter IV and any other relevant requirements of SOLAS 1974. A Record of Equipment for the Cargo Ship Safety Radio Certificate (Form R) shall be permanently attached.

SOLAS 1974, regulation I/12, as amended by the GMDSS amendments; 1988 SOLAS Protocol, regulation I/12

Cargo Ship Safety Certificate

A certificate called a Cargo Ship Safety Certificate may be issued after survey to a cargo ship which complies with the relevant requirements of chapters II-1, II-2, III, IV and V and other relevant requirements of SOLAS 1974 as modified by the 1988 SOLAS Protocol, as an alternative to the Cargo Ship Safety Construction Certificate, Cargo Ship Safety Equipment Certificate and Cargo Ship Safety Radio Certificate. A Record of Equipment for the Cargo Ship Safety Certificate (Form C) shall be permanently attached.

1988 SOLAS Protocol, regulation I/12

Ship Structure Access Manual

This regulation applies to oil tankers of 500 gross tonnage and over and bulk carriers, as defined in regulation IX/1, of 20,000 gross tonnage and over, constructed on or after 1 January 2006. A ship's means of access to carry out overall and close-up inspections and thickness measurements shall be described in a ship structure access manual approved by the Administration, an updated copy of which shall be kept on board.

SOLAS 1974, regulation II-1/3-6

Cargo Information

The shipper shall provide the master or his representative with appropriate information, confirmed in writing, on the cargo, in advance of loading. In bulk carriers, the density of the cargo shall be provided in the above information.

SOLAS 1974, regulations VI/2 and XII/10; MSC/Circ.663

Bulk Carrier Booklet

To enable the master to prevent excessive stress in the ship's structure, the ship loading and unloading solid bulk cargoes shall be provided with a booklet referred to in SOLAS regulation VI/7.2. The booklet shall be endorsed by the Administration or on its behalf to indicate that SOLAS regulations XII/4, 5, 6 and 7, as appropriate, are complied with. As an alternative to a separate booklet, the required information may be contained in the intact stability booklet.

SOLAS 1974, regulations VI/7 and XII/8; Code of Practice for the Safe Loading and Unloading of Bulk Carriers (BLU Code)

Document of authorization for the carriage of grain and grain loading manual

A document of authorization shall be issued for every ship loaded in accordance with the regulations of the International Code for the Safe Carriage of Grain in Bulk. The document shall accompany or be incorporated into the grain loading manual provided to enable the master to meet the stability requirements of the Code.

SOLAS 1974, regulation VI/9; International Code for the Safe Carriage of Grain in Bulk, section 3

Enhanced survey report file

Bulk carriers and oil tankers shall have a survey report file and supporting documents complying with paragraphs 6.2 and 6.3 of annex A/annex B, part A/part B, 2011 ESP Code.

SOLAS 1974, regulation XI-1/2; 2011 ESP Code (resolution A.1049(27), as amended)

Dedicated Clean Ballast Tank Operation Manual

Every product carrier of 40,000 tonnes deadweight and above delivered on or before 1 June 1982, operating with dedicated clean ballast tanks shall be provided with a Dedicated Clean Ballast Tank Operation Manual detailing the system and specifying operational procedures. Such a Manual shall be to the satisfaction of the Administration and shall contain all the information set out in the specifications referred to in subparagraph 8.2 of MARPOL Annex I regulation 18. If an alteration affecting the dedicated clean ballast tank system is made, the Operation Manual shall be revised accordingly.

MARPOL Annex I, regulation 18.8; resolution A.495(XII)

Condition Assessment Scheme (CAS) Statement of Compliance, CAS Final Report and Review Record

A Statement of Compliance shall be issued by the Administration to every oil tanker which has been surveyed in accordance with the requirements of the Condition Assessment Scheme (CAS) and found to be in compliance with these requirements. In addition, a copy of the CAS Final Report which was reviewed by the Administration for the issue of the Statement of Compliance and a copy of the relevant Review Record shall be placed on board to accompany the Statement of Compliance.

MARPOL Annex I, regulations 20 and 21; resolution MEPC.94(46), as amended by resolutions MEPC.99(48), MEPC.112(50), MEPC.131(53), MEPC.155(55), and MEPC.236(65)

Subdivision and stability information	
Every oil tanker to which regulation 28 of Annex I of MARPOL applies shall be provided in an approved form with information relative to loading and distribution of cargo necessary to ensure compliance with the provisions of this regulation and data on the ability of the ship to comply with damage stability criteria as determined by this regulation.	MARPOL Annex I, regulation 28
Record of oil discharge monitoring and control system for the last ballast voyage	
Subject to the provisions of paragraphs 4 and 5 of regulation 3 of MARPOL Annex I, every oil tanker of 150 gross tonnage and above shall be equipped with an oil discharge monitoring and control system approved by the Administration. The system shall be fitted with a recording device to provide a continuous record of the discharge in litres per nautical mile and total quantity discharged, or the oil content and rate of discharge. The record shall be identifiable as to time and date and shall be kept for at least three years.	MARPOL Annex I, regulation 31
Oil Discharge Monitoring and Control (ODMC) Operational Manual	
Every oil tanker fitted with an Oil Discharge Monitoring and Control system shall be provided with instructions as to the operation of the system in accordance with an operational manual approved by the Administration.	MARPOL Annex I, regulation 31; resolution A.496(XII); resolution A.586(14), as amended by resolution MEPC.24(22); resolution MEPC.108(49), as amended by MEPC.240(65)
Crude Oil Washing Operation and Equipment Manual (COW Manual)	
Every oil tanker operating with crude oil washing systems shall be provided with an Operations and Equipment Manual detailing the system and equipment and specifying operational procedures. Such a Manual shall be to the satisfaction of the Administration and shall contain all the information set out in the specifications referred to in regulation 33 of Annex I of MARPOL.	MARPOL Annex I, regulation 35; resolution MEPC.81(43)
STS Operation Plan and Records of STS Operations	
Any oil tanker involved in STS operations shall carry on board a plan prescribing how to conduct STS operations (STS operations plan) not later than the date of the first annual, intermediate or renewal survey of the ship to be carried out on or after 1 January 2011. Each oil tanker's STS operations plan shall be approved by the Administration. The STS operations plan shall be written in the working language of the ship. Records of STS operations shall be retained on board for three years and be readily available for inspection.	MARPOL Annex I, regulation 41
VOC Management Plan	
A tanker carrying crude oil, to which MARPOL Annex VI, regulation 15.1 applies, shall have on board and implement a VOC Management Plan.	MARPOL Annex VI, regulation 15.6

Document of approval for the stability instrument

All ships, subject to the IBC, IGC, BCH and GC Codes, should be fitted with a stability instrument capable of verifying compliance with intact and damage stability approved by the Administration, at the first scheduled renewal survey of the ship on or after 1 January 2016, but not later than 1 January 2021, having regard to the performance standards recommended by the Organization. The Administration should issue a document of approval for the stability instrument.

IBC Code para. 2.2.6; IGC Code para. 2.2.6; BCH Code para. 2.2.1.2; GC Code para. 2.2.4; 2008 IS Code; MSC.1/Circ.1229; MSC.1/Circ.1461.

Certificate of insurance or other financial security in respect of civil liability for oil pollution damage

A certificate attesting that insurance or other financial security is in force shall be issued to each ship carrying more than 2,000 tonnes of oil in bulk as cargo. It shall be issued or certified by the appropriate authority of the State of the ship's registry after determining that the requirements of article VII, paragraph 1, of the CLC Convention have been complied with.

CLC 1969, article VII

Certificate of insurance or other financial security in respect of civil liability for oil pollution damage

A certificate attesting that insurance or other financial security is in force in accordance with the provisions of the 1992 CLC Convention shall be issued to each ship carrying more than 2,000 tonnes of oil in bulk as cargo after the appropriate authority of a Contracting State has determined that the requirements of article VII, paragraph 1, of the Convention have been complied with. With respect to a ship registered in a Contracting State, such certificate shall be issued by the appropriate authority of the State of the ship's registry; with respect to a ship not registered in a Contracting State, it may be issued or certified by the appropriate authority of any Contracting State.

CLC 1992, article VII

4 In addition to the certificates listed in sections 1 and 3 above, where appropriate, any ship carrying noxious liquid chemical substances in bulk shall carry:

International Pollution Prevention Certificate for the Carriage of Noxious Liquid Substances in Bulk (NLS Certificate)

An international pollution prevention certificate for the carriage of noxious liquid substances in bulk (NLS Certificate) shall be issued, after survey in accordance with the provisions of regulation 8 of Annex II of MARPOL, to any ship carrying noxious liquid substances in bulk and which is engaged in voyages to ports or terminals under the jurisdiction of other Parties to MARPOL. In respect of chemical tankers, the Certificate of Fitness for the Carriage of Dangerous Chemicals in Bulk and the International Certificate of Fitness for the Carriage of Dangerous Chemicals in Bulk, issued under the provisions of the Bulk Chemical Code and International Bulk Chemical Code, respectively, shall have the same force and receive the same recognition as the NLS Certificate.

MARPOL Annex II, regulation 9

Cargo record book

Ships carrying noxious liquid substances in bulk shall be provided with a Cargo Record Book, whether as part of the ship's official log book or otherwise, in the form specified in appendix II to Annex II.

MARPOL Annex II, regulation 15.1

Procedures and Arrangements Manual (P & A Manual)

Every ship certified to carry noxious liquid substances in bulk shall have on board a Procedures and Arrangements Manual approved by the Administration.

MARPOL Annex II, regulation 14; resolution MEPC.18(22), as amended by resolution MEPC.62(35)

Shipboard Marine Pollution Emergency Plan for Noxious Liquid Substances

Every ship of 150 gross tonnage and above certified to carry noxious liquid substances in bulk shall carry on board a shipboard marine pollution emergency plan for noxious liquid substances approved by the Administration.

MARPOL Annex II, regulation 17; resolution MEPC.85(44), as amended by resolution MEPC.137(53)

5 In addition to the certificates listed in sections 1 and 3 above, where applicable, any chemical tanker shall carry:

Certificate of Fitness for the Carriage of Dangerous Chemicals in Bulk

A certificate called a Certificate of Fitness for the Carriage of Dangerous Chemicals in Bulk, the model form of which is set out in the appendix to the Bulk Chemical Code, should be issued after an initial or periodical survey to a chemical tanker engaged in international voyages which complies with the relevant requirements of the Code.

BCH Code, section 1.6

Note: The Code is mandatory under Annex II of MARPOL for chemical tankers constructed before 1 July 1986.

or

International Certificate of Fitness for the Carriage of Dangerous Chemicals in Bulk

A certificate called an International Certificate of Fitness for the Carriage of Dangerous Chemicals in Bulk, the model form of which is set out in the appendix to the International Bulk Chemical Code, should be issued after an initial or periodical survey to a chemical tanker engaged in international voyages, which complies with the relevant requirements of the Code.

IBC Code, section 1.5

Note: The Code is mandatory under both chapter VII of SOLAS 1974 and Annex II of MARPOL for chemical tankers constructed on or after 1 July 1986.

6 In addition to the certificates listed in sections 1 and 3 above, where applicable, any gas carrier shall carry:

Certificate of Fitness for the Carriage of Liquefied Gases in Bulk

A certificate called a Certificate of Fitness for the Carriage of Liquefied Gases in Bulk, the model form of which is set out in the appendix to the Gas Carrier Code, should be issued after an initial or periodical survey to a gas carrier which complies with the relevant requirements of the Code.

GC Code, section 1.6

International Certificate of Fitness for the Carriage of Liquefied Gases in Bulk

A certificate called an International Certificate of Fitness for the Carriage of Liquefied Gases in Bulk, the model form of which is set out in the appendix to the International Gas Carrier Code, should be issued after an initial or periodical survey to a gas carrier which complies with the relevant requirements of the Code.

IGC Code, section 1.4

Note: The Code is mandatory under chapter VII of SOLAS 1974 for gas carriers constructed on or after 1 July 1986.

7 In addition to the certificates listed in sections 1, and 2 or 3 above, where applicable, any high-speed craft shall carry:

High-Speed Craft Safety Certificate

A certificate called a High-Speed Craft Safety Certificate shall be issued after completion of an initial or renewal survey to a craft which complies with the requirements of the 1994 HSC Code or the 2000 HSC Code, as appropriate.

SOLAS 1974, regulation X/3;
1994 HSC Code, section 1.8;
2000 HSC Code, section 1.8

Permit to Operate High-Speed Craft

A certificate called a Permit to Operate High-Speed Craft shall be issued to a craft which complies with the requirements set out in paragraphs 1.2.2 to 1.2.7 of the 1994 HSC Code or the 2000 HSC Code, as appropriate.

1994 HSC Code, section 1.9;
2000 HSC Code, section 1.9

8 In addition to the certificates listed in sections 1, and 2 or 3 above, where applicable, any ship carrying dangerous goods shall carry:

Document of compliance with the special requirements for ships carrying dangerous goods

The Administration shall provide the ship with an appropriate document as evidence of compliance of construction and equipment with the requirements of regulation II-2/19 of SOLAS 1974. Certification for dangerous goods, except solid dangerous goods in bulk, is not required for those cargoes specified as class 6.2 and 7 and dangerous goods in limited quantities.

SOLAS 1974,
regulation II-2/19.4

9 In addition to the certificates listed in sections 1, and 2 or 3 above, where applicable, any ship carrying dangerous goods in packaged form shall carry:

Transport information

Transport information relating to the carriage of dangerous goods in packaged form and the container/vehicle packing certificate shall be in accordance with the relevant provisions of the IMDG Code and shall be made available to the person or organization designated by the port State authority

SOLAS 1974, regulation VII/4.1

Dangerous goods manifest or stowage plan

Each ship carrying dangerous goods in packaged form shall have a special list or manifest setting forth, in accordance with the classification set out in the IMDG Code, the dangerous goods on board and the location thereof. Each ship carrying dangerous goods in solid form in bulk shall have a list or manifest setting forth the dangerous goods on board and the location thereof. A detailed stowage plan, which identifies by class and sets out the location of all dangerous goods on board, may be used in place of such a special list or manifest. A copy of one of these documents shall be made available before departure to the person or organization designated by the port State authority.

SOLAS 1974, regulations VII/4.2 and VII/7-2.2; MARPOL Annex III, regulation 4

10 In addition to the certificates listed in sections 1, and 2 or 3 above, where applicable, any ship carrying INF cargo shall carry:

International Certificate of Fitness for the Carriage of INF Cargo

A ship carrying INF cargo shall comply with the requirements of the International Code for the Safe Carriage of Packaged Irradiated Nuclear Fuel, Plutonium and High Level Radioactive Wastes on Board Ships (INF Code) in addition to any other applicable requirements of the SOLAS regulations and shall be surveyed and be provided with the International Certificate of Fitness for the Carriage of INF Cargo.

SOLAS 1974, regulation VII/16; INF Code (resolution MSC.88(71), as amended), paragraph 1.3

11 In addition to the certificates listed in sections 1, and 2 or 3 above, where applicable, any nuclear ship shall carry:

Operating Manual for nuclear power plant

A fully detailed Operating Manual shall be prepared for the information and guidance of the operating personnel in their duties on all matters relating to the operation of the nuclear power plant having an important bearing on safety. The Administration, when satisfied, shall approve such Operating Manual and a copy shall be kept on board the ship. The Operating Manual shall always be kept up to date.

SOLAS 1974, regulation VIII/8

A Nuclear Cargo Ship Safety Certificate or Nuclear Passenger Ship Safety Certificate, in place of the Cargo Ship Safety Certificate or Passenger Ship Safety Certificate, as appropriate

Every nuclear powered ship shall be issued with the certificate required by SOLAS chapter VIII.

SOLAS 1974, regulation VIII/10

12 In addition to the certificates listed in sections 1, and 2 or 3 above, where applicable, any ship operating in Polar waters shall carry:

Polar Ship Certificate

Every ship to which the Polar Code applies shall have on board a valid Polar Ship Certificate. The certificate shall include a supplement recording equipment required by the Code.

Polar Code, part I-A Section 1.3

Polar water operational manual (PWOM)

Every ship to which the Polar Code applies shall have on board a Polar water operational manual (PWOM) as required in part I-A section 2.3 of the Code.

Polar Code, part I-A section 2.3

Other certificates and documents which are not mandatory

Special purpose ships

Special Purpose Ship Safety Certificate

In addition to SOLAS certificates as specified in paragraph 7 of the Preamble of the 1983 SPS Code and 2008 SPS Code, a Special Purpose Ship Safety Certificate should be issued after survey in accordance with the provisions of paragraph 1.6 of the 1983 SPS Code and 2008 SPS Code. The duration and validity of the certificate should be governed by the respective provisions for cargo ships in SOLAS 1974. If a certificate is issued for a special purpose ship of less than 500 gross tonnage, this certificate should indicate to what extent relaxations in accordance with 1.2 were accepted.

1983 SPS Code (resolution A.534(13), as amended); 2008 SPS Code (resolution MSC.266(84), as amended), SOLAS 1974, regulation I/12; 1988 SOLAS Protocol, regulation I/12

The 2008 SPS Code applies to every special purpose ship of not less than 500 gross tonnage certified on or after 13 May 2008.

Offshore support vessels

Offshore Supply Vessel Document of Compliance

The Document of Compliance should be issued after it has been satisfied that the vessel complies with the provisions of the Guidelines for the design and construction of Offshore Supply Vessels, 2006.

Resolution MSC.235(82), as amended by resolution MSC.335(90)

Certificate of Fitness for Offshore Support Vessels

When carrying such cargoes, offshore support vessels should carry a Certificate of Fitness issued under the "Guidelines for the Transport and Handling of Limited Amounts of Hazardous and Noxious Liquid Substances in Bulk on Offshore Support Vessels". If an offshore support vessel carries only noxious liquid substances, a suitably endorsed International Pollution Prevention Certificate for the Carriage of Noxious Liquid Substances in Bulk may be issued instead of the above Certificate of Fitness.

Resolution A.673(16), as amended by resolutions MSC.184(79), MSC.236(82) and MEPC.158(55); MARPOL Annex II, regulation 11.2

Diving systems

Diving System Safety Certificate

A certificate should be issued either by the Administration or any person or organization duly authorized by it after survey or inspection to a diving system which complies with the requirements of the Code of Safety for Diving Systems. In every case, the Administration should assume full responsibility for the certificate.

Resolution A. 831(19), as amended by resolution MSC.185(79), section 1.6

Passenger submersible craft

Safety Compliance Certificate for Passenger Submersible Craft

Applicable to submersible craft adapted to accommodate passengers and intended for underwater excursions with the pressure in the passenger compartment at or near one atmosphere.

MSC/Circ.981, as amended by MSC/Circ.1125

A Design and Construction Document issued by the Administration should be attached to the Safety Compliance Certificate.

Dynamically supported craft

Dynamically Supported Craft Construction and Equipment Certificate

To be issued after survey carried out in accordance with paragraph 1.5.1(a) of the Code of Safety for Dynamically Supported Craft.	DSC Code (resolution A.373(X), as amended) section 1.6

Mobile offshore drilling units

Mobile Offshore Drilling Unit Safety Certificate

To be issued after survey carried out in accordance with the provisions of the Code for the Construction and Equipment of Mobile Offshore Drilling Units, 1979, or, for units constructed on or after 1 May 1991, but before 1 January 2012, the Code for the Construction and Equipment of Drilling Units, 1989, or for units constructed on or after 1 January 2012, the Code for the Construction and Equipment of Drilling Units, 2009.	1979 MODU Code (resolution A.414(XI), as amended) section 1.6; 1989 MODU Code (resolution A.649(16), as amended) section 1.6; 2009 MODU Code (resolution A.1023(26), as amended) section 1.6

Wing-In-Ground (WIG) Craft

Wing-in-ground Craft Safety Certificate

A certificate called a WIG Craft Safety Certificate should be issued after completion of an initial or renewal survey to a craft, which complies with the provisions of the Interim Guidelines for WIG craft.	MSC/Circ.1054, as amended by MSC/Circ.1126, section 9

Permit to Operate WIG Craft

A permit to operate should be issued by the Administration to certify compliance with the provisions of the Interim Guidelines for WIG craft.	MSC/Circ.1054, as amended by MSC/Circ.1126, section 10

Noise levels

Noise Survey Report

Applicable to existing ships to which SOLAS II 1/3-12 does not apply. A noise survey report should be made for each ship in accordance with the Code on Noise Levels on board Ships.	Resolution A.468(XII), section 4.3

3

Prospective amendments to MARPOL Annexes

Prospective amendments to MARPOL Annex I

Resolution MEPC.276(70)
adopted on 28 October 2016

Amendments to the Annex of the International Convention for the Prevention of Pollution from Ships, 1973, as modified by the Protocol of 1978 relating thereto

Amendments to MARPOL Annex I

(Form B of the Supplement to the International Oil Pollution Prevention Certificate)

THE MARINE ENVIRONMENT PROTECTION COMMITTEE,

RECALLING Article 38(a) of the Convention on the International Maritime Organization concerning the functions of the Marine Environment Protection Committee conferred upon it by international conventions for the prevention and control of marine pollution from ships,

NOTING article 16 of the International Convention for the Prevention of Pollution from Ships, 1973, as modified by the Protocol of 1978 relating thereto (MARPOL), which specifies the amendment procedure and confers upon the appropriate body of the Organization the function of considering and adopting amendments thereto,

HAVING CONSIDERED, at its seventieth session, proposed amendments to appendix II of MARPOL Annex I concerning the Supplement to the International Oil Pollution Prevention Certificate,

1 ADOPTS, in accordance with article 16(2)(d) of MARPOL, amendments to appendix II of MARPOL Annex I, the text of which is set out in the annex to the present resolution;

2 DETERMINES, in accordance with article 16(2)(f)(iii) of MARPOL, that the amendments shall be deemed to have been accepted on 1 September 2017 unless prior to that date, not less than one third of the Parties or Parties the combined merchant fleets of which constitute not less than 50% of the gross tonnage of the world's merchant fleet, have communicated to the Organization their objection to the amendments;

3 INVITES the Parties to note that, in accordance with article 16(2)(g)(ii) of MARPOL, the said amendments shall enter into force on 1 March 2018 upon their acceptance in accordance with paragraph 2 above;

4 REQUESTS the Secretary-General, for the purposes of article 16(2)(e) of MARPOL, to transmit certified copies of the present resolution and the text of the amendments contained in the annex to all Parties to MARPOL;

5 REQUESTS FURTHER the Secretary-General to transmit copies of the present resolution and its annex to Members of the Organization which are not Parties to MARPOL.

Annex

Amendments to MARPOL Annex I
(Form B of the Supplement to the International Oil Pollution Prevention Certificate)

Annex I

Regulations for the prevention of pollution by oil

Appendix II
Form of IOPP Certificate and Supplements

Form B of the Supplement to the International Oil Pollution Prevention Certificate

RECORD OF CONSTRUCTION AND EQUIPMENT FOR OIL TANKERS

Section 1 – Particulars of ship

1 *Paragraphs 1.11.8 and 1.11.9 are deleted.*

Section 5 – Construction (regulations 18, 19, 20, 21, 22, 23, 26, 27, 28 and 33)

2 *Paragraph 5.1 is replaced with the following:*

"5.1 In accordance with the requirements of regulation 18, the ship is qualified as a segregated ballast tanker in compliance with regulation 18.9 .... ☐"

3 *Existing paragraphs 5.1.1 to 5.1.6 are deleted.*

4 *Paragraph 5.2 is replaced with the following:*

"5.2 Segregated ballast tanks (SBT) in compliance with regulation 18 are distributed as follows:

Tank	Volume (m^3)	Tank	Volume (m^3)
		Total volume .... m^3	

"

5 *Existing paragraphs 5.2.1 to 5.2.3, 5.3 and 5.3.1 to 5.3.5 are deleted.*

6 *Existing paragraphs 5.4 and 5.4.1 to 5.4.4 are renumbered as 5.3 and 5.3.1 to 5.3.4.*

7 *Existing paragraphs 5.5 and 5.5.1 to 5.5.2 are deleted.*

8 *All subsequent paragraphs in section 5 are renumbered accordingly.*

Prospective amendments to MARPOL Annex II

Resolution MEPC.270(69)
adopted on 22 April 2016

Amendments to the Annex of the International Convention for the Prevention of Pollution from Ships, 1973, as modified by the Protocol of 1978 relating thereto

Amendments to MARPOL Annex II

(Revised GESAMP Hazard Evaluation Procedure)

THE MARINE ENVIRONMENT PROTECTION COMMITTEE,

RECALLING Article 38(a) of the Convention on the International Maritime Organization concerning the functions of the Marine Environment Protection Committee conferred upon it by international conventions for the prevention and control of marine pollution from ships,

NOTING article 16 of the International Convention for the Prevention of Pollution from Ships, 1973, as modified by the Protocol of 1978 relating thereto (MARPOL), which specifies the amendment procedure and confers upon the appropriate body of the Organization the function of considering and adopting amendments thereto,

HAVING CONSIDERED, at its sixty-ninth session, proposed amendments to Appendix I of MARPOL Annex II concerning the abbreviated legend to the revised GESAMP Hazard Evaluation Procedure,

1 ADOPTS, in accordance with article 16(2)(d) of MARPOL, amendments to Appendix I of MARPOL Annex II, the text of which is set out in the annex to the present resolution;

2 DETERMINES, in accordance with article 16(2)(f)(iii) of MARPOL, that the amendments shall be deemed to have been accepted on 1 March 2017 unless prior to that date, not less than one third of the Parties or Parties the combined merchant fleets of which constitute not less than 50% of the gross tonnage of the world's merchant fleet, have communicated to the Organization their objection to the amendments;

3 INVITES the Parties to note that, in accordance with article 16(2)(g)(ii) of MARPOL, the said amendments shall enter into force on 1 September 2017 upon their acceptance in accordance with paragraph 2 above;

4 REQUESTS the Secretary-General, for the purposes of article 16(2)(e) of MARPOL, to transmit certified copies of the present resolution and the text of the amendments contained in the annex to all Parties to MARPOL;

5 REQUESTS FURTHER the Secretary-General to transmit copies of the present resolution and its annex to Members of the Organization which are not Parties to MARPOL.

Annex

Amendments to MARPOL Annex II
(Revised GESAMP Hazard Evaluation Procedure)

Annex II

Regulations for the control of pollution by noxious liquid substances in bulk

Appendix I
Guidelines for the categorization of noxious liquid substances

The tables under the title "Abbreviated legend to the revised GESAMP Hazard Evaluation Procedure" are replaced with the following:

The Revised GESAMP Hazard Evaluation Procedure

Columns A and B – Aquatic environment					
Rating	A			B	
	Bioaccumulation and biodegradation			Aquatic toxicity	
	A1 Bioaccumulation		A2 Biodegradation	B1 Acute toxicity	B2 Chronic toxicity
	log P_{OW}	BCF		$LC/EC/IC_{50}$ (mg/L)	NOEC (mg/L)
0	< 1 or > ca. 7	no measurable BCF	**R:** readily biodegradable **NR:** not readily biodegradable	> 1,000	> 1
1	≥ 1 – < 2	≥ 1 – < 10		> 100 – ≤ 1,000	> 0.1 – ≤ 1
2	≥ 2 – < 3	≥ 10 – < 100		> 10 – ≤ 100	> 0.01 – ≤ 0.1
3	≥ 3 – < 4	≥ 100 – < 500		> 1 – ≤ 10	> 0.001 – ≤ 0.01
4	≥ 4 – < 5	≥ 500 – < 4,000		> 0.1 – ≤ 1	≤ 0.001
5	≥ 5 – < ca. 7	> 4,000		> 0.01 – ≤ 0.1	
6				≤ 0.01	

<table>
<tr><th colspan="7">Columns C and D – Human health (toxic effects to mammals)</th></tr>
<tr><th rowspan="3">Rating</th><th colspan="3">C</th><th colspan="3">D</th></tr>
<tr><th colspan="3">Acute mammalian toxicity</th><th colspan="3">Irritation, corrosion and long-term health effects</th></tr>
<tr><th>C1
Oral toxicity
LD_{50}/ATE
(mg/kg)</th><th>C2
Dermal toxicity
LD_{50}/ATE
(mg/kg)</th><th>C3
Inhalation toxicity
LD_{50}/ATE
(mg/L)</th><th>D1
Skin irritation and corrosion</th><th>D2
Eye irritation and corrosion</th><th>D3
Long-term health effects</th></tr>
<tr><td>0</td><td>> 2,000</td><td>> 2,000</td><td>> 20</td><td>not irritating</td><td>not irritating</td><td rowspan="4">C: carcinogen
M: mutagenic
R: reprotoxic
Ss: sensitizing to skin
Sr: sensitizing to respiratory system
A: aspiration hazard
T: target organ toxicity
N: neurotoxic
I: immunotoxic</td></tr>
<tr><td>1</td><td>> 300 – ≤ 2,000</td><td>> 1,000 – ≤ 2,000</td><td>> 10 – ≤ 20</td><td>mildly irritating</td><td>mildly irritating</td></tr>
<tr><td>2</td><td>> 50 – ≤ 300</td><td>> 200 – ≤ 1,000</td><td>> 2 – ≤ 10</td><td>irritating</td><td>irritating</td></tr>
<tr><td>3</td><td>> 5 – ≤ 50</td><td>> 50 – ≤ 200</td><td>> 0.5 – ≤ 2</td><td>severely irritating or corrosive
3A Corr. (≤ 4 h)
3B Corr. (≤ 1 h)
3C Corr. (≤ 3 min)</td><td>severely irritating</td></tr>
<tr><td>4</td><td>≤ 5</td><td>≤ 50</td><td>≤ 0.5</td><td></td><td></td><td></td></tr>
</table>

<table>
<tr><th colspan="4">Column E – Interferences with other uses of the sea</th></tr>
<tr><th rowspan="2">E1
Tainting</th><th rowspan="2">E2
Physical effects on wildlife and benthic habitats</th><th colspan="2">E3
Interference with coastal amenities</th></tr>
<tr><th>Numerical rating</th><th>Description and action</th></tr>
<tr><td rowspan="2">NT: not tainting (tested)
T: tainting test positive</td><td rowspan="3">Fp: persistent floater
F: floater
S: sinking substances</td><td>0</td><td>no interference
no warning</td></tr>
<tr><td>1</td><td>slightly objectionable
warning, no closure of amenity</td></tr>
<tr><td></td><td>2</td><td>moderately objectionable
possible closure of amenity</td></tr>
<tr><td></td><td></td><td>3</td><td>highly objectionable
closure of amenity</td></tr>
</table>

Prospective amendments to MARPOL Annex IV

Resolution MEPC.274(69)
adopted on 22 April 2016

Amendments to the Annex of the Protocol of 1997 to amend the International Convention for the Prevention of Pollution from Ships, 1973, as modified by the Protocol of 1978 relating thereto

Amendments to MARPOL Annex IV

(Baltic Sea Special Area and Form of ISPP Certificate)

THE MARINE ENVIRONMENT PROTECTION COMMITTEE,

RECALLING Article 38(a) of the Convention on the International Maritime Organization concerning the functions of the Marine Environment Protection Committee conferred upon it by international conventions for the prevention and control of marine pollution from ships,

NOTING article 16 of the International Convention for the Prevention of Pollution from Ships, 1973, as modified by the Protocol of 1978 relating thereto (MARPOL), which specifies the amendment procedure and confers upon the appropriate body of the Organization the function of considering and adopting amendments thereto,

HAVING CONSIDERED, at its sixty-ninth session, proposed amendments to regulations 1 and 11 and to the appendix to MARPOL Annex IV,

1 ADOPTS, in accordance with article 16(2)(d) of MARPOL, amendments to regulations 1 and 11 of MARPOL Annex IV concerning the Baltic Sea Special Area and to the appendix to MARPOL Annex IV concerning the Form of the International Sewage Pollution Prevention Certificate, the texts of which are set out in the annex to the present resolution;

2 DETERMINES, in accordance with article 16(2)(f)(iii) of MARPOL, that the amendments shall be deemed to have been accepted on 1 March 2017, unless prior to that date, not less than one third of the Parties or Parties the combined merchant fleets of which constitute not less than 50% of the gross tonnage of the world's merchant fleet, have communicated to the Organization their objection to the amendments;

3 INVITES the Parties to note that, in accordance with article 16(2)(g)(ii) of MARPOL, the said amendments shall enter into force on 1 September 2017 upon their acceptance in accordance with paragraph 2 above;

4 REQUESTS the Secretary-General, for the purposes of article 16(2)(e) of MARPOL, to transmit certified copies of the present resolution and the text of the amendments contained in the annex to all Parties to MARPOL;

5 REQUESTS FURTHER the Secretary-General to transmit copies of the present resolution and its annex to Members of the Organization which are not Parties to MARPOL.

Annex

Amendments to MARPOL Annex IV
Regulations for the prevention of pollution by sewage from ships

Chapter 1 – General

Regulation 1
Definitions

1 *Paragraph 10 is replaced by the following:*

"**10** A *passenger ship* means a ship which carries more than twelve passengers.

For the application of regulation 11.3, a new passenger ship is a passenger ship:

.1 for which the building contract is placed, or in the absence of a building contract, the keel of which is laid, or which is in similar stage of construction, on or after 1 June 2019; or

.2 the delivery of which is on or after 1 June 2021.

An *existing passenger ship* is a passenger ship which is not a new passenger ship."

Chapter 3 – Equipment and control of discharge

Regulation 11
Discharge of sewage

2 *Paragraph 3 is replaced by the following:*

"*B* *Discharge of sewage from passenger ships within a special area*

3 Subject to the provisions of regulation 3 of this Annex, the discharge of sewage from a passenger ship within a special area* shall be prohibited:

.1 for new passenger ships, on a date determined by the Organization pursuant to regulation 13.2 of this Annex, but in no event prior to 1 June 2019; and

.2 for existing passenger ships, on a date determined by the Organization pursuant to regulation 13.2 of this Annex, but in no event prior to 1 June 2021,

except when the following conditions are satisfied:

> the ship has in operation an approved sewage treatment plant which has been certified by the Administration to meet the operational requirements referred to in regulation 9.2.1 of this Annex, and the effluent shall not produce visible floating solids nor cause discoloration of the surrounding water."

* Refer to Establishment of the date on which regulation 11.3 of MARPOL Annex IV in respect of the Baltic Sea Special Area shall take effect (resolution MEPC.275(69)).

Appendix
Form of International Sewage Pollution Prevention Certificate

INTERNATIONAL SEWAGE POLLUTION PREVENTION CERTIFICATE

3 *The final paragraph under section 1.1 is replaced by the following:*

"The sewage treatment plant is certified by the Administration to meet the effluent standards as provided for in the Guidelines on implementation of effluent standards and performance test for sewage treatment plants, adopted by resolution MEPC.227(64), as amended, including/excluding* the standards of section 4.2 thereof."

* Delete as appropriate.

Prospective amendments to MARPOL Annex V

Resolution MEPC.277(70)
adopted on 28 October 2016

Amendments to the Annex of the International Convention for the Prevention of Pollution from Ships, 1973, as modified by the Protocol of 1978 relating thereto

Amendments to MARPOL Annex V

(HME substances and Form of Garbage Record Book)

THE MARINE ENVIRONMENT PROTECTION COMMITTEE,

RECALLING Article 38(a) of the Convention on the International Maritime Organization concerning the functions of the Marine Environment Protection Committee conferred upon it by international conventions for the prevention and control of marine pollution from ships,

NOTING article 16 of the International Convention for the Prevention of Pollution from Ships, 1973, as modified by the Protocol of 1978 relating thereto (MARPOL), which specifies the amendment procedure and confers upon the appropriate body of the Organization the function of considering and adopting amendments thereto,

HAVING CONSIDERED, at its seventieth session, proposed amendments to MARPOL Annex V concerning substances that are harmful to the marine environment (HME) and Form of Garbage Record Book,

1 ADOPTS, in accordance with article 16(2)(d) of MARPOL, amendments to MARPOL Annex V, the text of which is set out in the annex to the present resolution;

2 DETERMINES, in accordance with article 16(2)(f)(iii) of MARPOL, that the amendments shall be deemed to have been accepted on 1 September 2017 unless prior to that date, not less than one third of the Parties or Parties the combined merchant fleets of which constitute not less than 50% of the gross tonnage of the world's merchant fleet, have communicated to the Organization their objection to the amendments;

3 INVITES the Parties to note that, in accordance with article 16(2)(g)(ii) of MARPOL, the said amendments shall enter into force on 1 March 2018 upon their acceptance in accordance with paragraph 2 above;

4 REQUESTS the Secretary-General, for the purposes of article 16(2)(e) of MARPOL, to transmit certified copies of the present resolution and the text of the amendments contained in the annex to all Parties to MARPOL;

5 REQUESTS FURTHER the Secretary-General to transmit copies of the present resolution and its annex to Members of the Organization which are not Parties to MARPOL.

Annex

Amendments to MARPOL Annex V
(HME substances and Form of Garbage Record Book)

Annex V

Regulations for the prevention of pollution by garbage from ships

Regulation 4
Discharge of garbage outside special areas

1 *In the second sentence of paragraph 1.3, the words* "taking into account guidelines developed by the Organization" *are replaced with the words* "in accordance with the criteria set out in appendix I of this Annex".

2 *A new paragraph 3 is added as follows:*

"**3** Solid bulk cargoes as defined in regulation VI/1-1.2 of the International Convention for the Safety of Life at Sea (SOLAS), 1974, as amended, other than grain, shall be classified in accordance with appendix I of this Annex, and declared by the shipper as to whether or not they are harmful to the marine environment.*

* For ships engaged in international voyages, reference is made to section 4.2.3 of the International Maritime Solid Bulk Cargoes (IMSBC) Code; for ships not engaged in international voyages, other means of declaration may be used, as determined by the Administration."

3 *The existing paragraph 3 is renumbered as paragraph 4.*

Regulation 6
Discharge of garbage within special areas

4 *Paragraph 1.2.1 is replaced with the following:*

"**.1** Cargo residues contained in hold washing water do not include any substances classified as harmful to the marine environment according to the criteria set out in appendix I of this Annex;"

5 *A new paragraph 1.2.2 is added as follows:*

"**.2** Solid bulk cargoes as defined in regulation VI/1-1.2 of the International Convention for the Safety of Life at Sea (SOLAS), 1974, as amended, other than grain, shall be classified in accordance with appendix I of this Annex, and declared by the shipper as to whether or not they are harmful to the marine environment;*

* For ships engaged in international voyages, reference is made to section 4.2.3 of the International Maritime Solid Bulk Cargoes (IMSBC) Code; for ships not engaged in international voyages, other means of declaration may be used, as determined by the Administration."

6 *A new paragraph 1.2.3 is added as follows:*

"**.3** Cleaning agents or additives contained in hold washing water do not include any substances classified as harmful to the marine environment taking into account guidelines developed by the Organization;"

7 *The existing paragraphs 1.2.2 to 1.2.4 are renumbered as paragraphs 1.2.4 to 1.2.6. The renumbered paragraph 1.2.6 is amended to read as follows:*

"**.6** Where the conditions of subparagraphs .2.1 to .2.5 of this paragraph have been fulfilled, discharge of cargo hold washing water containing residues shall be made as far as practicable from the nearest land or the nearest ice shelf and not less than 12 nautical miles from the nearest land or the nearest ice shelf."

Regulation 10
Placards, garbage management plans and garbage record-keeping

8 *In the chapeau of paragraph 3, the words* "the appendix" *are replaced with the words* "appendix II".

9 *Paragraph 3.2 is replaced with the following:*

"**.2** The entry for each discharge into the sea under regulations 4, 5, 6 or section 5.2 of chapter 5 of part II-A of the Polar Code shall include date and time, position of the ship (latitude and longitude), category of the garbage and the estimated amount (in cubic metres) discharged. For discharge of cargo residues the discharge start and stop positions shall be recorded in addition to the foregoing."

10 *After the existing paragraph 3.2, new paragraphs 3.3 and 3.4 are inserted as follows:*

"**.3** The entry for each completed incineration shall include date and time and position of the ship (latitude and longitude) at the start and stop of incineration, categories of garbage incinerated and the estimated amount incinerated for each category in cubic metres.

.4 The entry for each discharge to a port reception facility or another ship shall include date and time of discharge, port or facility or name of ship, categories of garbage discharged, and the estimated amount discharged for each category in cubic metres."

11 *The existing paragraph 3.3 is renumbered as 3.5 and between the words* "Book" *and* "shall"; *the words* "along with receipts obtained from reception facilities" *are inserted.*

12 *The existing paragraph 3.4 is renumbered as 3.6 and is replaced with the following:*

"**.6** In the event of any discharge or accidental loss referred to in regulation 7 of this Annex an entry shall be made in the Garbage Record Book, or in the case of any ship of less than 400 gross tonnage, an entry shall be made in the ship's official logbook of the date and time of occurrence, port or position of the ship at time of occurrence (latitude, longitude and water depth if known), the reason for the discharge or loss, details of the items discharged or lost, categories of garbage discharged or lost, estimated amount for each category in cubic metres, reasonable precautions taken to prevent or minimize such discharge or accidental loss and general remarks."

13 *A new appendix I is added as follows and the existing appendix is renumbered as appendix II:*

"Appendix I
Criteria for the classification of solid bulk cargoes as harmful to the marine environment

For the purpose of this Annex, cargo residues are considered to be harmful to the marine environment (HME) if they are residues of solid bulk cargoes which are classified according to the criteria of the United Nations Globally Harmonized System of Classification and Labelling of Chemicals (GHS) meeting the following parameters:*

.1 Acute Aquatic Toxicity Category 1; and/or

.2 Chronic Aquatic Toxicity Category 1 or 2; and/or

.3 Carcinogenicity Category 1A or 1B combined with not being rapidly degradable and having high bioaccumulation; and/or

.4 Mutagenicity† Category 1A or 1B combined with not being rapidly degradable and having high bioaccumulation; and/or

.5 Reproductive Toxicity† Category 1A or 1B combined with not being rapidly degradable and having high bioaccumulation; and/or

.6 Specific Target Organ Toxicity Repeated Exposure† Category 1 combined with not being rapidly degradable and having high bioaccumulation; and/or

.7 Solid bulk cargoes containing or consisting of synthetic polymers, rubber, plastics, or plastic feedstock pellets (this includes materials that are shredded, milled, chopped or macerated or similar materials)."

Appendix II
Form of Garbage Record Book

14 *Section 3 of the renumbered appendix II is replaced with the following:*

"3 Description of the garbage

Garbage is to be grouped into categories for the purposes of recording in parts I and II of the Garbage Record Book (or ship's official logbook) as follows:

Part I

A Plastics
B Food wastes
C Domestic Wastes
D Cooking oil
E Incinerator ashes
F Operational wastes
G Animal carcasses
H Fishing gear
I E-waste

* The criteria are based on UN GHS. For specific products (e.g. metals and inorganic metal compounds) guidance available in UN GHS, annexes 9 and 10 is essential for proper interpretation of the criteria and classification and should be followed.

† Products that are classified for Carcinogenicity, Mutagenicity, Reproductive Toxicity or Specific Target Organ Toxicity Repeated Exposure for oral and dermal hazards or without specification of the exposure route in the hazard statement.

Part II

J Cargo residues (non-HME)

K Cargo residues (HME)"

15 *The Record of Garbage Discharges in the renumbered appendix II is replaced with the following:*

"RECORD OF GARBAGE DISCHARGES

PART I
For all garbage other than cargo residues as defined in regulation 1.2 (Definitions)

(All ships)

Ship's name . . . .

Distinctive number or letters . . . .

IMO number . . . .

Garbage categories

A Plastics
B Food wastes
C Domestic wastes
D Cooking oil
E Incinerator ashes
F Operational wastes
G Animal Carcasses
H Fishing gear
I E-waste

Discharges under MARPOL Annex V regulations 4 (Discharge of garbage outside special areas), 5 (Special requirements for discharge of garbage from fixed or floating platforms) or 6 (Discharge of garbage within special areas) or chapter 5 of part II-A of the Polar Code

Date/time	Position of the ship (latitude/longitude) or port if discharged ashore or name of ship if discharged to another ship	Category	Estimated amount discharged		Estimated amount incinerated (m^3)	Remarks: (e.g. start/stop time and position of incineration; general remarks)	Certification/signature
			Into sea (m^3)	To reception facilities or to another ship (m^3)			

Exceptional discharge or loss of garbage under regulation 7 (Exceptions)

Date/ time	Port or position of the ship (latitude/longitude and water depth if known)	Category	Estimated amount lost or discharged (m^3)	Remarks on the reason for the discharge or loss and general remarks (e.g. reasonable precautions taken to prevent or minimize such discharge or accidental loss and general remarks)	Certification/ signature

Master's signature. . . . . Date . . . .

PART II

For all cargo residues as defined in regulation 1.2 (Definitions)

(Ships that carry solid bulk cargoes))

Ship's name . . . .

Distinctive number or letters . . . .

IMO number . . . .

Garbage categories

J Cargo residues (non-HME)

K Cargo residues (HME)

Discharges under regulations 4 (Discharge of garbage outside special areas) and 6 (Discharge of garbage within special areas)

Date/ time	Position of the ship (latitude/ longitude) or port if discharged ashore	Category	Estimated amount discharged		Start and stop positions of the ship for discharges into the sea	Certification/ signature
			Into sea (m^3)	To reception facilities or to another ship (m^3)		

Master's signature. . . . . Date . . . .''

Consolidated text of MARPOL Annex V, including amendments adopted by resolution MEPC.277(70)

Regulations for the prevention of pollution by garbage from ships*

Chapter 1 – General

Regulation 1
Definitions

For the purposes of this Annex:

1 *Animal carcasses* means the bodies of any animals that are carried on board as cargo and that die or are euthanized during the voyage.

2 *Cargo residues* means the remnants of any cargo which are not covered by other Annexes to the present Convention and which remain on the deck or in holds following loading or unloading, including loading and unloading excess or spillage, whether in wet or dry condition or entrained in wash water but does not include cargo dust remaining on the deck after sweeping or dust on the external surfaces of the ship.

3 *Cooking oil* means any type of edible oil or animal fat used or intended to be used for the preparation or cooking of food, but does not include the food itself that is prepared using these oils.

4 *Domestic wastes* means all types of wastes not covered by other Annexes that are generated in the accommodation spaces on board the ship. Domestic wastes does not include grey water.

5 *En route* means that the ship is underway at sea on a course or courses, including deviation from the shortest direct route, which as far as practicable for navigational purposes, will cause any discharge to be spread over as great an area of the sea as is reasonable and practicable.

6 *Fishing gear* means any physical device or part thereof or combination of items that may be placed on or in the water or on the sea-bed with the intended purpose of capturing, or controlling for subsequent capture or harvesting, marine or fresh water organisms.

7 *Fixed or floating platforms* means fixed or floating structures located at sea which are engaged in the exploration, exploitation or associated offshore processing of sea-bed mineral resources.

8 *Food wastes* means any spoiled or unspoiled food substances and includes fruits, vegetables, dairy products, poultry, meat products and food scraps generated aboard ship.

* Refer to 2017 Guidelines for implementation of MARPOL Annex V (resolution MEPC.295(71)).

9 *Garbage* means all kinds of food wastes, domestic wastes and operational wastes, all plastics, cargo residues, incinerator ashes, cooking oil, fishing gear, and animal carcasses generated during the normal operation of the ship and liable to be disposed of continuously or periodically except those substances which are defined or listed in other Annexes to the present Convention. Garbage does not include fresh fish and parts thereof generated as a result of fishing activities undertaken during the voyage, or as a result of aquaculture activities which involve the transport of fish including shellfish for placement in the aquaculture facility and the transport of harvested fish including shellfish from such facilities to shore for processing.

10 *Incinerator ashes* means ash and clinkers resulting from shipboard incinerators used for the incineration of garbage.

11 *Nearest land.* The term "from the nearest land" means from the baseline from which the territorial sea of the territory in question is established in accordance with international law, except that, for the purposes of the present Annex, "from the nearest land" off the north-eastern coast of Australia shall mean from a line drawn from a point on the coast of Australia in:

latitude 11°00′ S, longitude 142°08′ E
to a point in latitude 10°35′ S, longitude 141°55′ E,
thence to a point latitude 10°00′ S, longitude 142°00′ E,
thence to a point latitude 09°10′ S, longitude 143°52′ E,
thence to a point latitude 09°00′ S, longitude 144°30′ E,
thence to a point latitude 10°41′ S, longitude 145°00′ E,
thence to a point latitude 13°00′ S, longitude 145°00′ E,
thence to a point latitude 15°00′ S, longitude 146°00′ E,
thence to a point latitude 17°30′ S, longitude 147°00′ E,
thence to a point latitude 21°00′ S, longitude 152°55′ E,
thence to a point latitude 24°30′ S, longitude 154°00′ E,
thence to a point on the coast of Australia in
latitude 24°42′ S, longitude 153°15′ E.

12 *Operational wastes* means all solid wastes (including slurries) not covered by other Annexes that are collected on board during normal maintenance or operations of a ship, or used for cargo stowage and handling. Operational wastes also includes cleaning agents and additives contained in cargo hold and external wash water. Operational wastes does not include grey water, bilge water, or other similar discharges essential to the operation of a ship, taking into account the guidelines developed by the Organization.

13 *Plastic* means a solid material which contains as an essential ingredient one or more high molecular mass polymers and which is formed (shaped) during either manufacture of the polymer or the fabrication into a finished product by heat and/or pressure. Plastics have material properties ranging from hard and brittle to soft and elastic. For the purposes of this annex, "all plastics" means all garbage that consists of or includes plastic in any form, including synthetic ropes, synthetic fishing nets, plastic garbage bags and incinerator ashes from plastic products.

14 *Special area* means a sea area where for recognized technical reasons in relation to its oceanographic and ecological condition and to the particular character of its traffic the adoption of special mandatory methods for the prevention of sea pollution by garbage is required.

For the purposes of this Annex the special areas are the Mediterranean Sea area, the Baltic Sea area, the Black Sea area, the Red Sea area, the Gulfs area, the North Sea area, the Antarctic area and the Wider Caribbean Region, which are defined as follows:

.1 The Mediterranean Sea area means the Mediterranean Sea proper including the gulfs and seas therein with the boundary between the Mediterranean and the Black Sea constituted by the 41° N parallel and bounded to the west by the Straits of Gibraltar at the meridian 5°36′ W.

.2 The Baltic Sea area means the Baltic Sea proper with the Gulf of Bothnia and the Gulf of Finland and the entrance to the Baltic Sea bounded by the parallel of the Skaw in the Skagerrak at 57°44.8′ N.

.3 The Black Sea area means the Black Sea proper with the boundary between the Mediterranean and the Black Sea constituted by the parallel 41° N.

.4 The Red Sea area means the Red Sea proper including the Gulfs of Suez and Aqaba bounded at the south by the rhumb line between Ras si Ane (12°28.5′ N, 43°19.6′ E) and Husn Murad (12°40.4′ N, 43°30.2′ E).

.5 The Gulfs area means the sea area located north-west of the rhumb line between Ras al Hadd (22°30′ N, 59°48′ E) and Ras al Fasteh (25°04′ N, 61°25′ E).

.6 The North Sea area means the North Sea proper including seas therein with the boundary between:

.1 the North Sea southwards of latitude 62°N and eastwards of longitude 4° W;

.2 the Skagerrak, the southern limit of which is determined east of the Skaw by latitude 57°44.8′ N; and

.3 the English Channel and its approaches eastwards of longitude 5° W and northwards of latitude 48°30′ N.

.7 The Antarctic area means the sea area south of latitude 60° S.

.8 The Wider Caribbean Region means the Gulf of Mexico and Caribbean Sea proper including the bays and seas therein and that portion of the Atlantic Ocean within the boundary constituted by the 30° N parallel from Florida eastward to 77°30′ W meridian, thence a rhumb line to the intersection of 20° N parallel and 59° W meridian, thence a rhumb line to the intersection of 7°20′ N parallel and 50° W meridian, thence a rhumb line drawn southwesterly to the eastern boundary of French Guiana.

15 *Audit* means a systematic, independent and documented process for obtaining audit evidence and evaluating it objectively to determine the extent to which audit criteria are fulfilled.

16 *Audit Scheme* means the IMO Member State Audit Scheme established by the Organization and taking into account the guidelines developed by the Organization.*

17 *Code for Implementation* means the IMO Instruments Implementation Code (III Code) adopted by the Organization by resolution A.1070(28).

18 *Audit Standard* means the Code for Implementation.

Regulation 2
Application

Unless expressly provided otherwise, the provisions of this Annex shall apply to all ships.

Regulation 3
General prohibition on discharge of garbage into the sea

1 Discharge of all garbage into the sea is prohibited, except as provided otherwise in regulations 4, 5, 6 and 7 of this Annex and section 5.2 of part II-A of the Polar Code, as defined in regulation 13.1 of this Annex.

2 Except as provided in regulation 7 of this Annex, discharge into the sea of all plastics, including but not limited to synthetic ropes, synthetic fishing nets, plastic garbage bags and incinerator ashes from plastic products is prohibited.

3 Except as provided in regulation 7 of this Annex, the discharge into the sea of cooking oil is prohibited.

* Refer to the Framework and Procedures for the IMO Member State Audit Scheme (resolution A.1067(28)).

Regulation 4

Discharge of garbage outside special areas

1 Discharge of the following garbage into the sea outside special areas shall only be permitted while the ship is en route and as far as practicable from the nearest land, but in any case not less than:

.1 3 nautical miles from the nearest land for food wastes which have been passed through a comminuter or grinder. Such comminuted or ground food wastes shall be capable of passing through a screen with openings no greater than 25 mm.

.2 12 nautical miles from the nearest land for food wastes that have not been treated in accordance with subparagraph .1 above.

.3 12 nautical miles from the nearest land for cargo residues that cannot be recovered using commonly available methods for unloading. These cargo residues shall not contain any substances classified as harmful to the marine environment, in accordance with the criteria set out in appendix I of this Annex.

.4 For animal carcasses, discharge shall occur as far from the nearest land as possible, taking into account the guidelines developed by the Organization.

2 Cleaning agents or additives contained in cargo hold, deck and external surfaces wash water may be discharged into the sea, but these substances must not be harmful to the marine environment, taking into account guidelines developed by the Organization.

3 Solid bulk cargoes as defined in regulation VI/1-1.2 of the International Convention for the Safety of Life at Sea (SOLAS), 1974, as amended, other than grain, shall be classified in accordance with appendix I of this Annex, and declared by the shipper as to whether or not they are harmful to the marine environment.*

4 When garbage is mixed with or contaminated by other substances prohibited from discharge or having different discharge requirements, the more stringent requirements shall apply.

Regulation 5

Special requirements for discharge of garbage from fixed or floating platforms

1 Subject to the provisions of paragraph 2 of this regulation, the discharge into the sea of any garbage is prohibited from fixed or floating platforms and from all other ships when alongside or within 500 m of such platforms.

2 Food wastes may be discharged into the sea from fixed or floating platforms located more than 12 nautical miles from the nearest land and from all other ships when alongside or within 500 m of such platforms, but only when the wastes have been passed through a comminuter or grinder. Such comminuted or ground food wastes shall be capable of passing through a screen with openings no greater than 25 mm.

Regulation 6

Discharge of garbage within special areas

1 Discharge of the following garbage into the sea within special areas shall only be permitted while the ship is en route and as follows:

.1 Discharge into the sea of food wastes as far as practicable from the nearest land, but not less than 12 nautical miles from the nearest land or the nearest ice shelf. Food wastes shall be comminuted or ground and shall be capable of passing through a screen with openings no greater than 25 mm. Food wastes shall not be contaminated by any other garbage type. Discharge of introduced avian products, including poultry and poultry parts, is not permitted in the Antarctic area unless it has been treated to be made sterile.

* For ships engaged in international voyages, reference is made to section 4.2.3 of the International Maritime Solid Bulk Cargoes (IMSBC) Code; for ships not engaged in international voyages, other means of declaration may be used, as determined by the Administration.

.2 Discharge of cargo residues that cannot be recovered using commonly available methods for unloading, where all the following conditions are satisfied:

.1 cargo residues contained in hold washing water do not include any substances classified as harmful to the marine environment according to the criteria set out in appendix I of this Annex;

.2 solid bulk cargoes as defined in regulation VI/1-1.2 of the International Convention for the Safety of Life at Sea (SOLAS), 1974, as amended, other than grain, shall be classified in accordance with appendix I of this Annex, and declared by the shipper as to whether or not they are harmful to the marine environment;*

.3 cleaning agents or additives contained in hold washing water do not include any substances classified as harmful to the marine environment taking into account guidelines developed by the Organization;

.4 both the port of departure and the next port of destination are within the special area and the ship will not transit outside the special area between those ports;

.5 no adequate reception facilities are available at those ports taking into account guidelines developed by the Organization; and

.6 where the conditions of subparagraphs .2.1 to .2.5 of this paragraph have been fulfilled, discharge of cargo hold washing water containing residues shall be made as far as practicable from the nearest land or the nearest ice shelf and not less than 12 nautical miles from the nearest land or the nearest ice shelf.

2 Cleaning agents or additives contained in deck and external surfaces wash water may be discharged into the sea, but only if these substances are not harmful to the marine environment, taking into account guidelines developed by the Organization.

3 The following rules (in addition to the rules in paragraph 1 of this regulation) apply with respect to the Antarctic area:

.1 Each Party at whose ports ships depart en route to or arrive from the Antarctic area undertakes to ensure that as soon as practicable adequate facilities are provided for the reception of all garbage from all ships, without causing undue delay, and according to the needs of the ships using them.

.2 Each Party shall ensure that all ships entitled to fly its flag, before entering the Antarctic area, have sufficient capacity on board for the retention of all garbage, while operating in the area and have concluded arrangements to discharge such garbage at a reception facility after leaving the area.

4 When garbage is mixed with or contaminated by other substances prohibited from discharge or having different discharge requirements, the more stringent requirements shall apply.

Regulation 7
Exceptions

1 Regulations 3, 4, 5 and 6 of this Annex and section 5.2 of chapter 5 of part II-A of the Polar Code shall not apply to:

.1 the discharge of garbage from a ship necessary for the purpose of securing the safety of a ship and those on board or saving life at sea; or

.2 the accidental loss of garbage resulting from damage to a ship or its equipment, provided that all reasonable precautions have been taken before and after the occurrence of the damage, to prevent or minimize the accidental loss; or

.3 the accidental loss of fishing gear from a ship provided that all reasonable precautions have been taken to prevent such loss; or

* For ships engaged in international voyages, reference is made to section 4.2.3 of the International Maritime Solid Bulk Cargoes (IMSBC) Code; for ships not engaged in international voyages, other means of declaration may be used, as determined by the Administration.

.4 the discharge of fishing gear from a ship for the protection of the marine environment or for the safety of that ship or its crew.

2 Exception of en route:

.1 The en route requirements of regulations 4 and 6 of this Annex and chapter 5 of part II-A of the Polar Code shall not apply to the discharge of food wastes where it is clear the retention on board of these food wastes presents an imminent health risk to the people on board.

Regulation 8
*Reception facilities**

1 Each Party undertakes to ensure the provision of adequate facilities at ports and terminals for the reception of garbage without causing undue delay to ships, and according to the needs of the ships using them.

2 Reception facilities within special areas

.1 Each Party, the coastline of which borders a special area, undertakes to ensure that as soon as possible, in all ports and terminals within the special area, adequate reception facilities are provided, taking into account the needs of ships operating in these areas.

.2 Each Party concerned shall notify the Organization of the measures taken pursuant to paragraph 2.1 of this regulation. Upon receipt of sufficient notifications the Organization shall establish a date from which the requirements of regulation 6 of this Annex in respect of the area in question are to take effect. The Organization shall notify all Parties of the date so established no less than 12 months in advance of that date. Until the date so established, ships that are navigating in a special area shall comply with the requirements of regulation 4 of this Annex as regards discharges outside special areas.

3 Small Island Developing States may satisfy the requirements in paragraphs 1 and 2.1 of this regulation through regional arrangements when, because of those States' unique circumstances, such arrangements are the only practical means to satisfy these requirements. Parties participating in a regional arrangement shall develop a Regional Reception Facilities Plan, taking into account the guidelines developed by the Organization.†

The Government of each Party participating in the Arrangement shall consult with the Organization for circulation to the Parties of the present Convention:

.1 how the Regional Reception Facilities Plan takes into account the Guidelines;

.2 particulars of the identified Regional Ships Waste Reception Centres; and

.3 particulars of those ports with only limited facilities.

4 Each Party shall notify the Organization for transmission to the Contracting Parties concerned of all cases where the facilities provided under this regulation are alleged to be inadequate.

Regulation 9
Port State control on operational requirements‡

1 A ship when in a port or an offshore terminal of another Party is subject to inspection by officers duly authorized by such Party concerning operational requirements under this Annex, where there are clear grounds for believing that the master or crew are not familiar with essential shipboard procedures relating to the prevention of pollution by garbage.

---

* Refer to Consolidated guidance for port reception facility providers and users (MEPC.1/Circ.834).

† Refer to 2012 Guidelines for the development of a regional reception facilities plan (resolution MEPC.221(63)).

‡ Refer to Procedures for port State control, 2011 (resolution A.1052(27)).

2 In the circumstances given in paragraph 1 of this regulation, the Party shall take such steps as will ensure that the ship shall not sail until the situation has been brought to order in accordance with the requirements of this Annex.

3 Procedures relating to the port State control prescribed in article 5 of the present Convention shall apply to this regulation.

4 Nothing in this regulation shall be construed to limit the rights and obligations of a Party carrying out control over operational requirements specifically provided for in the present Convention.

Regulation 10
Placards, garbage management plans and garbage record-keeping*

1 **.1** Every ship of 12 m or more in length overall and fixed or floating platforms shall display placards which notify the crew and passengers of the discharge requirements of regulations 3, 4, 5 and 6 of this Annex and section 5.2 of part II-A of the Polar Code, as applicable.

.2 The placards shall be written in the working language of the ship's crew and, for ships engaged in voyages to ports or offshore terminals under the jurisdiction of other Parties to the Convention, shall also be in English, French or Spanish.

2 Every ship of 100 gross tonnage and above, and every ship which is certified to carry 15 or more persons, and fixed or floating platforms shall carry a garbage management plan which the crew shall follow. This plan shall provide written procedures for minimizing, collecting, storing, processing and disposing of garbage, including the use of the equipment on board. It shall also designate the person or persons in charge of carrying out the plan. Such a plan shall be based on the guidelines developed by the Organization* and written in the working language of the crew.

3 Every ship of 400 gross tonnage and above and every ship which is certified to carry 15 or more persons engaged in voyages to ports or offshore terminals under the jurisdiction of another Party to the Convention and every fixed or floating platform shall be provided with a Garbage Record Book. The Garbage Record Book, whether as a part of the ship's official logbook or otherwise, shall be in the form specified in appendix II to this Annex:

.1 Each discharge into the sea or to a reception facility, or a completed incineration, shall be promptly recorded in the Garbage Record Book and signed for on the date of the discharge or incineration by the officer in charge. Each completed page of the Garbage Record Book shall be signed by the master of the ship. The entries in the Garbage Record Book shall be at least in English, French or Spanish. Where the entries are also made in an official language of the State whose flag the ship is entitled to fly, the entries in that language shall prevail in case of a dispute or discrepancy.

.2 The entry for each discharge into the sea under regulations 4, 5, 6 or section 5.2 of chapter 5 of part II-A of the Polar Code shall include date and time, position of the ship (latitude and longitude), category of the garbage and the estimated amount (in cubic metres) discharged. For discharge of cargo residues the discharge start and stop positions shall be recorded in addition to the foregoing.

.3 The entry for each completed incineration shall include date and time and position of the ship (latitude and longitude) at the start and stop of incineration, categories of garbage incinerated and the estimated amount incinerated for each category in cubic metres.

.4 The entry for each discharge to a port reception facility or another ship shall include date and time of discharge, port or facility or name of ship, categories of garbage discharged, and the estimated amount discharged for each category in cubic metres.

* Refer to 2012 Guidelines for the development of garbage management plans (resolution MEPC.220(63), as amended).

.5 The Garbage Record Book along with receipts obtained from reception facilities shall be kept on board the ship or the fixed or floating platform, and in such a place as to be readily available for inspection at all reasonable times. This document shall be preserved for a period of at least two years from the date of the last entry made in it.

.6 In the event of any discharge or accidental loss referred to in regulation 7 of this Annex an entry shall be made in the Garbage Record Book, or in the case of any ship of less than 400 gross tonnage, an entry shall be made in the ship's official logbook of the date and time of occurrence, port or position of the ship at time of occurrence (latitude, longitude and water depth if known), the reason for the discharge or loss, details of the items discharged or lost, categories of garbage discharged or lost, estimated amount for each category in cubic metres, reasonable precautions taken to prevent or minimize such discharge or accidental loss and general remarks.

4 The Administration may waive the requirements for Garbage Record Books for:

.1 any ship engaged on voyages of 1 h or less in duration which is certified to carry 15 or more persons; or

.2 fixed or floating platforms.

5 The competent authority of the Government of a Party to the Convention may inspect the Garbage Record Books or ship's official logbook on board any ship to which this regulation applies while the ship is in its ports or offshore terminals and may make a copy of any entry in those books, and may require the master of the ship to certify that the copy is a true copy of such an entry. Any copy so made, which has been certified by the master of the ship as a true copy of an entry in the ship's Garbage Record Book or ship's official logbook, shall be admissible in any judicial proceedings as evidence of the facts stated in the entry. The inspection of a Garbage Record Book or ship's official logbook and the taking of a certified copy by the competent authority under this paragraph shall be performed as expeditiously as possible without causing the ship to be unduly delayed.

6 The accidental loss or discharge of fishing gear as provided for in regulations 7.1.3 and 7.1.4 which poses a significant threat to the marine environment or navigation shall be reported to the State whose flag the ship is entitled to fly, and, where the loss or discharge occurs within waters subject to the jurisdiction of a coastal State, also to that coastal State.

Chapter 2 – Verification of compliance with the provisions of this Annex

Regulation 11
Application

Parties shall use the provisions of the Code for Implementation in the execution of their obligations and responsibilities contained in this Annex.

Regulation 12
Verification of compliance

1 Every Party shall be subject to periodic audits by the Organization in accordance with the audit standard to verify compliance with and implementation of this Annex.

2 The Secretary-General of the Organization shall have responsibility for administering the Audit Scheme, based on the guidelines developed by the Organization.*

* Refer to the Framework and Procedures for the IMO Member State Audit Scheme (resolution A.1067(28)).

3 Every Party shall have responsibility for facilitating the conduct of the audit and implementation of a programme of actions to address the findings, based on the guidelines developed by the Organization.*

4 Audit of all Parties shall be:

.1 based on an overall schedule developed by the Secretary-General of the Organization, taking into account the guidelines developed by the Organization; and

.2 conducted at periodic intervals, taking into account the guidelines developed by the Organization.

Chapter 3 – International Code for Ships Operating in Polar Waters

Regulation 13
Definitions

For the purpose of this Annex,

1 *Polar Code* means the International Code for Ships Operating in Polar Waters, consisting of an introduction, part I-A and part II-A and parts I-B and II-B, as adopted by resolutions MSC.385(94) and MEPC.264(68), as may be amended, provided that:

.1 amendments to the environment-related provisions of the introduction and chapter 5 of part II-A of the Polar Code are adopted, brought into force and take effect in accordance with the provisions of article 16 of the present Convention concerning the amendment procedures applicable to an appendix to an annex; and

.2 amendments to part II-B of the Polar Code are adopted by the Marine Environment Protection Committee in accordance with its Rules of Procedure.

2 *Arctic waters* means those waters which are located north of a line from the latitude 58°00′.0 N and longitude 042°00′.0 W to latitude 64°37′.0 N, longitude 035°27′.0 W and thence by a rhumb line to latitude 67°03′.9 N, longitude 026°33′.4 W and thence by a rhumb line to the latitude 70°49′.56 N and longitude 008°59′.61 W (Sørkapp, Jan Mayen) and by the southern shore of Jan Mayen to 73°31′.6 N and 019°01′.0 E by the Island of Bjørnøya, and thence by a great circle line to the latitude 68°38′.29 N and longitude 043°23′.08 E (Cap Kanin Nos) and thence by the northern shore of the Asian Continent eastward to the Bering Strait and thence from the Bering Strait westward to latitude 60° N as far as Il'pyrskiy and following the 60th North parallel eastward as far as and including Etolin Strait and thence by the northern shore of the North American continent as far south as latitude 60° N and thence eastward along parallel of latitude 60° N, to longitude 056°37′.1 W and thence to the latitude 58°00′.0 N, longitude 042°00′.0 W.

3 *Polar waters* means Arctic waters and/or the Antarctic area.

Regulation 14
Application and requirements

1 This chapter applies to all ships to which this Annex applies, operating in polar waters.

2 Unless expressly provided otherwise, any ship covered by paragraph 1 of this regulation shall comply with the environment-related provisions of the introduction and with chapter 5 of part II-A of the Polar Code, in addition to any other applicable requirements of this Annex.

3 In applying chapter 5 of part II-A of the Polar Code, consideration should be given to the additional guidance in part II-B of the Polar Code.

Appendices to Annex V

Appendix I

Criteria for the classification of solid bulk cargoes as harmful to the marine environment

For the purpose of this Annex, cargo residues are considered to be harmful to the marine environment (HME) if they are residues of solid bulk cargoes which are classified according to the criteria of the United Nations Globally Harmonized System of Classification and Labelling of Chemicals (GHS) meeting the following parameters:*

.1 Acute Aquatic Toxicity Category 1; and/or

.2 Chronic Aquatic Toxicity Category 1 or 2; and/or

.3 Carcinogenicity† Category 1A or 1B combined with not being rapidly degradable and having high bioaccumulation; and/or

.4 Mutagenicity† Category 1A or 1B combined with not being rapidly degradable and having high bioaccumulation; and/or

.5 Reproductive Toxicity† Category 1A or 1B combined with not being rapidly degradable and having high bioaccumulation; and/or

.6 Specific Target Organ Toxicity Repeated Exposure† Category 1 combined with not being rapidly degradable and having high bioaccumulation; and/or

.7 Solid bulk cargoes containing or consisting of synthetic polymers, rubber, plastics, or plastic feedstock pellets (this includes materials that are shredded, milled, chopped or macerated or similar materials).

* The criteria are based on UN GHS. For specific products (e.g. metals and inorganic metal compounds) guidance available in UN GHS, annexes 9 and 10 is essential for proper interpretation of the criteria and classification and should be followed.

† Products that are classified for Carcinogenicity, Mutagenicity, Reproductive Toxicity or Specific Target Organ Toxicity Repeated Exposure for oral and dermal hazards or without specification of the exposure route in the hazard statement.

Appendix II
Form of Garbage Record Book

GARBAGE RECORD BOOK

Name of ship. . . . .

Distinctive number or letters. . . . .

IMO Number. . . . .

Period . . . . from: . . . . to. . . . .

1 Introduction

In accordance with regulation 10 of Annex V of the International Convention for the Prevention of Pollution from Ships, 1973, as modified by the Protocol of 1978 (MARPOL), a record is to be kept of each discharge operation or completed incineration. This includes discharges into the sea, to reception facilities, or to other ships, as well as the accidental loss of garbage.

2 Garbage and garbage management

Garbage means all kinds of food wastes, domestic wastes and operational wastes, all plastics, cargo residues, incinerator ashes, cooking oil, fishing gear, and animal carcasses generated during the normal operation of the ship and liable to be disposed of continuously or periodically except those substances which are defined or listed in other Annexes to the present Convention. Garbage does not include fresh fish and parts thereof generated as a result of fishing activities undertaken during the voyage, or as a result of aquaculture activities which involve the transport of fish including shellfish for placement in the aquaculture facility and the transport of harvested fish including shellfish from such facilities to shore for processing.

The Guidelines for the Implementation of MARPOL Annex V should also be referred to for relevant information.

3 Description of the garbage

Garbage is to be grouped into categories for the purposes of recording in parts I and II of the Garbage Record Book (or ship's official logbook) as follows:

Part I

A Plastics
B Food wastes
C Domestic wastes
D Cooking oil
E Incinerator ashes
F Operational wastes
G Animal carcasses
H Fishing gear
I E-waste

Part II

J Cargo residues (non-HME)
K Cargo residues (HME)

4 Entries in the Garbage Record Book

4.1 Entries in the Garbage Record Book shall be made on each of the following occasions:

4.1.1 When garbage is discharged to a reception facility ashore or to other ships:

.1 Date and time of discharge
.2 Port or facility, or name of ship
.3 Categories of garbage discharged
.4 Estimated amount discharged for each category in cubic metres
.5 Signature of officer in charge of the operation.

4.1.2 When garbage is incinerated:

.1 Date and time of start and stop of incineration

.2 Position of the ship (latitude and longitude) at the start and stop of incineration

.3 Categories of garbage incinerated

.4 Estimated amount incinerated in cubic metres

.5 Signature of the officer in charge of the operation.

4.1.3 When garbage is discharged into the sea in accordance with regulations 4, 5 or 6 of MARPOL Annex V or chapter 5 of part II-A of the Polar Code:

.1 Date and time of discharge

.2 Position of the ship (latitude and longitude). Note: for cargo residue discharges, include discharge start and stop positions.

.3 Category of garbage discharged

.4 Estimated amount discharged for each category in cubic metres

.5 Signature of the officer in charge of the operation.

4.1.4 Accidental or other exceptional discharges or loss of garbage into the sea, including in accordance with regulation 7 of MARPOL Annex V:

.1 Date and time of occurrence

.2 Port or position of the ship at time of occurrence (latitude, longitude and water depth if known)

.3 Categories of garbage discharged or lost

.4 Estimated amount for each category in cubic metres

.5 The reason for the discharge or loss and general remarks.

4.2 Amount of garbage

The amount of garbage on board should be estimated in cubic metres, if possible separately according to category. The Garbage Record Book contains many references to estimated amount of garbage. It is recognized that the accuracy of estimating amounts of garbage is left to interpretation. Volume estimates will differ before and after processing. Some processing procedures may not allow for a usable estimate of volume, e.g. the continuous processing of food waste. Such factors should be taken into consideration when making and interpreting entries made in a record..

RECORD OF GARBAGE DISCHARGES

PART I

For all garbage other than cargo residues as defined in regulation 1.2 (Definitions)

(All ships)

Ship's name . . . .

Distinctive number or letters . . . .

IMO number . . . .

Garbage categories

- A Plastics
- B Food wastes
- C Domestic wastes
- D Cooking oil
- E Incinerator ashes
- F Operational wastes
- G Animal Carcasses
- H Fishing gear
- I E-waste

Discharges under MARPOL Annex V regulations 4 (Discharge of garbage outside special areas), 5 (Special requirements for discharge of garbage from fixed or floating platforms) or 6 (Discharge of garbage within special areas) or chapter 5 of part II-A of the Polar Code

Date/ time	Position of the ship (latitude/ longitude) or port if discharged ashore or name of ship if discharged to another ship	Category	Estimated amount discharged		Estimated amount incinerated (m^3)	Remarks: (e.g. start/ stop time and position of incineration; general remarks)	Certification/ signature
			Into sea (m^3)	To reception facilities or to another ship (m^3)			

Exceptional discharge or loss of garbage under regulation 7 (Exceptions)

Date/ time	Port or position of the ship (latitude/longitude and water depth if known)	Category	Estimated amount lost or discharged (m^3)	Remarks on the reason for the discharge or loss and general remarks (e.g. reasonable precautions taken to prevent or minimize such discharge or accidental loss and general remarks)	Certification/ signature

Master's signature . . . . Date . . . .

PART II

For all cargo residues as defined in regulation 1.2 (Definitions)

(Ships that carry solid bulk cargoes))

Ship's name ....

Distinctive number or letters ....

IMO number ....

Garbage categories

J Cargo residues (non-HME)

K Cargo residues (HME)

Discharges under regulations 4 (Discharge of garbage outside special areas) and 6 (Discharge of garbage within special areas)

Date/ time	Position of the ship (latitude/ longitude) or port if discharged ashore	Category	Estimated amount discharged		Start and stop positions of the ship for discharges into the sea	Certification/ signature
			Into sea (m^3)	To reception facilities or to another ship (m^3)		

Master's signature .... Date ....

Prospective amendments to MARPOL Annex VI

Resolution MEPC.271(69)
adopted on 22 April 2016

Amendments to the Annex of the Protocol of 1997 to amend the International Convention for the Prevention of Pollution from Ships, 1973, as modified by the Protocol of 1978 relating thereto

Amendments to regulation 13 of MARPOL Annex VI

(Record requirements for operational compliance with NO_x Tier III emission control areas)

THE MARINE ENVIRONMENT PROTECTION COMMITTEE,

RECALLING article 38(a) of the Convention on the International Maritime Organization concerning the functions of the Marine Environment Protection Committee conferred upon it by international conventions for the prevention and control of marine pollution from ships,

NOTING article 16 of the International Convention for the Prevention of Pollution from Ships, 1973, as modified by the Protocols of 1978 and 1997 relating thereto (MARPOL), which specifies the amendment procedure and confers upon the appropriate body of the Organization the function of considering and adopting amendments thereto,

HAVING CONSIDERED, at its sixty-ninth session, draft amendments to MARPOL Annex VI, related to record requirements for operational compliance with NOX Tier III emission control areas,

1 ADOPTS, in accordance with article 16(2)(d) of MARPOL, amendments to regulation 13 of MARPOL Annex VI, the text of which is set out in the annex to the present resolution;

2 DETERMINES, in accordance with article 16(2)(f)(iii) of MARPOL, that the amendments shall be deemed to have been accepted on 1 March 2017, unless prior to that date, not less than one third of the Parties or Parties the combined merchant fleets of which constitute not less than 50% of the gross tonnage of the world's merchant fleet, have communicated to the Organization their objection to the amendments;

3 INVITES the Parties to note that, in accordance with article 16(2)(g)(ii) of MARPOL, the said amendments shall enter into force on 1 September 2017 upon their acceptance in accordance with paragraph 2 above;

4 REQUESTS the Secretary-General, for the purposes of article 16(2)(e) of MARPOL, to transmit certified copies of the present resolution and the text of the amendments contained in the annex to all Parties to MARPOL;

5 REQUESTS FURTHER the Secretary-General to transmit copies of the present resolution and its annex to the Members of the Organization which are not Parties MARPOL.

Annex

Amendments to MARPOL Annex VI
(Record requirements for operational compliance
with NO_x Tier III emission control areas)

Annex VI

Regulation for the prevention of air pollution from ships

Chapter 3 – Requirements for control of emissions from ships

Regulation 13

Nitrogen oxides (NO_x)

1 A new paragraph 5.3 is added after existing paragraph 5.2, as follows:

"**5.3** The tier and on/off status of marine diesel engines installed on board a ship to which paragraph 5.1 of this regulation applies which are certified to both Tier II and Tier III or which are certified to Tier II only shall be recorded in such logbook as prescribed by the Administration at entry into and exit from an emission control area designated under paragraph 6 of this regulation, or when the on/off status changes within such an area, together with the date, time and position of the ship."

2 In paragraph 5.1.1, the symbol "NO_x" is replaced with the symbol "NO_2".

Prospective amendments to MARPOL Annex VI

Resolution MEPC.278(70)
adopted on 28 October 2016

Amendments to the Annex of the Protocol of 1997 to amend the International Convention for the Prevention of Pollution from Ships, 1973, as modified by the Protocol of 1978 relating thereto

Amendments to MARPOL Annex VI

(Data collection system for fuel oil consumption of ships)

THE MARINE ENVIRONMENT PROTECTION COMMITTEE,

RECALLING Article 38(a) of the Convention on the International Maritime Organization concerning the functions of the Marine Environment Protection Committee conferred upon it by international conventions for the prevention and control of marine pollution from ships,

NOTING article 16 of the International Convention for the Prevention of Pollution from Ships, 1973, as modified by the Protocols of 1978 and 1997 relating thereto (MARPOL), which specifies the amendment procedure and confers upon the appropriate body of the Organization the function of considering and adopting amendments thereto,

HAVING CONSIDERED, at its seventieth session, proposed amendments to MARPOL Annex VI concerning the data collection system for fuel oil consumption,

1 ADOPTS, in accordance with article 16(2)(d) of MARPOL, amendments to MARPOL Annex VI, the text of which is set out in the annex to the present resolution;

2 DETERMINES, in accordance with article 16(2)(f)(iii) of MARPOL, that the amendments shall be deemed to have been accepted on 1 September 2017 unless prior to that date, not less than one third of the Parties or Parties the combined merchant fleets of which constitute not less than 50% of the gross tonnage of the world's merchant fleet, have communicated to the Organization their objection to the amendments;

3 INVITES the Parties to note that, in accordance with article 16(2)(g)(ii) of MARPOL, the said amendments shall enter into force on 1 March 2018 upon their acceptance in accordance with paragraph 2 above;

4 INVITES FURTHER the Parties to consider the application of the aforesaid amendments to Annex VI of MARPOL as soon as possible to ships entitled to fly their flag;

5 ENCOURAGES the Organization to establish as soon as possible the IMO Ship Fuel Oil Consumption Database;

6 REQUESTS the Secretary-General, for the purposes of article 16(2)(e) of MARPOL, to transmit certified copies of the present resolution and the text of the amendments contained in the annex to all Parties to MARPOL;

7 REQUESTS FURTHER the Secretary-General to transmit copies of the present resolution and its annex to Members of the Organization which are not Parties to MARPOL.

Annex

Amendments to MARPOL Annex VI
(Data collection system for fuel oil consumption of ships)

Annex VI

Regulations for the prevention of air pollution from ships

Regulation 1
Application

1 *The reference to* "regulations 3, 5, 6, 13, 15, 16, 18, 19, 20, 21 and 22" *is replaced with* "regulations 3, 5, 6, 13, 15, 16, 18, 19, 20, 21, 22 and 22A".

Regulation 2
Definitions

2 *After existing paragraph 47, new paragraphs 48, 49 and 50 are added as follows:*

"**48** Calendar year means the period from 1 January until 31 December inclusive.

49 Company means the owner of the ship or any other organization or person such as the manager, or the bareboat charterer, who has assumed the responsibility for operation of the ship from the owner of the ship and who on assuming such responsibility has agreed to take over all the duties and responsibilities imposed by the International Management Code for the Safe Operation of Ships and for Pollution Prevention, as amended.

50 Distance travelled means distance travelled over ground."

Regulation 3
Exceptions and exemptions

3 *In the chapeau of paragraph 2, between existing sentences 2 and 3, a new sentence is added as follows:*

"A permit issued under this regulation shall not exempt a ship from the reporting requirement under regulation 22A and shall not alter the type and scope of data required to be reported under regulation 22A."

Regulation 5
Surveys

4 *At the end of paragraph 4.3, after the words* "on board", *new text is added as follows:*

"and for a ship to which regulation 22A applies, has been revised appropriately to reflect a major conversion in those cases where the major conversion affects data collection methodology and/or reporting processes"

and the word "and" *following the semicolon at the end of the paragraph is deleted.*

5 *In paragraph 4.4, the full stop at the end of the paragraph is replaced by* "; and".

6 *After the existing paragraph 4.4, a new paragraph 4.5 is added as follows:*

> "**.5** The Administration shall ensure that for each ship to which regulation 22A applies, the SEEMP complies with regulation 22.2 of this Annex. This shall be done prior to collecting data under regulation 22A of this Annex in order to ensure the methodology and processes are in place prior to the beginning of the ship's first reporting period. Confirmation of compliance shall be provided to and retained on board the ship."

Regulation 6
Issue or endorsement of Certificates and Statements of Compliance related to fuel oil consumption reporting

7 *In the title of regulation 6, the words* "and Statements of Compliance related to fuel oil consumption reporting" *are inserted following the word* "Certificates".

8 *After existing paragraph 5, new paragraphs 6 and 7 are added as follows:*

"Statement of Compliance – Fuel Oil Consumption Reporting

6 Upon receipt of reported data pursuant to regulation 22A.3 of this Annex, the Administration or any organization duly authorized by it* shall determine whether the data has been reported in accordance with regulation 22A of this Annex and, if so, issue a Statement of Compliance related to fuel oil consumption to the ship no later than five months from the beginning of the calendar year. In every case, the Administration assumes full responsibility for this Statement of Compliance.

7 Upon receipt of reported data pursuant to regulations 22A.4, 22A.5 or 22A.6 of this Annex, the Administration or any organization duly authorized by it* shall promptly determine whether the data has been reported in accordance with regulation 22A and, if so, issue a Statement of Compliance related to fuel oil consumption to the ship at that time. In every case, the Administration assumes full responsibility for this Statement of Compliance.

* Refer to Guidelines for the authorization of organizations acting on behalf of the Administration (resolution A.739(18), as may be amended), and Specifications on the survey and certification functions of recognized organizations acting on behalf of the Administration (resolution A.789(19), as may be amended)."

Regulation 8
Form of Certificates and Statements of Compliance related to fuel oil consumption reporting

9 *In the title of regulation 8, the words* "and Statements of Compliance related to fuel oil consumption reporting" *are inserted following the word* "Certificates".

10 *After existing paragraph 2, a new paragraph 3 is added as follows:*

"Statement of Compliance – Fuel Oil Consumption Reporting

3 The Statement of Compliance pursuant to regulations 6.6 and 6.7 of this Annex shall be drawn up in a form corresponding to the model given in appendix X to this Annex and shall be at least in English, French, or Spanish. If an official language of the issuing Party is also used, this shall prevail in case of a dispute or discrepancy."

Regulation 9

Duration and validity of Certificates and Statements of Compliance related to fuel oil consumption reporting

11 *In the title of regulation 9, the words* "and Statements of Compliance related to fuel oil consumption reporting" *are inserted following the word* "Certificates".

12 *After existing paragraph 11, a new paragraph 12 is added as follows:*

"Statement of Compliance – Fuel Oil Consumption Reporting

12 The Statement of Compliance pursuant to regulation 6.6 of this Annex shall be valid for the calendar year in which it is issued and for the first five months of the following calendar year. The Statement of Compliance pursuant to regulation 6.7 of this Annex shall be valid for the calendar year in which it is issued, for the following calendar year, and for the first five months of the subsequent calendar year. All Statements of Compliance shall be kept on board for at least the period of their validity."

Regulation 10

Port State control on operational requirements

13 *In paragraph 5, the words* "Statement of Compliance related to fuel oil consumption reporting and" *are inserted before the words* "International Energy Efficiency Certificate".

Regulation 22

Ship Energy Efficiency Management Plan (SEEMP)

14 *After existing paragraph 1, a new paragraph 2 is inserted as follows and the existing paragraph 2 is renumbered as paragraph 3:*

"**2** On or before 31 December 2018, in the case of a ship of 5,000 gross tonnage and above, the SEEMP shall include a description of the methodology that will be used to collect the data required by regulation 22A.1 of this Annex and the processes that will be used to report the data to the ship's Administration."

15 *After existing regulation 22, a new 22A is inserted as follows:*

"Regulation 22A

Collection and reporting of ship fuel oil consumption data

1 From calendar year 2019, each ship of 5,000 gross tonnage and above shall collect the data specified in appendix IX to this Annex, for that and each subsequent calendar year or portion thereof, as appropriate, according to the methodology included in the SEEMP.

2 Except as provided for in paragraphs 4, 5 and 6 of this regulation, at the end of each calendar year, the ship shall aggregate the data collected in that calendar year or portion thereof, as appropriate.

3 Except as provided for in paragraphs 4, 5 and 6 of this regulation, within three months after the end of each calendar year, the ship shall report to its Administration or any organization duly authorized by it,* the aggregated value for each datum specified in appendix IX to this Annex, via electronic communication and using a standardized format to be developed by the Organization.†

* Refer to the Guidelines for the authorization of organizations acting on behalf of the Administration, adopted by the Organization by resolution A.739(18), as may be amended by the Organization, and the Specifications on the survey and certification functions of recognized organizations acting on behalf of the Administration, adopted by the Organization by resolution A.789(19), as may be amended by the Organization.

† Refer to 2016 Guidelines for the development of a Ship Energy Efficiency Management Plan (SEEMP Guidelines) (resolution MEPC.282(70)).

4 In the event of the transfer of a ship from one Administration to another, the ship shall on the day of completion of the transfer or as close as practical thereto report to the losing Administration or any organization duly authorized by it,* the aggregated data for the period of the calendar year corresponding to that Administration, as specified in appendix IX to this Annex and, upon prior request of that Administration, the disaggregated data.

5 In the event of a change from one Company to another, the ship shall on the day of completion of the change or as close as practical thereto report to its Administration or any organization duly authorized by it,* the aggregated data for the portion of the calendar year corresponding to the Company, as specified in appendix IX to this Annex and, upon request of its Administration, the disaggregated data.†

6 In the event of change from one Administration to another and from one Company to another concurrently, paragraph 4 of this regulation shall apply.

7 The data shall be verified according to procedures established by the Administration, taking into account guidelines to be developed by the Organization.

8 Except as provided for in paragraphs 4, 5 and 6 of this regulation, the disaggregated data that underlies the reported data noted in appendix IX to this Annex for the previous calendar year shall be readily accessible for a period of not less than 12 months from the end of that calendar year and be made available to the Administration upon request.

9 The Administration shall ensure that the reported data noted in appendix IX to this Annex by its registered ships of 5,000 gross tonnage and above are transferred to the IMO Ship Fuel Oil Consumption Database via electronic communication and using a standardized format to be developed by the Organization not later than one month after issuing the Statements of Compliance of these ships.

10 On the basis of the reported data submitted to the IMO Ship Fuel Oil Consumption Database, the Secretary-General of the Organization shall produce an annual report to the Marine Environment Protection Committee summarizing the data collected, the status of missing data, and such other relevant information as may be requested by the Committee.

11 The Secretary-General of the Organization shall maintain an anonymized database such that identification of a specific ship will not be possible. Parties shall have access to the anonymized data strictly for their analysis and consideration.

12 The IMO Ship Fuel Oil Consumption Database shall be undertaken and managed by the Secretary-General of the Organization, pursuant to guidelines to be developed by the Organization."

* Refer to Guidelines for the authorization of organizations acting on behalf of the Administration (resolution A.739(18), as may be amended), and Specifications on the survey and certification functions of recognized organizations acting on behalf of the Administration (resolution A.789(19), as may be amended).

† Refer to 2016 Guidelines for the development of a Ship Energy Efficiency Management Plan (SEEMP Guidelines) (resolution MEPC.282(70)).

16 *After existing appendix VIII, new appendices IX and X are inserted as follows:*

Appendix IX
Information to be submitted to the IMO Ship Fuel Oil Consumption Database

Identity of the ship

IMO number . . . .

Period of calendar year for which the data is submitted

Start date (dd/mm/yyyy). . . . .

End date (dd/mm/yyyy). . . . .

Technical characteristics of the ship

Ship type, as defined in regulation 2 of this Annex or other (to be stated) . . . .

Gross tonnage (GT)* . . . .

Net tonnage (NT)† . . . .

Deadweight tonnage (DWT)‡ . . . .

Power output (rated power)§ of main and auxiliary reciprocating internal combustion engines over 130 kW (to be stated in kW). . . . .

EEDI¶ (if applicable). . . . .

Ice class** . . . .

Fuel oil consumption, by fuel oil type in metric tonnes and methods used for collecting fuel oil consumption data . . . .

Distance travelled . . . .

Hours underway. . . . .

---

* Gross tonnage should be calculated in accordance with the International Convention on Tonnage Measurement of Ships, 1969.

† Net tonnage should be calculated in accordance with the International Convention on Tonnage Measurement of Ships, 1969. If not applicable, note "N/A".

‡ DWT means the difference in tonnes between the displacement of a ship in water of relative density of 1,025 kg/m^3 at the summer load draught and the lightweight of the ship. The summer load draught should be taken as the maximum summer draught as certified in the stability booklet approved by the Administration or an organization recognized by it. If not applicable, note "N/A".

§ Rated power means the maximum continuous rated power as specified on the nameplate of the engine.

¶ As defined in the 2014 Guidelines on the method of calculation of the Attained Energy Efficiency Design Index (EEDI) for new ships (resolution MEPC.245(66), as amended) or other (to be stated).

** Ice class should be consistent with the definition set out in the International Code for ships operating in polar waters (Polar Code) (resolutions MEPC.264(68) and MSC.385(94)). If not applicable, note "N/A".

Appendix X
Form of Statement of Compliance – Fuel Oil Consumption Reporting

STATEMENT OF COMPLIANCE – FUEL OIL CONSUMPTION REPORTING

Issued under the provisions of the Protocol of 1997, as amended, to amend the International Convention for the Prevention of Pollution by Ships, 1973, as modified by the Protocol of 1978 related thereto (hereinafter referred to as "the Convention") under the authority of the Government of:

. . . .

(full designation of the Party)

by . . . .

(full designation of the competent person or organization authorized under the provisions of the Convention)

Particulars of ship[*]

Name of ship . . . .

Distinctive number or letters . . . .

IMO Number[†] . . . .

Port of registry . . . .

Gross tonnage . . . .

THIS IS TO DECLARE:

1. That the ship has submitted to this Administration the data required by regulation 22A of Annex VI of the Convention, covering ship operations from (dd/mm/yyyy) through (dd/mm/yyyy); and
2. The data was collected and reported in accordance with the methodology and processes set out in the ship's SEEMP that was in effect over the period from (dd/mm/yyyy) through (dd/mm/yyyy).

This Statement of Compliance is valid until (dd/mm/yyyy). . . . .

Issued at: . . . .

(place of issue of Statement)

Date (dd/mm/yyyy) . . . .

(date of issue)   *(signature of duly authorized official issuing the Statement)*

(seal or stamp of the authority, as appropriate)

[*] Alternatively, the particulars of the ship may be placed horizontally in boxes.

[†] In accordance with the IMO Ship Identification Number Scheme, adopted by the Organization by resolution A.1078(28).

Consolidated text of MARPOL Annex VI, including amendments adopted by resolutions MEPC.271(69) and MEPC.278(70)*

Regulations for the prevention of air pollution from ships

Chapter 1 – General

Regulation 1
Application

The provisions of this Annex shall apply to all ships, except where expressly provided otherwise in regulations 3, 5, 6, 13, 15, 16, 18, 19, 20, 21, 22 and 22A of this Annex.

Regulation 2
Definitions

For the purpose of this Annex:

1 *Annex* means Annex VI to the International Convention for the Prevention of Pollution from Ships, 1973 (MARPOL), as modified by the Protocol of 1978 relating thereto, and as modified by the Protocol of 1997, as amended by the Organization, provided that such amendments are adopted and brought into force in accordance with the provisions of article 16 of the present Convention.

2 *A similar stage of construction* means the stage at which:

- **.1** construction identifiable with a specific ship begins; and
- **.2** assembly of that ship has commenced comprising at least 50 tonnes or one per cent of the estimated mass of all structural material, whichever is less.

3 *Anniversary date* means the day and the month of each year that will correspond to the date of expiry of the International Air Pollution Prevention Certificate.

4 *Auxiliary control device* means a system, function or control strategy installed on a marine diesel engine that is used to protect the engine and/or its ancillary equipment against operating conditions that could result in damage or failure, or that is used to facilitate the starting of the engine. An auxiliary control device may also be a strategy or measure that has been satisfactorily demonstrated not to be a defeat device.

5 *Continuous feeding* is defined as the process whereby waste is fed into a combustion chamber without human assistance while the incinerator is in normal operating conditions with the combustion chamber operative temperature between 850°C and 1,200°C.

---

* The original MARPOL Annex VI entered into force on 19 May 2005. The revised MARPOL Annex VI adopted by resolution MEPC. 176(58) entered into force on 1 July 2010. The amendments thereto, adopted by resolutions MEPC.190(60), MEPC.194(61), MEPC.202(62), MEPC.203(62), MEPC.217(63), MEPC.247(66), MEPC.251(66) and MEPC.258(67), have entered into force. Additionally, further amendments thereto, adopted by resolution MEPC.271(69), are expected to enter into force on 1 September 2017, and those adopted by resolution MEPC.278(70) are expected to enter into force on 1 March 2018.

6 *Defeat device* means a device that measures, senses or responds to operating variables (e.g. engine speed, temperature, intake pressure or any other parameter) for the purpose of activating, modulating, delaying or deactivating the operation of any component or the function of the emission control system such that the effectiveness of the emission control system is reduced under conditions encountered during normal operation, unless the use of such a device is substantially included in the applied emission certification test procedures.

7 *Emission* means any release of substances, subject to control by this Annex, from ships into the atmosphere or sea.

8 *Emission control area* means an area where the adoption of special mandatory measures for emissions from ships is required to prevent, reduce and control air pollution from NO_x or SO_x and particulate matter or all three types of emissions and their attendant adverse impacts on human health and the environment. Emission control areas shall include those listed in, or designated under, regulations 13 and 14 of this Annex.

9 *Fuel oil* means any fuel delivered to and intended for combustion purposes for propulsion or operation on board a ship, including gas, distillate and residual fuels.

10 *Gross tonnage* means the gross tonnage calculated in accordance with the tonnage measurement regulations contained in Annex I to the International Convention on Tonnage Measurements of Ships, 1969, or any successor Convention.

11 *Installations* in relation to regulation 12 of this Annex means the installation of systems, equipment, including portable fire-extinguishing units, insulation, or other material on a ship, but excludes the repair or recharge of previously installed systems, equipment, insulation or other material, or the recharge of portable fire-extinguishing units.

12 *Installed* means a marine diesel engine that is or is intended to be fitted on a ship, including a portable auxiliary marine diesel engine, only if its fuelling, cooling or exhaust system is an integral part of the ship. A fuelling system is considered integral to the ship only if it is permanently affixed to the ship. This definition includes a marine diesel engine that is used to supplement or augment the installed power capacity of the ship and is intended to be an integral part of the ship.

13 *Irrational emission control strategy* means any strategy or measure that, when the ship is operated under normal conditions of use, reduces the effectiveness of an emission control system to a level below that expected on the applicable emission test procedures.

14 *Marine diesel engine* means any reciprocating internal combustion engine operating on liquid or dual fuel, to which regulation 13 of this Annex applies, including booster/compound systems if applied. In addition, a gas-fuelled engine installed on a ship constructed on or after 1 March 2016 or a gas-fuelled additional or non-identical replacement engine installed on or after that date is also considered as a marine diesel engine.

15 *NO_x Technical Code* means the Technical Code on Control of Emission of Nitrogen Oxides from Marine Diesel Engines adopted by resolution 2 of the 1997 MARPOL Conference, as amended by the Organization, provided that such amendments are adopted and brought into force in accordance with the provisions of article 16 of the present Convention.

16 *Ozone-depleting substances* means controlled substances defined in paragraph (4) of article 1 of the Montreal Protocol on Substances that Deplete the Ozone Layer, 1987, listed in Annexes A, B, C or E to the said Protocol in force at the time of application or interpretation of this Annex.

Ozone-depleting substances that may be found on board ship include, but are not limited to:

Halon 1211	Bromochlorodifluoromethane
Halon 1301	Bromotrifluoromethane
Halon 2402	1,2-Dibromo-1,1,2,2-tetraflouroethane (also known as Halon 114B2)
CFC-11	Trichlorofluoromethane
CFC-12	Dichlorodifluoromethane

CFC-113	1,1,2-Trichloro-1,2,2-trifluoroethane
CFC-114	1,2-Dichloro-1,1,2,2-tetrafluoroethane
CFC-115	Chloropentafluoroethane

17 *Shipboard incineration* means the incineration of wastes or other matter on board a ship, if such wastes or other matter were generated during the normal operation of that ship.

18 *Shipboard incinerator* means a shipboard facility designed for the primary purpose of incineration.

19 *Ships constructed* means ships the keels of which are laid or that are at a similar stage of construction.

20 *Sludge oil* means sludge from the fuel oil or lubricating oil separators, waste lubricating oil from main or auxiliary machinery, or waste oil from bilge water separators, oil filtering equipment or drip trays.

21 *Tanker* in relation to regulation 15 of this Annex means an oil tanker as defined in regulation 1 of Annex I of the present Convention or a chemical tanker as defined in regulation 1 of Annex II of the present Convention.

For the purpose of chapter 4:

22 *Existing ship* means a ship which is not a new ship.

23 *New ship* means a ship:

.1 for which the building contract is placed on or after 1 January 2013; or

.2 in the absence of a building contract, the keel of which is laid or which is at a similar stage of construction on or after 1 July 2013; or

.3 the delivery of which is on or after 1 July 2015.

24 *Major conversion* means in relation to chapter 4 of this Annex a conversion of a ship:

.1 which substantially alters the dimensions, carrying capacity or engine power of the ship; or

.2 which changes the type of the ship; or

.3 the intent of which in the opinion of the Administration is substantially to prolong the life of the ship; or

.4 which otherwise so alters the ship that, if it were a new ship, it would become subject to relevant provisions of the present Convention not applicable to it as an existing ship; or

.5 which substantially alters the energy efficiency of the ship and includes any modifications that could cause the ship to exceed the applicable required EEDI as set out in regulation 21 of this Annex.

25 *Bulk carrier* means a ship which is intended primarily to carry dry cargo in bulk, including such types as ore carriers as defined in regulation 1 of chapter XII of SOLAS 74 (as amended) but excluding combination carriers.

26 *Gas carrier* in relation to chapter 4 of this Annex means a cargo ship, other than an LNG carrier as defined in paragraph 38 of this regulation, constructed or adapted and used for the carriage in bulk of any liquefied gas.

27 *Tanker* in relation to chapter 4 of this Annex means an oil tanker as defined in regulation 1 of Annex I of the present Convention or a chemical tanker or an NLS tanker as defined in regulation 1 of Annex II of the present Convention.

28 *Containership* means a ship designed exclusively for the carriage of containers in holds and on deck.

29 *General cargo ship* means a ship with a multi-deck or single deck hull designed primarily for the carriage of general cargo. This definition excludes specialized dry cargo ships, which are not included in the calculation of reference lines for general cargo ships, namely livestock carrier, barge carrier, heavy load carrier, yacht carrier, nuclear fuel carrier.

30 *Refrigerated cargo carrier* means a ship designed exclusively for the carriage of refrigerated cargoes in holds.

31 *Combination carrier* means a ship designed to load 100% deadweight with both liquid and dry cargo in bulk.

32 *Passenger ship* means a ship which carries more than 12 passengers.

33 *Ro-ro cargo ship (vehicle carrier)* means a multi deck roll-on-roll-off cargo ship designed for the carriage of empty cars and trucks.

34 *Ro-ro cargo ship* means a ship designed for the carriage of roll-on-roll-off cargo transportation units.

35 *Ro-ro passenger ship* means a passenger ship with roll-on-roll-off cargo spaces.

36 *Attained EEDI* is the EEDI value achieved by an individual ship in accordance with regulation 20 of this Annex.

37 *Required EEDI* is the maximum value of attained EEDI that is allowed by regulation 21 of this Annex for the specific ship type and size.

38 *LNG carrier* in relation to chapter 4 of this Annex means a cargo ship constructed or adapted and used for the carriage in bulk of liquefied natural gas (LNG).

39 *Cruise passenger ship* in relation to chapter 4 of this Annex means a passenger ship not having a cargo deck, designed exclusively for commercial transportation of passengers in overnight accommodations on a sea voyage.

40 *Conventional propulsion* in relation to chapter 4 of this Annex means a method of propulsion where a main reciprocating internal combustion engine(s) is the prime mover and coupled to a propulsion shaft either directly or through a gear box.

41 *Non-conventional propulsion* in relation to chapter 4 of this Annex means a method of propulsion, other than conventional propulsion, including diesel-electric propulsion, turbine propulsion, and hybrid propulsion systems.

42 *Cargo ship having ice-breaking capability* in relation to chapter 4 of this Annex means a cargo ship which is designed to break level ice independently with a speed of at least 2 knots when the level ice thickness is 1.0 m or more having ice bending strength of at least 500 kPa.

43 *A ship delivered on or after 1 September 2019* means a ship:

.1 for which the building contract is placed on or after 1 September 2015; or

.2 in the absence of a building contract, the keel of which is laid, or which is at a similar stage of construction, on or after 1 March 2016; or

.3 the delivery of which is on or after 1 September 2019.

For the purposes of this Annex:

44 *Audit* means a systematic, independent and documented process for obtaining audit evidence and evaluating it objectively to determine the extent to which audit criteria are fulfilled.

45 *Audit Scheme* means the IMO Member State Audit Scheme established by the Organization and taking into account the guidelines developed by the Organization.*

46 *Code for Implementation* means the IMO Instruments Implementation Code (III Code) adopted by the Organization by resolution A.1070(28).

47 *Audit Standard* means the Code for Implementation.

For the purpose of chapter 4:

48 *Calendar year* means the period from 1 January until 31 December inclusive.

49 *Company* means the owner of the ship or any other organization or person such as the manager, or the bareboat charterer, who has assumed the responsibility for operation of the ship from the owner of the ship and who on assuming such responsibility has agreed to take over all the duties and responsibilities imposed by the International Management Code for the Safe Operation of Ships and for Pollution Prevention, as amended.

50 *Distance travelled* means distance travelled over ground.

Regulation 3
Exceptions and exemptions

General

1 Regulations of this Annex shall not apply to:

.1 any emission necessary for the purpose of securing the safety of a ship or saving life at sea; or

.2 any emission resulting from damage to a ship or its equipment:

.2.1 provided that all reasonable precautions have been taken after the occurrence of the damage or discovery of the emission for the purpose of preventing or minimizing the emission; and

.2.2 except if the owner or the master acted either with intent to cause damage, or recklessly and with knowledge that damage would probably result.

Trials for ship emission reduction and control technology research

2 The Administration of a Party may, in cooperation with other Administrations as appropriate, issue an exemption from specific provisions of this Annex for a ship to conduct trials for the development of ship emission reduction and control technologies and engine design programmes. Such an exemption shall only be provided if the applications of specific provisions of the Annex or the revised NO_x Technical Code 2008 could impede research into the development of such technologies or programmes. A permit issued under this regulation shall not exempt a ship from the reporting requirement under regulation 22A and shall not alter the type and scope of data required to be reported under regulation 22A. A permit for such an exemption shall only be provided to the minimum number of ships necessary and be subject to the following provisions:

.1 for marine diesel engines with a per cylinder displacement up to 30 L, the duration of the sea trial shall not exceed 18 months. If additional time is required, a permitting Administration or Administrations may permit a renewal for one additional 18-month period; or

.2 for marine diesel engines with a per cylinder displacement at or above 30 L, the duration of the ship trial shall not exceed five years and shall require a progress review by the permitting Administration or Administrations at each intermediate survey. A permit may be withdrawn based on this review if the testing has not adhered to the conditions of the permit or if it is determined that the technology or programme is not likely to produce effective results in the reduction and control of ship emissions. If the reviewing Administration or Administrations determine that additional time is required to conduct a test of a particular technology or programme, a permit may be renewed for an additional time period not to exceed five years.

* Refer to the Framework and Procedures for the IMO Member State Audit Scheme (resolution A.1067(28)).

Emissions from sea-bed mineral activities

3.1 Emissions directly arising from the exploration, exploitation and associated offshore processing of sea-bed mineral resources are, consistent with article 2(3)(b)(ii) of the present Convention, exempt from the provisions of this Annex. Such emissions include the following:

- **.1** emissions resulting from the incineration of substances that are solely and directly the result of exploration, exploitation and associated offshore processing of sea-bed mineral resources, including but not limited to the flaring of hydrocarbons and the burning of cuttings, muds, and/or stimulation fluids during well completion and testing operations, and flaring arising from upset conditions;
- **.2** the release of gases and volatile compounds entrained in drilling fluids and cuttings;
- **.3** emissions associated solely and directly with the treatment, handling or storage of sea-bed minerals; and
- **.4** emissions from marine diesel engines that are solely dedicated to the exploration, exploitation and associated offshore processing of sea-bed mineral resources.

3.2 The requirements of regulation 18 of this Annex shall not apply to the use of hydrocarbons that are produced and subsequently used on site as fuel, when approved by the Administration.

Regulation 4
Equivalents

1 The Administration of a Party may allow any fitting, material, appliance or apparatus to be fitted in a ship or other procedures, alternative fuel oils, or compliance methods used as an alternative to that required by this Annex if such fitting, material, appliance or apparatus or other procedures, alternative fuel oils, or compliance methods are at least as effective in terms of emissions reductions as that required by this Annex, including any of the standards set forth in regulations 13 and 14.

2 The Administration of a Party that allows a fitting, material, appliance or apparatus or other procedures, alternative fuel oils, or compliance methods used as an alternative to that required by this Annex shall communicate to the Organization for circulation to the Parties particulars thereof, for their information and appropriate action, if any.

3 The Administration of a Party should take into account any relevant guidelines developed by the Organization* pertaining to the equivalents provided for in this regulation.

4 The Administration of a Party that allows the use of an equivalent as set forth in paragraph 1 of this regulation shall endeavour not to impair or damage its environment, human health, property or resources or those of other States.

Chapter 2 – Survey, certification and means of control

Regulation 5
Surveys

1 Every ship of 400 gross tonnage and above and every fixed and floating drilling rig and other platforms shall, to ensure compliance with the requirements of chapter 3 of this Annex, be subject to the surveys specified below:

- **.1** An initial survey before the ship is put into service or before the certificate required under regulation 6 of this Annex is issued for the first time. This survey shall be such as to ensure that the equipment, systems, fittings, arrangements and material fully comply with the applicable requirements of chapter 3 of this Annex;

---

* Refer to 2015 Guidelines for exhaust gas cleaning systems (resolution MEPC.259(68)).

.2 A renewal survey at intervals specified by the Administration, but not exceeding five years, except where regulation 9.2, 9.5, 9.6 or 9.7 of this Annex is applicable. The renewal survey shall be such as to ensure that the equipment, systems, fittings, arrangements and material fully comply with applicable requirements of chapter 3 of this Annex;

.3 An intermediate survey within three months before or after the second anniversary date or within three months before or after the third anniversary date of the certificate which shall take the place of one of the annual surveys specified in paragraph 1.4 of this regulation. The intermediate survey shall be such as to ensure that the equipment and arrangements fully comply with the applicable requirements of chapter 3 of this Annex and are in good working order. Such intermediate surveys shall be endorsed on the IAPP Certificate issued under regulation 6 or 7 of this Annex;

.4 An annual survey within three months before or after each anniversary date of the certificate, including a general inspection of the equipment, systems, fittings, arrangements and material referred to in paragraph 1.1 of this regulation to ensure that they have been maintained in accordance with paragraph 5 of this regulation and that they remain satisfactory for the service for which the ship is intended. Such annual surveys shall be endorsed on the IAPP Certificate issued under regulation 6 or 7 of this Annex; and

.5 An additional survey either general or partial, according to the circumstances, shall be made whenever any important repairs or renewals are made as prescribed in paragraph 5 of this regulation or after a repair resulting from investigations prescribed in paragraph 6 of this regulation. The survey shall be such as to ensure that the necessary repairs or renewals have been effectively made, that the material and workmanship of such repairs or renewals are in all respects satisfactory and that the ship complies in all respects with the requirements of chapter 3 of this Annex.

2 In the case of ships of less than 400 gross tonnage, the Administration may establish appropriate measures in order to ensure that the applicable provisions of chapter 3 of this Annex are complied with.

3 Surveys of ships as regards the enforcement of the provisions of this Annex shall be carried out by officers of the Administration.

.1 The Administration may, however, entrust the surveys either to surveyors nominated for the purpose or to organizations recognized by it. Such organizations shall comply with the guidelines adopted by the Organization;*

.2 The survey of marine diesel engines and equipment for compliance with regulation 13 of this Annex shall be conducted in accordance with the revised NO_x Technical Code 2008;

.3 When a nominated surveyor or recognized organization determines that the condition of the equipment does not correspond substantially with the particulars of the certificate, it shall ensure that corrective action is taken and shall in due course notify the Administration. If such corrective action is not taken, the certificate shall be withdrawn by the Administration. If the ship is in a port of another Party, the appropriate authorities of the port State shall also be notified immediately. When an officer of the Administration, a nominated surveyor or recognized organization has notified the appropriate authorities of the port State, the Government of the port State concerned shall give such officer, surveyor or organization any necessary assistance to carry out their obligations under this regulation; and

.4 In every case, the Administration concerned shall fully guarantee the completeness and efficiency of the survey and shall undertake to ensure the necessary arrangements to satisfy this obligation.

* Refer to Guidelines for the authorization of organizations acting on behalf of the Administration (resolution A.739(18), as amended by resolution MSC.208(81)), and Specifications on the survey and certification functions of recognized organizations acting on behalf of the Administration (resolution A.789(19), as may be amended). Refer also to Survey Guidelines under the Harmonized System of Survey and Certification for the revised MARPOL Annex VI (resolution MEPC.180(59)).

4 Ships to which chapter 4 of this Annex applies shall also be subject to the surveys specified below, taking into account the guidelines adopted by the Organization:*

.1 An initial survey before a new ship is put in service and before the International Energy Efficiency Certificate is issued. The survey shall verify that the ship's attained EEDI is in accordance with the requirements in chapter 4 of this Annex, and that the SEEMP required by regulation 22 of this Annex is on board;

.2 A general or partial survey, according to the circumstances, after a major conversion of a new ship to which this regulation applies. The survey shall ensure that the attained EEDI is recalculated as necessary and meets the requirement of regulation 21 of this Annex, with the reduction factor applicable to the ship type and size of the converted ship in the phase corresponding to the date of contract or keel laying or delivery determined for the original ship in accordance with regulation 2.23 of this Annex;

.3 In cases where the major conversion of a new or existing ship is so extensive that the ship is regarded by the Administration as a newly constructed ship, the Administration shall determine the necessity of an initial survey on attained EEDI. Such a survey, if determined necessary, shall ensure that the attained EEDI is calculated and meets the requirement of regulation 21 of this Annex, with the reduction factor applicable corresponding to the ship type and size of the converted ship at the date of the contract of the conversion, or in the absence of a contract, the commencement date of the conversion. The survey shall also verify that the SEEMP required by regulation 22 of this Annex is on board and, for a ship to which regulation 22A applies, has been revised appropriately to reflect a major conversion in those cases where the major conversion affects data collection methodology and/or reporting processes;

.4 For existing ships, the verification of the requirement to have a SEEMP on board according to regulation 22 of this Annex shall take place at the first intermediate or renewal survey identified in paragraph 1 of this regulation, whichever is the first, on or after 1 January 2013; and

.5 The Administration shall ensure that for each ship to which regulation 22A applies, the SEEMP complies with regulation 22.2 of this Annex. This shall be done prior to collecting data under regulation 22A of this Annex in order to ensure the methodology and processes are in place prior to the beginning of the ship's first reporting period. Confirmation of compliance shall be provided to and retained on board the ship.

5 The equipment shall be maintained to conform with the provisions of this Annex and no changes shall be made in the equipment, systems, fittings, arrangements or material covered by the survey, without the express approval of the Administration. The direct replacement of such equipment and fittings with equipment and fittings that conform with the provisions of this Annex is permitted.

6 Whenever an accident occurs to a ship or a defect is discovered that substantially affects the efficiency or completeness of its equipment covered by this Annex, the master or owner of the ship shall report at the earliest opportunity to the Administration, a nominated surveyor or recognized organization responsible for issuing the relevant certificate.

Regulation 6

Issue or endorsement of Certificates and Statements of Compliance related to fuel oil consumption reporting

International Air Pollution Prevention Certificate

1 An International Air Pollution Prevention Certificate shall be issued, after an initial or renewal survey in accordance with the provisions of regulation 5 of this Annex, to:

.1 any ship of 400 gross tonnage and above engaged in voyages to ports or offshore terminals under the jurisdiction of other Parties; and

* Refer to 2014 Guidelines on survey and certification of the Energy Efficiency Design Index (resolution MEPC.254(67), as amended by resolution MEPC.261(68)).

.2 platforms and drilling rigs engaged in voyages to waters under the sovereignty or jurisdiction of other Parties.

2 A ship constructed before the date this Annex enters into force for that particular ship's Administration, shall be issued with an International Air Pollution Prevention Certificate in accordance with paragraph 1 of this regulation no later than the first scheduled dry-docking after the date of such entry into force, but in no case later than three years after this date.

3 Such certificate shall be issued or endorsed either by the Administration or by any person or organization duly authorized by it.* In every case, the Administration assumes full responsibility for the certificate.

International Energy Efficiency Certificate

4 An International Energy Efficiency Certificate for the ship shall be issued after a survey in accordance with the provisions of regulation 5.4 of this Annex to any ship of 400 gross tonnage and above before that ship may engage in voyages to ports or offshore terminals under the jurisdiction of other Parties.

5 The certificate shall be issued or endorsed either by the Administration or any organization duly authorized by it.* In every case, the Administration assumes full responsibility for the certificate.

Statement of Compliance – Fuel Oil Consumption Reporting

6 Upon receipt of reported data pursuant to regulation 22A.3 of this Annex, the Administration or any organization duly authorized by it shall determine whether the data has been reported in accordance with regulation 22A of this Annex and, if so, issue a Statement of Compliance related to fuel oil consumption to the ship no later than five months from the beginning of the calendar year. In every case, the Administration assumes full responsibility for this Statement of Compliance.

7 Upon receipt of reported data pursuant to regulations 22A.4, 22A.5 or 22A.6 of this Annex, the Administration or any organization duly authorized by it* shall promptly determine whether the data has been reported in accordance with regulation 22A and, if so, issue a Statement of Compliance related to fuel oil consumption to the ship at that time. In every case, the Administration assumes full responsibility for this Statement of Compliance.

Regulation 7
Issue of a Certificate by another Party

1 A Party may, at the request of the Administration, cause a ship to be surveyed and, if satisfied that the provisions of this Annex are complied with, shall issue or authorize the issuance of an International Air Pollution Prevention Certificate or an International Energy Efficiency Certificate to the ship, and where appropriate, endorse or authorize the endorsement of such certificates on the ship, in accordance with this Annex.

2 A copy of the certificate and a copy of the survey report shall be transmitted as soon as possible to the requesting Administration.

3 A certificate so issued shall contain a statement to the effect that it has been issued at the request of the Administration and it shall have the same force and receive the same recognition as a certificate issued under regulation 6 of this Annex.

4 No International Air Pollution Prevention Certificate or an International Energy Efficiency Certificate shall be issued to a ship which is entitled to fly the flag of a State which is not a Party.

---

* Refer to Guidelines for the authorization of organizations acting on behalf of the Administration (resolution A.739(18), as amended by resolution MSC.208(81)), and Specifications on the survey and certification functions of recognized organizations acting on behalf of the Administration (resolution A.789(19), as may be amended).

Regulation 8
Form of Certificates and Statements of Compliance related to fuel oil consumption reporting

International Air Pollution Prevention Certificate

1 The International Air Pollution Prevention (IAPP) Certificate shall be drawn up in a form corresponding to the model given in appendix I to this Annex and shall be at least in English, French or Spanish. If an official language of the issuing country is also used, this shall prevail in case of a dispute or discrepancy.

International Energy Efficiency Certificate

2 The International Energy Efficiency Certificate shall be drawn up in a form corresponding to the model given in appendix VIII to this Annex and shall be at least in English, French or Spanish. If an official language of the issuing Party is also used, this shall prevail in case of a dispute or discrepancy.

Statement of Compliance – Fuel Oil Consumption Reporting

3 The Statement of Compliance pursuant to regulations 6.6 and 6.7 of this Annex shall be drawn up in a form corresponding to the model given in appendix X to this Annex and shall be at least in English, French or Spanish. If an official language of the issuing Party is also used, this shall prevail in case of a dispute or discrepancy.

Regulation 9
Duration and validity of Certificates and Statements of Compliance related to fuel oil consumption reporting

International Air Pollution Prevention Certificate

1 An International Air Pollution Prevention (IAPP) Certificate shall be issued for a period specified by the Administration, which shall not exceed five years.

2 Notwithstanding the requirements of paragraph 1 of this regulation:

.1 when the renewal survey is completed within three months before the expiry date of the existing certificate, the new certificate shall be valid from the date of completion of the renewal survey to a date not exceeding five years from the date of expiry of the existing certificate;

.2 when the renewal survey is completed after the expiry date of the existing certificate, the new certificate shall be valid from the date of completion of the renewal survey to a date not exceeding five years from the date of expiry of the existing certificate; and

.3 when the renewal survey is completed more than three months before the expiry date of the existing certificate, the new certificate shall be valid from the date of completion of the renewal survey to a date not exceeding five years from the date of completion of the renewal survey.

3 If a certificate is issued for a period of less than five years, the Administration may extend the validity of the certificate beyond the expiry date to the maximum period specified in paragraph 1 of this regulation, provided that the surveys referred to in regulations 5.1.3 and 5.1.4 of this Annex applicable when a certificate is issued for a period of five years are carried out as appropriate.

4 If a renewal survey has been completed and a new certificate cannot be issued or placed on board the ship before the expiry date of the existing certificate, the person or organization authorized by the Administration may endorse the existing certificate and such a certificate shall be accepted as valid for a further period that shall not exceed five months from the expiry date.

5 If a ship, at the time when a certificate expires, is not in a port in which it is to be surveyed, the Administration may extend the period of validity of the certificate, but this extension shall be granted only for the purpose of allowing the ship to complete its voyage to the port in which it is to be surveyed, and then only in cases where it appears proper and reasonable to do so. No certificate shall be extended for a period longer than three months, and a ship to which an extension is granted shall not, on its arrival in the port in which it is to be surveyed, be entitled by virtue of such extension to leave that port without having a new certificate. When the renewal survey is completed, the new certificate shall be valid to a date not exceeding five years from the date of expiry of the existing certificate before the extension was granted.

6 A certificate issued to a ship engaged on short voyages that has not been extended under the foregoing provisions of this regulation may be extended by the Administration for a period of grace of up to one month from the date of expiry stated on it. When the renewal survey is completed, the new certificate shall be valid to a date not exceeding five years from the date of expiry of the existing certificate before the extension was granted.

7 In special circumstances, as determined by the Administration, a new certificate need not be dated from the date of expiry of the existing certificate as required by paragraph 2.1, 5 or 6 of this regulation. In these special circumstances, the new certificate shall be valid to a date not exceeding five years from the date of completion of the renewal survey.

8 If an annual or intermediate survey is completed before the period specified in regulation 5 of this Annex, then:

.1 the anniversary date shown on the certificate shall be amended by endorsement to a date that shall not be more than three months later than the date on which the survey was completed;

.2 the subsequent annual or intermediate survey required by regulation 5 of this Annex shall be completed at the intervals prescribed by that regulation using the new anniversary date; and

.3 the expiry date may remain unchanged, provided one or more annual or intermediate surveys, as appropriate, are carried out so that the maximum intervals between the surveys prescribed by regulation 5 of this Annex are not exceeded.

9 A certificate issued under regulation 6 or 7 of this Annex shall cease to be valid in any of the following cases:

.1 if the relevant surveys are not completed within the periods specified under regulation 5.1 of this Annex;

.2 if the certificate is not endorsed in accordance with regulation 5.1.3 or 5.1.4 of this Annex; and

.3 upon transfer of the ship to the flag of another State. A new certificate shall only be issued when the Government issuing the new certificate is fully satisfied that the ship is in compliance with the requirements of regulation 5.4 of this Annex. In the case of a transfer between Parties, if requested within three months after the transfer has taken place, the Government of the Party whose flag the ship was formerly entitled to fly shall, as soon as possible, transmit to the Administration copies of the certificate carried by the ship before the transfer and, if available, copies of the relevant survey reports.

International Energy Efficiency Certificate

10 The International Energy Efficiency Certificate shall be valid throughout the life of the ship subject to the provisions of paragraph 11 below.

11 An International Energy Efficiency Certificate issued under this Annex shall cease to be valid in any of the following cases:

.1 if the ship is withdrawn from service or if a new certificate is issued following major conversion of the ship; or

.2 upon transfer of the ship to the flag of another State. A new certificate shall only be issued when the Government issuing the new certificate is fully satisfied that the ship is in compliance with the requirements of chapter 4 of this Annex. In the case of a transfer between Parties, if requested within three months after the transfer has taken place, the Government of the Party whose flag the ship was formerly entitled to fly shall, as soon as possible, transmit to the Administration copies of the certificate carried by the ship before the transfer and, if available, copies of the relevant survey reports.

Statement of Compliance – Fuel Oil Consumption Reporting

12 The Statement of Compliance pursuant to regulation 6.6 of this Annex shall be valid for the calendar year in which it is issued and for the first five months of the following calendar year. The Statement of Compliance pursuant to regulation 6.7 of this Annex shall be valid for the calendar year in which it is issued, for the following calendar year, and for the first five months of the subsequent calendar year. All Statements of Compliance shall be kept on board for at least the period of their validity.

Regulation 10
Port State control on operational requirements

1 A ship, when in a port or an offshore terminal under the jurisdiction of another Party, is subject to inspection by officers duly authorized by such Party concerning operational requirements under this Annex,* where there are clear grounds for believing that the master or crew are not familiar with essential shipboard procedures relating to the prevention of air pollution from ships.

2 In the circumstances given in paragraph 1 of this regulation, the Party shall take such steps as to ensure that the ship shall not sail until the situation has been brought to order in accordance with the requirements of this Annex.

3 Procedures relating to the port State control prescribed in article 5 of the present Convention shall apply to this regulation.

4 Nothing in this regulation shall be construed to limit the rights and obligations of a Party carrying out control over operational requirements specifically provided for in the present Convention.

5 In relation to chapter 4 of this Annex, any port State inspection shall be limited to verifying, when appropriate, that there is a valid Statement of Compliance related to fuel oil consumption reporting and International Energy Efficiency Certificate on board, in accordance with article 5 of the Convention.

Regulation 11
Detection of violations and enforcement

1 Parties shall cooperate in the detection of violations and the enforcement of the provisions of this Annex, using all appropriate and practicable measures of detection and environmental monitoring, adequate procedures for reporting and accumulation of evidence.

2 A ship to which this Annex applies may, in any port or offshore terminal of a Party, be subject to inspection by officers appointed or authorized by that Party for the purpose of verifying whether the ship has emitted any of the substances covered by this Annex in violation of the provision of this Annex. If an inspection indicates a violation of this Annex, a report shall be forwarded to the Administration for any appropriate action.

3 Any Party shall furnish to the Administration evidence, if any, that the ship has emitted any of the substances covered by this Annex in violation of the provisions of this Annex. If it is practicable to do so, the competent authority of the former Party shall notify the master of the ship of the alleged violation.

* Refer to Procedures for port State control (resolution A.1052(27)). Refer also to 2009 Guidelines for port State control under the revised MARPOL Annex VI (resolution MEPC.181(59)).

4 Upon receiving such evidence, the Administration so informed shall investigate the matter, and may request the other Party to furnish further or better evidence of the alleged contravention. If the Administration is satisfied that sufficient evidence is available to enable proceedings to be brought in respect of the alleged violation, it shall cause such proceedings to be taken in accordance with its law as soon as possible. The Administration shall promptly inform the Party that has reported the alleged violation, as well as the Organization, of the action taken.

5 A Party may also inspect a ship to which this Annex applies when it enters the ports or offshore terminals under its jurisdiction, if a request for an investigation is received from any Party together with sufficient evidence that the ship has emitted any of the substances covered by the Annex in any place in violation of this Annex. The report of such investigation shall be sent to the Party requesting it and to the Administration so that the appropriate action may be taken under the present Convention.

6 The international law concerning the prevention, reduction and control of pollution of the marine environment from ships, including that law relating to enforcement and safeguards, in force at the time of application or interpretation of this Annex, applies, mutatis mutandis, to the rules and standards set forth in this Annex.

Chapter 3 – Requirements for control of emissions from ships

Regulation 12
Ozone-depleting substances

1 This regulation does not apply to permanently sealed equipment where there are no refrigerant charging connections or potentially removable components containing ozone-depleting substances.

2 Subject to the provisions of regulation 3.1, any deliberate emissions of ozone-depleting substances shall be prohibited. Deliberate emissions include emissions occurring in the course of maintaining, servicing, repairing or disposing of systems or equipment, except that deliberate emissions do not include minimal releases associated with the recapture or recycling of an ozone-depleting substance. Emissions arising from leaks of an ozone-depleting substance, whether or not the leaks are deliberate, may be regulated by Parties.

3.1 Installations that contain ozone-depleting substances, other than hydrochlorofluorocarbons, shall be prohibited:

.1 on ships constructed on or after 19 May 2005; or

.2 in the case of ships constructed before 19 May 2005, which have a contractual delivery date of the equipment to the ship on or after 19 May 2005 or, in the absence of a contractual delivery date, the actual delivery of the equipment to the ship on or after 19 May 2005.

3.2 Installations that contain hydrochlorofluorocarbons shall be prohibited:

.1 on ships constructed on or after 1 January 2020; or

.2 in the case of ships constructed before 1 January 2020, which have a contractual delivery date of the equipment to the ship on or after 1 January 2020 or, in the absence of a contractual delivery date, the actual delivery of the equipment to the ship on or after 1 January 2020.

4 The substances referred to in this regulation, and equipment containing such substances, shall be delivered to appropriate reception facilities when removed from ships.

5 Each ship subject to regulation 6.1 shall maintain a list of equipment containing ozone-depleting substances.*

* See appendix I, Supplement to International Air Pollution Prevention Certificate (IAPP Certificate), section 2.1.

6 Each ship subject to regulation 6.1 that has rechargeable systems that contain ozone-depleting substances shall maintain an *ozone-depleting substances record book*. This record book may form part of an existing logbook or electronic recording system as approved by the Administration.

7 Entries in the ozone-depleting substances record book shall be recorded in terms of mass (kg) of substance and shall be completed without delay on each occasion, in respect of the following:

.1 recharge, full or partial, of equipment containing ozone-depleting substances;

.2 repair or maintenance of equipment containing ozone-depleting substances;

.3 discharge of ozone-depleting substances to the atmosphere:

.3.1 deliberate; and

.3.2 non-deliberate;

.4 discharge of ozone-depleting substances to land-based reception facilities; and

.5 supply of ozone-depleting substances to the ship.

Regulation 13
Nitrogen oxides (NO_x)

Application

1.1 This regulation shall apply to:

.1 each marine diesel engine with a power output of more than 130 kW installed on a ship; and

.2 each marine diesel engine with a power output of more than 130 kW that undergoes a major conversion on or after 1 January 2000 except when demonstrated to the satisfaction of the Administration that such engine is an identical replacement to the engine that it is replacing and is otherwise not covered under paragraph 1.1.1 of this regulation.

1.2 This regulation does not apply to:

.1 a marine diesel engine intended to be used solely for emergencies, or solely to power any device or equipment intended to be used solely for emergencies on the ship on which it is installed, or a marine diesel engine installed in lifeboats intended to be used solely for emergencies; and

.2 a marine diesel engine installed on a ship solely engaged in voyages within waters subject to the sovereignty or jurisdiction of the State the flag of which the ship is entitled to fly, provided that such engine is subject to an alternative NO_x control measure established by the Administration.

1.3 Notwithstanding the provisions of paragraph 1.1 of this regulation, the Administration may provide an exclusion from the application of this regulation for any marine diesel engine that is installed on a ship constructed, or for any marine diesel engine that undergoes a major conversion, before 19 May 2005, provided that the ship on which the engine is installed is solely engaged in voyages to ports or offshore terminals within the State the flag of which the ship is entitled to fly.

Major conversion

2.1 For the purpose of this regulation, *major conversion* means a modification on or after 1 January 2000 of a marine diesel engine that has not already been certified to the standards set forth in paragraph 3, 4, or 5.1.1 of this regulation where:

.1 the engine is replaced by a marine diesel engine or an additional marine diesel engine is installed, or

.2 any substantial modification, as defined in the revised NO_x Technical Code 2008, is made to the engine, or

.3 the maximum continuous rating of the engine is increased by more than 10% compared to the maximum continuous rating of the original certification of the engine.

2.2 For a major conversion involving the replacement of a marine diesel engine with a non-identical marine diesel engine, or the installation of an additional marine diesel engine, the standards in this regulation at the time of the replacement or addition of the engine shall apply. In the case of replacement engines only, if it is not possible for such a replacement engine to meet the standards set forth in paragraph 5.1.1 of this regulation (Tier III, as applicable), then that replacement engine shall meet the standards set forth in paragraph 4 of this regulation (Tier II), taking into account guidelines developed by the Organization.*

2.3 A marine diesel engine referred to in paragraph 2.1.2 or 2.1.3 of this regulation shall meet the following standards:

.1 for ships constructed prior to 1 January 2000, the standards set forth in paragraph 3 of this regulation shall apply; and

.2 for ships constructed on or after 1 January 2000, the standards in force at the time the ship was constructed shall apply.

Tier I†

3 Subject to regulation 3 of this Annex, the operation of a marine diesel engine that is installed on a ship constructed on or after 1 January 2000 and prior to 1 January 2011 is prohibited, except when the emission of nitrogen oxides (calculated as the total weighted emission of NO_2) from the engine is within the following limits, where n = rated engine speed (crankshaft revolutions per minute):

.1 17.0 g/kWh when n is less than 130 rpm;

.2 $45 \cdot n^{(-0.2)}$ g/kWh when n is 130 or more but less than 2,000 rpm;

.3 9.8 g/kWh when n is 2,000 rpm or more.

Tier II

4 Subject to regulation 3 of this Annex, the operation of a marine diesel engine that is installed on a ship constructed on or after 1 January 2011 is prohibited, except when the emission of nitrogen oxides (calculated as the total weighted emission of NO_2) from the engine is within the following limits, where n = rated engine speed (crankshaft revolutions per minute):

.1 14.4 g/kWh when n is less than 130 rpm;

.2 $44 \cdot n^{(-0.23)}$ g/kWh when n is 130 or more but less than 2,000 rpm;

.3 7.7 g/kWh when n is 2,000 rpm or more.

Tier III

5.1 Subject to regulation 3 of this Annex, in an emission control area designated for Tier III NO_x control under paragraph 6 of this regulation, the operation of a marine diesel engine that is installed on a ship:

.1 is prohibited except when the emission of nitrogen oxides (calculated as the total weighted emission of NO_2) from the engine is within the following limits, where n = rated engine speed (crankshaft revolutions per minute):

.1.1 3.4 g/kWh when n is less than 130 rpm;

.1.2 $9 \cdot n^{(-0.2)}$ g/kWh when n is 130 or more but less than 2,000 rpm;

.1.3 2.0 g/kWh when n is 2,000 rpm or more;

when:

.2 that ship is constructed on or after 1 January 2016 and is operating in the North American Emission Control Area or the United States Caribbean Sea Emission Control Area;

---

* Refer to 2013 Guidelines as required by regulation 13.2.2 of MARPOL Annex VI in respect of non-identical replacement engines not required to meet the Tier III limit (resolution MEPC.230(65)).

† Refer to Guidelines for the application of the NO_x Technical Code relative to certification and amendments of Tier I engines (MEPC.1/Circ.679).

when:

.3 that ship is operating in an emission control area designated for Tier III NO_x control under paragraph 6 of this regulation, other than an emission control area described in paragraph 5.1.2 of this regulation, and is constructed on or after the date of adoption of such an emission control area, or a later date as may be specified in the amendment designating the NO_x Tier III emission control area, whichever is later.

5.2 The standards set forth in paragraph 5.1.1 of this regulation shall not apply to:

.1 a marine diesel engine installed on a ship with a length (*L*), as defined in regulation 1.19 of Annex I to the present Convention, of less than 24 m when it has been specifically designed, and is used solely, for recreational purposes; or

.2 a marine diesel engine installed on a ship with a combined nameplate diesel engine propulsion power of less than 750 kW if it is demonstrated, to the satisfaction of the Administration, that the ship cannot comply with the standards set forth in paragraph 5.1.1 of this regulation because of design or construction limitations of the ship; or

.3 a marine diesel engine installed on a ship constructed prior to 1 January 2021 of less than 500 gross tonnage, with a length (*L*), as defined in regulation 1.19 of Annex I to the present convention, of 24 m or over when it has been specifically designed, and is used solely, for recreational purposes.

5.3 The tier and on/off status of marine diesel engines installed on board a ship to which paragraph 5.1 of this regulation applies which are certified to both Tier II and Tier III or which are certified to Tier II only shall be recorded in such logbook as prescribed by the Administration at entry into and exit from an emission control area designated under paragraph 6 of this regulation, or when the on/off status changes within such an area, together with the date, time and position of the ship.

Emission control area

6 For the purposes of this regulation, emission control areas shall be:

.1 the North American area, which means the area described by the coordinates provided in appendix VII to this Annex;

.2 the United States Caribbean sea area, which means the area described by the coordinates provided in appendix VII to this Annex; and

.3 any other sea area, including any port area, designated by the Organization in accordance with the criteria and procedures set forth in appendix III to this Annex.

Marine diesel engines installed on a ship constructed prior to 1 January 2000

7.1 Notwithstanding paragraph 1.1.1 of this regulation, a marine diesel engine with a power output of more than 5,000 kW and a per cylinder displacement at or above 90 L installed on a ship constructed on or after 1 January 1990 but prior to 1 January 2000 shall comply with the emission limits set forth in paragraph 7.4 of this regulation, provided that an approved method* for that engine has been certified by an Administration of a Party and notification of such certification has been submitted to the Organization by the certifying Administration.† Compliance with this paragraph shall be demonstrated through one of the following:

.1 installation of the certified approved method, as confirmed by a survey using the verification procedure specified in the approved method file, including appropriate notation on the ship's International Air Pollution Prevention Certificate of the presence of the approved method; or

.2 certification of the engine confirming that it operates within the limits set forth in paragraph 3, 4, or 5.1.1 of this regulation and an appropriate notation of the engine certification on the ship's International Air Pollution Prevention Certificate.

---

* Refer to 2014 Guidelines on the approved method process (resolution MEPC.243(66)).

† Refer to 2014 Guidelines in respect of the information to be submitted by an Administration to the Organization covering the certification of an approved method as required under regulation 13.7.1 of MARPOL Annex VI (resolution MEPC.242(66)).

7.2 Paragraph 7.1 of this regulation shall apply no later than the first renewal survey that occurs 12 months or more after deposit of the notification in paragraph 7.1. If a shipowner of a ship on which an approved method is to be installed can demonstrate to the satisfaction of the Administration that the approved method was not commercially available despite best efforts to obtain it, then that approved method shall be installed on the ship no later than the next annual survey of that ship that falls after the approved method is commercially available.

7.3 With regard to a marine diesel engine with a power output of more than 5,000 kW and a per cylinder displacement at or above 90 L installed on a ship constructed on or after 1 January 1990, but prior to 1 January 2000, the International Air Pollution Prevention Certificate shall, for a marine diesel engine to which paragraph 7.1 of this regulation applies, indicate one of the following:

.1 an approved method has been applied pursuant to paragraph 7.1.1 of this regulation;

.2 the engine has been certified pursuant to paragraph 7.1.2 of this regulation;

.3 an approved method is not yet commercially available as described in paragraph 7.2 of this regulation; or

.4 an approved method is not applicable.

7.4 Subject to regulation 3 of this Annex, the operation of a marine diesel engine described in paragraph 7.1 of this regulation is prohibited, except when the emission of nitrogen oxides (calculated as the total weighted emission of NO_2) from the engine is within the following limits, where n = rated engine speed (crankshaft revolutions per minute):

.1 17.0 g/kWh when n is less than 130 rpm;

.2 $45 \cdot n^{(-0.2)}$ g/kWh when n is 130 or more but less than 2,000 rpm; and

.3 9.8 g/kWh when n is 2,000 rpm or more.

7.5 Certification of an approved method shall be in accordance with chapter 7 of the revised NO_x Technical Code 2008 and shall include verification:

.1 by the designer of the base marine diesel engine to which the approved method applies that the calculated effect of the approved method will not decrease engine rating by more than 1.0%, increase fuel consumption by more than 2.0% as measured according to the appropriate test cycle set forth in the revised NO_x Technical Code 2008, or adversely affect engine durability or reliability; and

.2 that the cost of the approved method is not excessive, which is determined by a comparison of the amount of NO_x reduced by the approved method to achieve the standard set forth in paragraph 7.4 of this regulation and the cost of purchasing and installing such approved method.*

Certification

8 The revised NO_x Technical Code 2008 shall be applied in the certification, testing and measurement procedures for the standards set forth in this regulation.

9 The procedures for determining NO_x emissions set out in the revised NO_x Technical Code 2008 are intended to be representative of the normal operation of the engine. Defeat devices and irrational emission control strategies undermine this intention and shall not be allowed. This regulation shall not prevent the use of auxiliary control devices that are used to protect the engine and/or its ancillary equipment against operating conditions that could result in damage or failure or that are used to facilitate the starting of the engine.

* The cost of an approved method shall not exceed 375 Special Drawing Rights/metric tonne NO_x calculated in accordance with the cost-effectiveness (Ce) formula below:

$$Ce = \frac{\text{Cost of approved method} \cdot 10^6}{\text{Power (kW)} \cdot 0.768 \cdot 6{,}000 \text{ (hours/year)} \cdot 5 \text{ (years)} \cdot \Delta NO_x \text{ (g/kWh)}}$$

Refer to Definitions for the cost-effectiveness formula in regulation 13.7.5 of the revised MARPOL Annex VI (MEPC.1/Circ.678).

Regulation 14
Sulphur oxides (SO_x) and particulate matter

General requirements

1 The sulphur content of any fuel oil used on board ships shall not exceed the following limits:

.1 4.50% m/m prior to 1 January 2012;

.2 3.50% m/m on and after 1 January 2012; and

.3 0.50% m/m on and after 1 January 2020.

2 The worldwide average sulphur content of residual fuel oil supplied for use on board ships shall be monitored taking into account guidelines developed by the Organization.*

Requirements within emission control areas

3 For the purpose of this regulation, emission control areas shall include:

.1 the Baltic Sea area as defined in regulation 1.11.2 of Annex I and the North Sea as defined in regulation 1.14.6 of Annex V;

.2 the North American area as described by the coordinates provided in appendix VII to this Annex;

.3 the United States Caribbean Sea area as described by the coordinates provided in appendix VII to this Annex; and

.4 any other sea area, including any port area, designated by the Organization in accordance with the criteria and procedures set forth in appendix III to this Annex.

4 While ships are operating within an emission control area, the sulphur content of fuel oil used on board ships shall not exceed the following limits:

.1 1.50% m/m prior to 1 July 2010;

.2 1.00% m/m on and after 1 July 2010; and

.3 0.10% m/m on and after 1 January 2015.

.4 Prior to 1 January 2020, the sulphur content of fuel oil referred to in paragraph 4 of this regulation shall not apply to ships operating in the North American area or the United States Caribbean Sea area defined in paragraph 3, built on or before 1 August 2011 that are powered by propulsion boilers that were not originally designed for continued operation on marine distillate fuel or natural gas.

5 The sulphur content of fuel oil referred to in paragraph 1 and paragraph 4 of this regulation shall be documented by its supplier as required by regulation 18 of this Annex.

6 Those ships using separate fuel oils to comply with paragraph 4 of this regulation and entering or leaving an emission control area set forth in paragraph 3 of this regulation shall carry a written procedure showing how the fuel oil changeover is to be done, allowing sufficient time for the fuel oil service system to be fully flushed of all fuel oils exceeding the applicable sulphur content specified in paragraph 4 of this regulation prior to entry into an emission control area. The volume of low sulphur fuel oils in each tank as well as the date, time and position of the ship when any fuel oil changeover operation is completed prior to the entry into an emission control area or commenced after exit from such an area shall be recorded in such logbook as prescribed by the Administration.

7 During the first 12 months immediately following entry into force of an amendment designating a specific emission control area under paragraph 3 of this regulation, ships operating in that emission control area are exempt from the requirements in paragraphs 4 and 6 of this regulation and from the requirements of paragraph 5 of this regulation insofar as they relate to paragraph 4 of this regulation.

---

* Refer to 2010 Guidelines for monitoring the worldwide average sulphur content of fuel oils supplied for use on board ships (resolution MEPC.192(61), as amended by resolution MEPC.273(69)).

Review provision*

8 A review of the standard set forth in paragraph 1.3 of this regulation shall be completed by 2018 to determine the availability of fuel oil to comply with the fuel oil standard set forth in that paragraph and shall take into account the following elements:

.1 the global market supply and demand for fuel oil to comply with paragraph 1.3 of this regulation that exist at the time that the review is conducted;

.2 an analysis of the trends in fuel oil markets; and

.3 any other relevant issue.

9 The Organization shall establish a group of experts, comprising representatives with the appropriate expertise in the fuel oil market and appropriate maritime, environmental, scientific and legal expertise, to conduct the review referred to in paragraph 8 of this regulation. The group of experts shall develop the appropriate information to inform the decision to be taken by the Parties.

10 The Parties, based on the information developed by the group of experts, may decide whether it is possible for ships to comply with the date in paragraph 1.3 of this regulation. If a decision is taken that it is not possible for ships to comply, then the standard in that paragraph shall become effective on 1 January 2025.

Regulation 15

Volatile organic compounds (VOCs)

1 If the emissions of VOCs from a tanker are to be regulated in a port or ports or a terminal or terminals under the jurisdiction of a Party, they shall be regulated in accordance with the provisions of this regulation.

2 A Party regulating tankers for VOC emissions shall submit a notification to the Organization.† This notification shall include information on the size of tankers to be controlled, the cargoes requiring vapour emission control systems and the effective date of such control. The notification shall be submitted at least six months before the effective date.

3 A Party that designates ports or terminals at which VOC emissions from tankers are to be regulated shall ensure that vapour emission control systems, approved by that Party taking into account the safety standards for such systems developed by the Organization,‡ are provided in any designated port and terminal and are operated safely and in a manner so as to avoid undue delay to a ship.

4 The Organization shall circulate a list of the ports and terminals designated by Parties to other Parties and Member States of the Organization for their information.

5 A tanker to which paragraph 1 of this regulation applies shall be provided with a vapour emission collection system approved by the Administration taking into account the safety standards for such systems developed by the Organization,† and shall use this system during the loading of relevant cargoes. A port or terminal that has installed vapour emission control systems in accordance with this regulation may accept tankers that are not fitted with vapour collection systems for a period of three years after the effective date identified in paragraph 2 of this regulation.

* Refer to Effective date of implementation of the fuel oil standard in regulation 14.1.3 of MARPOL Annex VI (resolution MEPC.280(70)).

† Refer to Notification to the Organization on ports or terminals where volatile organic compounds (VOCs) emissions are to be regulated (MEPC.1/Circ.509).

‡ Refer to Standards for vapour emission control systems (MSC/Circ.585).

6 A tanker carrying crude oil shall have on board and implement a VOC management plan approved by the Administration.* Such a plan shall be prepared taking into account the guidelines developed by the Organization. The plan shall be specific to each ship and shall at least:

.1 provide written procedures for minimizing VOC emissions during the loading, sea passage and discharge of cargo;

.2 give consideration to the additional VOC generated by crude oil washing;

.3 identify a person responsible for implementing the plan; and

.4 for ships on international voyages, be written in the working language of the master and officers and, if the working language of the master and officers is not English, French or Spanish, include a translation into one of these languages.

7 This regulation shall also apply to gas carriers only if the types of loading and containment systems allow safe retention of non-methane VOCs on board or their safe return ashore.†

Regulation 16
Shipboard incineration

1 Except as provided in paragraph 4 of this regulation, shipboard incineration shall be allowed only in a shipboard incinerator.

2 Shipboard incineration of the following substances shall be prohibited:

.1 residues of cargoes subject to Annex I, II or III or related contaminated packing materials;

.2 polychlorinated biphenyls (PCBs);

.3 garbage, as defined by Annex V, containing more than traces of heavy metals;

.4 refined petroleum products containing halogen compounds;

.5 sewage sludge and sludge oil either of which is not generated on board the ship; and

.6 exhaust gas cleaning system residues.

3 Shipboard incineration of polyvinyl chlorides (PVCs) shall be prohibited, except in shipboard incinerators for which IMO Type Approval Certificates‡ have been issued.

4 Shipboard incineration of sewage sludge and sludge oil generated during normal operation of a ship may also take place in the main or auxiliary power plant or boilers, but in those cases, shall not take place inside ports, harbours and estuaries.

5 Nothing in this regulation neither:

.1 affects the prohibition in, or other requirements of, the Convention on the Prevention of Marine Pollution by Dumping of Wastes and Other Matter, 1972, as amended, and the 1996 Protocol thereto, nor

.2 precludes the development, installation and operation of alternative design shipboard thermal waste treatment devices that meet or exceed the requirements of this regulation.

---

* Refer to Guidelines for the development of a VOC management plan (resolution MEPC.185(59)). Refer also to Technical information on systems and operation to assist development of VOC management plans (MEPC.1/Circ.680), and Technical information on a vapour pressure control system to facilitate the development and update of VOC management plans (MEPC.1/Circ.719).

† Refer to the International Code for the construction and equipment of ships carrying liquefied gases in bulk (resolution MSC.370(93)).

‡ Type Approval Certificates issued in accordance with Revised guidelines for the implementation of Annex V of MARPOL 73/78 (resolution MEPC.59(33), as amended by resolution MEPC.92(45)), or Standard specification for shipboard incinerators (resolution MEPC.76(40), as amended by resolution MEPC.93(45)), or 2014 Standard specification for shipboard incinerators (resolution MEPC 244(66)).

6.1 Except as provided in paragraph 6.2 of this regulation, each incinerator on a ship constructed on or after 1 January 2000 or incinerator that is installed on board a ship on or after 1 January 2000 shall meet the requirements contained in appendix IV to this Annex. Each incinerator subject to this paragraph shall be approved by the Administration taking into account the standard specification for shipboard incinerators developed by the Organization;* or

6.2 The Administration may allow exclusion from the application of paragraph 6.1 of this regulation to any incinerator installed on board a ship before 19 May 2005, provided that the ship is solely engaged in voyages within waters subject to the sovereignty or jurisdiction of the State the flag of which the ship is entitled to fly.

7 Incinerators installed in accordance with the requirements of paragraph 6.1 of this regulation shall be provided with a manufacturer's operating manual, which is to be retained with the unit and which shall specify how to operate the incinerator within the limits described in paragraph 2 of appendix IV of this Annex.

8 Personnel responsible for the operation of an incinerator installed in accordance with the requirements of paragraph 6.1 of this regulation shall be trained to implement the guidance provided in the manufacturer's operating manual as required by paragraph 7 of this regulation.

9 For incinerators installed in accordance with the requirements of paragraph 6.1 of this regulation the combustion chamber gas outlet temperature shall be monitored at all times the unit is in operation. Where that incinerator is of the continuous-feed type, waste shall not be fed into the unit when the combustion chamber gas outlet temperature is below 850°C. Where that incinerator is of the batch-loaded type, the unit shall be designed so that the combustion chamber gas outlet temperature shall reach 600°C within five minutes after start-up and will thereafter stabilize at a temperature not less than 850°C.

Regulation 17
Reception facilities

1 Each Party undertakes to ensure the provision of facilities adequate to meet the:

.1 needs of ships using its repair ports for the reception of ozone-depleting substances and equipment containing such substances when removed from ships;

.2 needs of ships using its ports, terminals or repair ports for the reception of exhaust gas cleaning residues from an exhaust gas cleaning system;

without causing undue delay to ships, and

.3 needs in ship-breaking facilities for the reception of ozone-depleting substances and equipment containing such substances when removed from ships.

2 Small Island Developing States† may satisfy the requirements in paragraph 1 of this regulation through regional arrangements when, because of those States' unique circumstances, such arrangements are the only practical means to satisfy these requirements. Parties participating in a regional arrangement shall develop a Regional Reception Facilities Plan, taking into account the guidelines developed by the Organization.‡

The Government of each Party participating in the arrangement shall consult with the Organization for circulation to the Parties of the present Convention:

.1 how the Regional Reception Facilities Plan takes into account the Guidelines;

.2 particulars of the identified Regional Ships Waste Reception Centres; and

.3 particulars of those ports with only limited facilities.

---

* Refer to 2014 Standard specification for shipboard incinerators (resolution MEPC.244(66)), or Standard specification for shipboard incinerators (resolution MEPC.76(40), as amended by resolution MEPC.93(45)), and Type approval of shipboard incinerators (MEPC.1/Circ.793).

† Refer to 2012 Guidelines for the development of a regional reception facilities plan (resolution MEPC.221(63)).

‡ Refer to 2011 Guidelines for reception facilities under MARPOL Annex VI (resolution MEPC.199(62)).

3 If a particular port or terminal of a Party is, taking into account the guidelines to be developed by the Organization, remotely located from, or lacking in, the industrial infrastructure necessary to manage and process those substances referred to in paragraph 1 of this regulation and therefore cannot accept such substances, then the Party shall inform the Organization of any such port or terminal so that this information may be circulated to all Parties and Member States of the Organization for their information and any appropriate action. Each Party that has provided the Organization with such information shall also notify the Organization of its ports and terminals where reception facilities are available to manage and process such substances.

4 Each Party shall notify the Organization for transmission to the Members of the Organization of all cases where the facilities provided under this regulation are unavailable or alleged to be inadequate.

Regulation 18
Fuel oil availability and quality

Fuel oil availability

1 Each Party shall take all reasonable steps to promote the availability of fuel oils that comply with this Annex and inform the Organization of the availability of compliant fuel oils in its ports and terminals.

2.1 If a ship is found by a Party not to be in compliance with the standards for compliant fuel oils set forth in this Annex, the competent authority of the Party is entitled to require the ship to:

.1 present a record of the actions taken to attempt to achieve compliance; and

.2 provide evidence that it attempted to purchase compliant fuel oil in accordance with its voyage plan and, if it was not made available where planned, that attempts were made to locate alternative sources for such fuel oil and that despite best efforts to obtain compliant fuel oil, no such fuel oil was made available for purchase.

2.2 The ship should not be required to deviate from its intended voyage or to delay unduly the voyage in order to achieve compliance.

2.3 If a ship provides the information set forth in paragraph 2.1 of this regulation, a Party shall take into account all relevant circumstances and the evidence presented to determine the appropriate action to take, including not taking control measures.

2.4 A ship shall notify its Administration and the competent authority of the relevant port of destination when it cannot purchase compliant fuel oil.

2.5 A Party shall notify the Organization when a ship has presented evidence of the non-availability of compliant fuel oil.

Fuel oil quality

3 Fuel oil for combustion purposes delivered to and used on board ships to which this Annex applies shall meet the following requirements:

.1 except as provided in paragraph 3.2 of this regulation:

.1.1 the fuel oil shall be blends of hydrocarbons derived from petroleum refining. This shall not preclude the incorporation of small amounts of additives intended to improve some aspects of performance;

.1.2 the fuel oil shall be free from inorganic acid; and

.1.3 the fuel oil shall not include any added substance or chemical waste that:

.1.3.1 jeopardizes the safety of ships or adversely affects the performance of the machinery, or

.1.3.2 is harmful to personnel, or

.1.3.3 contributes overall to additional air pollution.

.2 fuel oil for combustion purposes derived by methods other than petroleum refining shall not:

.2.1 exceed the applicable sulphur content set forth in regulation 14 of this Annex;

.2.2 cause an engine to exceed the applicable NO_x emission limit set forth in paragraphs 3, 4, 5.1.1 and 7.4 of regulation 13;

.2.3 contain inorganic acid; or

.2.4.1 jeopardize the safety of ships or adversely affect the performance of the machinery, or

.2.4.2 be harmful to personnel, or

.2.4.3 contribute overall to additional air pollution.

4 This regulation does not apply to coal in its solid form or nuclear fuels. Paragraphs 5, 6, 7.1, 7.2, 8.1, 8.2, 9.2, 9.3, and 9.4 of this regulation do not apply to gas fuels such as liquefied natural gas, compressed natural gas or liquefied petroleum gas. The sulphur content of gas fuels delivered to a ship specifically for combustion purposes on board that ship shall be documented by the supplier.

5 For each ship subject to regulations 5 and 6 of this Annex, details of fuel oil for combustion purposes delivered to and used on board shall be recorded by means of a bunker delivery note that shall contain at least the information specified in appendix V to this Annex.

6 The bunker delivery note shall be kept on board the ship in such a place as to be readily available for inspection at all reasonable times. It shall be retained for a period of three years after the fuel oil has been delivered on board.

7.1 The competent authority of a Party may inspect the bunker delivery notes on board any ship to which this Annex applies while the ship is in its port or offshore terminal, may make a copy of each delivery note, and may require the master or person in charge of the ship to certify that each copy is a true copy of such bunker delivery note. The competent authority may also verify the contents of each note through consultations with the port where the note was issued.

7.2 The inspection of the bunker delivery notes and the taking of certified copies by the competent authority under paragraph 7.1 shall be performed as expeditiously as possible without causing the ship to be unduly delayed.

8.1 The bunker delivery note shall be accompanied by a representative sample of the fuel oil delivered taking into account guidelines developed by the Organization.* The sample is to be sealed and signed by the supplier's representative and the master or officer in charge of the bunker operation on completion of bunkering operations and retained under the ship's control until the fuel oil is substantially consumed, but in any case for a period of not less than 12 months from the time of delivery.

8.2 If an Administration requires the representative sample to be analysed, it shall be done in accordance with the verification procedure set forth in appendix VI to determine whether the fuel oil meets the requirements of this Annex.

9 Parties undertake to ensure that appropriate authorities designated by them:

.1 maintain a register of local suppliers of fuel oil;

.2 require local suppliers to provide the bunker delivery note and sample as required by this regulation, certified by the fuel oil supplier that the fuel oil meets the requirements of regulations 14 and 18 of this Annex;

.3 require local suppliers to retain a copy of the bunker delivery note for at least three years for inspection and verification by the port State as necessary;

.4 take action as appropriate against fuel oil suppliers that have been found to deliver fuel oil that does not comply with that stated on the bunker delivery note;

* Refer to 2009 Guidelines for the sampling of fuel oil for determination of compliance with the revised MARPOL Annex VI (resolution MEPC.182(59)).

.5 inform the Administration of any ship receiving fuel oil found to be non-compliant with the requirements of regulation 14 or 18 of this Annex; and

.6 inform the Organization for transmission to Parties and Member States of the Organization of all cases where fuel oil suppliers have failed to meet the requirements specified in regulations 14 or 18 of this Annex.

10 In connection with port State inspections carried out by Parties, the Parties further undertake to:

.1 inform the Party or non-Party under whose jurisdiction a bunker delivery note was issued of cases of delivery of non-compliant fuel oil, giving all relevant information; and

.2 ensure that remedial action as appropriate is taken to bring non-compliant fuel oil discovered into compliance.

11 For every ship of 400 gross tonnage and above on scheduled services with frequent and regular port calls, an Administration may decide after application and consultation with affected States that compliance with paragraph 6 of this regulation may be documented in an alternative manner that gives similar certainty of compliance with regulations 14 and 18 of this Annex.

Chapter 4 – Regulations on energy efficiency for ships

Regulation 19
Application

1 This chapter shall apply to all ships of 400 gross tonnage and above.

2 The provisions of this chapter shall not apply to:

.1 ships solely engaged in voyages within waters subject to the sovereignty or jurisdiction of the State the flag of which the ship is entitled to fly. However, each Party should ensure, by the adoption of appropriate measures, that such ships are constructed and act in a manner consistent with the requirements of chapter 4 of this Annex, so far as is reasonable and practicable.

.2 ships not propelled by mechanical means, and platforms including FPSOs and FSUs and drilling rigs, regardless of their propulsion.

3 Regulations 20 and 21 of this Annex shall not apply to ships which have non-conventional propulsion, except that regulations 20 and 21 shall apply to cruise passenger ships having non-conventional propulsion and LNG carriers having conventional or non-conventional propulsion, delivered on or after 1 September 2019, as defined in paragraph 43 of regulation 2. Regulations 20 and 21 shall not apply to cargo ships having ice-breaking capability.

4 Notwithstanding the provisions of paragraph 1 of this regulation, the Administration may waive the requirement for a ship of 400 gross tonnage and above from complying with regulations 20 and 21 of this Annex.

5 The provision of paragraph 4 of this regulation shall not apply to ships of 400 gross tonnage and above:

.1 for which the building contract is placed on or after 1 January 2017; or

.2 in the absence of a building contract, the keel of which is laid or which is at a similar stage of construction on or after 1 July 2017; or

.3 the delivery of which is on or after 1 July 2019; or

.4 in cases of a major conversion of a new or existing ship, as defined in regulation 2.24 of this Annex, on or after 1 January 2017, and in which regulations 5.4.2 and 5.4.3 of this Annex apply.

6 The Administration of a Party to the present Convention which allows application of paragraph 4, or suspends, withdraws or declines the application of that paragraph, to a ship entitled to fly its flag shall forthwith communicate to the Organization for circulation to the Parties to the present Protocol particulars thereof, for their information.

Regulation 20
Attained Energy Efficiency Design Index (attained EEDI)

1 The attained EEDI shall be calculated for:

.1 each new ship;

.2 each new ship which has undergone a major conversion; and

.3 each new or existing ship which has undergone a major conversion, that is so extensive that the ship is regarded by the Administration as a newly-constructed ship, which falls into one or more of the categories in regulations 2.25 to 2.35, 2.38 and 2.39 of this Annex. The attained EEDI shall be specific to each ship and shall indicate the estimated performance of the ship in terms of energy efficiency, and be accompanied by the EEDI technical file that contains the information necessary for the calculation of the attained EEDI and that shows the process of calculation. The attained EEDI shall be verified, based on the EEDI technical file, either by the Administration or by any organization duly authorized by it.*

2 The attained EEDI shall be calculated taking into account guidelines† developed by the Organization.

Regulation 21
Required EEDI

1 For each:

.1 new ship,

.2 new ship which has undergone a major conversion, and

.3 new or existing ship which has undergone a major conversion that is so extensive that the ship is regarded by the Administration as a newly-constructed ship, which falls into one of the categories in regulations 2.25 to 2.31, 2.33 to 2.35, 2.38 and 2.39 and to which this chapter is applicable, the attained EEDI shall be as follows:

$$\text{Attained EEDI} \leq \text{Required EEDI} = \left(1 - \frac{X}{100}\right) \cdot \text{Reference line value}$$

where X is the reduction factor specified in table 1 for the required EEDI compared to the EEDI reference line.

2 For each new and existing ship that has undergone a major conversion which is so extensive that the ship is regarded by the Administration as a newly constructed ship, the attained EEDI shall be calculated and meet the requirement of paragraph 21.1 with the reduction factor applicable corresponding to the ship type and size of the converted ship at the date of the contract of the conversion, or in the absence of a contract, the commencement date of the conversion.

---

* Refer to Guidelines for the authorization of organizations acting on behalf of the Administration (resolution A.739(18), as amended by resolution MSC.208(81)), and Specifications on the survey and certification functions of recognized organizations acting on behalf of the Administration (resolution A.789(19), as may be amended).

† Refer to 2014 Guidelines on the method of calculation of the Energy Efficiency Design Index for new ships (resolution MEPC.245(66), as amended by resolutions MEPC.263(68) and MEPC.281(70)).

Table 1 – *Reduction factors (in percentage) for the EEDI relative to the EEDI reference line*

Ship type	Size	Phase 0 1 Jan 2013 – 31 Dec 2014	Phase 1 1 Jan 2015 – 31 Dec 2019	Phase 2 1 Jan 2020 – 31 Dec 2024	Phase 3 1 Jan 2025 and onwards
Bulk carrier	20,000 DWT and above	0	10	20	30
	10,000 – 20,000 DWT	n/a	0–10*	0–20*	0–30*
Gas carrier	10,000 DWT and above	0	10	20	30
	2,000 – 10,000 DWT	n/a	0–10*	0–20*	0–30*
Tanker	20,000 DWT and above	0	10	20	30
	4,000 – 20,000 DWT	n/a	0–10*	0–20*	0–30*
Containership	15,000 DWT and above	0	10	20	30
	10,000 – 15,000 DWT	n/a	0–10*	0–20*	0–30*
General cargo ships	15,000 DWT and above	0	10	15	30
	3,000 – 15,000 DWT	n/a	0–10*	0–15*	0–30*
Refrigerated cargo carrier	5,000 DWT and above	0	10	15	30
	3,000 – 5,000 DWT	n/a	0–10*	0–15*	0–30*
Combination carrier	20,000 DWT and above	0	10	20	30
	4,000 – 20,000 DWT	n/a	0–10*	0–20*	0–30*
LNG carrier***	10,000 DWT and above	n/a	10**	20	30
Ro-ro cargo ship (vehicle carrier)***	10,000 DWT and above	n/a	5**	15	30
Ro-ro cargo ship***	2,000 DWT and above	n/a	5**	20	30
	1,000 – 2,000 DWT	n/a	0–5* **	0–20*	0–30*
Ro-ro passenger ship***	1,000 DWT and above	n/a	5**	20	30
	250 – 1,000 DWT	n/a	0–5* **	0–20*	0–30*
Cruise passenger ship*** having non-conventional propulsion	85,000 GT and above	n/a	5**	20	30
	25,000 – 85,000 GT	n/a	0–5* **	0–20*	0–30*

* Reduction factor to be linearly interpolated between the two values dependent upon vessel size. The lower value of the reduction factor is to be applied to the smaller ship size.

** Phase 1 commences for those ships on 1 September 2015.

*** Reduction factor applies to those ships delivered on or after 1 September 2019, as defined in paragraph 43 of regulation 2.

Note: n/a means that no required EEDI applies.

3 The reference line values shall be calculated as follows:

$$\text{Reference line value} = a \cdot b^{-c}$$

where a, b and c are the parameters given in table 2.

Table 2 – *Parameters for determination of reference values for the different ship types*

Ship type defined in regulation 2	a	b	c
2.25 Bulk carrier	961.79	DWT of the ship	0.477
2.26 Gas carrier	1,120.00	DWT of the ship	0.456
2.27 Tanker	1,218.80	DWT of the ship	0.488
2.28 Containership	174.22	DWT of the ship	0.201
2.29 General cargo ship	107.48	DWT of the ship	0.216
2.30 Refrigerated cargo carrier	227.01	DWT of the ship	0.244
2.31 Combination carrier	1,219.00	DWT of the ship	0.488
2.33 Ro-ro cargo ship (vehicle carrier)	$(DWT/GT)^{-0.7} \cdot 780.36$ where DWT/GT < 0.3 1,812.63 where DWT/GT ≥ 0.3	DWT of the ship	0.471
2.34 Ro-ro cargo ship	1,405.15	DWT of the ship	0.498
2.35 Ro-ro passenger ship	752.16	DWT of the ship	0.381
2.38 LNG carrier	2,253.7	DWT of the ship	0.474
2.39 Cruise passenger ship having non-conventional propulsion	170.84	GT of the ship	0.214

4 If the design of a ship allows it to fall into more than one of the above ship type definitions specified in table 2, the required EEDI for the ship shall be the most stringent (the lowest) required EEDI.

5 For each ship to which this regulation applies, the installed propulsion power shall not be less than the propulsion power needed to maintain the manoeuvrability of the ship under adverse conditions as defined in the guidelines to be developed by the Organization.*

6 At the beginning of phase 1 and at the midpoint of phase 2, the Organization shall review the status of technological developments and, if proven necessary, amend the time periods, the EEDI reference line parameters for relevant ship types and reduction rates set out in this regulation.

Regulation 22

Ship Energy Efficiency Management Plan (SEEMP)

1 Each ship shall keep on board a ship specific Ship Energy Efficiency Management Plan (SEEMP). This may form part of the ship's Safety Management System (SMS).

2 On or before 31 December 2018, in the case of a ship of 5,000 gross tonnage and above, the SEEMP shall include a description of the methodology that will be used to collect the data required by regulation 22A.1 of this Annex and the processes that will be used to report the data to the ship's Administration.

3 The SEEMP shall be developed taking into account guidelines adopted by the Organization.†

* Refer to 2013 Interim Guidelines for determining minimum propulsion power to maintain the manoeuvrability of ships in adverse conditions (resolution MEPC.232(65), as amended by resolutions MEPC.255(67) and MEPC.262(68)).

† Refer to 2016 Guidelines for the development of a Ship Energy Efficiency Management Plan (SEEMP) (resolution MEPC.282(70)).

Regulation 22A
Collection and reporting of ship fuel oil consumption data

1 From calendar year 2019, each ship of 5,000 gross tonnage and above shall collect the data specified in appendix IX to this Annex, for that and each subsequent calendar year or portion thereof, as appropriate according to the methodology included in the SEEMP.

2 Except as provided for in paragraphs 4, 5 and 6 of this regulation, at the end of each calendar year, the ship shall aggregate the data collected in that calendar year or portion thereof, as appropriate.

3 Except as provided for in paragraphs 4, 5 and 6 of this regulation, within three months after the end of each calendar year, the ship shall report to its Administration or any organization duly authorized by it,* the aggregated value for each datum specified in appendix IX to this Annex, via electronic communication and using a standardized format to be developed by the Organization.†

4 In the event of the transfer of a ship from one Administration to another, the ship shall on the day of completion of the transfer or as close as practical thereto report to the losing Administration or any organization duly authorized by it,* the aggregated data for the period of the calendar year corresponding to that Administration, as specified in appendix IX to this Annex and, upon prior request of that Administration, the disaggregated data.

5 In the event of a change from one Company to another, the ship shall on the day of completion of the change or as close as practical thereto report to its Administration or any organization duly authorized by it,* the aggregated data for the portion of the calendar year corresponding to the Company, as specified in appendix IX to this Annex and, upon request of its Administration, the disaggregated data.

6 In the event of change from one Administration to another and from one Company to another concurrently, paragraph 4 of this regulation shall apply.

7 The data shall be verified according to procedures established by the Administration, taking into account guidelines to be developed by the Organization.

8 Except as provided for in paragraphs 4, 5 and 6 of this regulation, the disaggregated data that underlies the reported data noted in appendix IX to this Annex for the previous calendar year shall be readily accessible for a period of not less than 12 months from the end of that calendar year and be made available to the Administration upon request.

9 The Administration shall ensure that the reported data noted in appendix IX to this Annex by its registered ships of 5,000 gross tonnage and above are transferred to the IMO Ship Fuel Oil Consumption Database via electronic communication and using a standardized format to be developed by the Organization not later than one month after issuing the Statements of Compliance of these ships.

10 On the basis of the reported data submitted to the IMO Ship Fuel Oil Consumption Database, the Secretary-General of the Organization shall produce an annual report to the Marine Environment Protection Committee summarizing the data collected, the status of missing data, and such other relevant information as may be requested by the Committee.

11 The Secretary-General of the Organization shall maintain an anonymized database such that identification of a specific ship will not be possible. Parties shall have access to the anonymized data strictly for their analysis and consideration.

12 The IMO Ship Fuel Oil Consumption Database shall be undertaken and managed by the Secretary-General of the Organization, pursuant to guidelines to be developed by the Organization.

* Refer to Guidelines for the authorization of organizations acting on behalf of the Administration (resolution A.739(18)), and Specifications on the survey and certification functions of recognized organizations acting on behalf of the Administration (resolution A.789(19)).

† Refer to 2016 Guidelines for the development of a Ship Energy Efficiency Management Plan (SEEMP Guidelines) (resolution MEPC.282(70)).

Regulation 23
*Promotion of technical cooperation and transfer of technology relating to the improvement of energy efficiency of ships**

1 Administrations shall, in cooperation with the Organization and other international bodies, promote and provide, as appropriate, support directly or through the Organization to States, especially developing States, that request technical assistance.

2 The Administration of a Party shall cooperate actively with other Parties, subject to its national laws, regulations and policies, to promote the development and transfer of technology and exchange of information to States which request technical assistance, particularly developing States, in respect of the implementation of measures to fulfil the requirements of chapter 4 of this Annex, in particular regulations 19.4 to 19.6.

Chapter 5 – Verification of compliance with the provisions of this Annex

Regulation 24
Application

Parties shall use the provisions of the Code for Implementation in the execution of their obligations and responsibilities contained in this Annex.

Regulation 25
Verification of compliance

1 Every Party shall be subject to periodic audits by the Organization in accordance with the audit standard to verify compliance with and implementation of this Annex.

2 The Secretary-General of the Organization shall have responsibility for administering the Audit Scheme, based on the guidelines developed by the Organization.†

3 Every Party shall have responsibility for facilitating the conduct of the audit and implementation of a programme of actions to address the findings, based on the guidelines developed by the Organization.*

4 Audit of all Parties shall be:

.1 based on an overall schedule developed by the Secretary-General of the Organization, taking into account the guidelines developed by the Organization;* and

.2 conducted at periodic intervals, taking into account the guidelines developed by the Organization.*

---

* Refer to Promotion of technical cooperation and transfer of technology relating to the improvement of energy efficiency of ships (resolution MEPC.229(65)), and Model agreement between governments on technological cooperation for the implementation of the regulations in chapter 4 of MARPOL Annex VI (MEPC.1/Circ.861).

† Refer to Framework and Procedures for the IMO Member State Audit Scheme (resolution A.1067(28)).

Appendices to Annex VI

Appendix I
Form of International Air Pollution Prevention (IAPP) Certificate (regulation 8)

INTERNATIONAL AIR POLLUTION PREVENTION CERTIFICATE

Issued under the provisions of the Protocol of 1997, as amended, to amend the International Convention for the Prevention of Pollution from Ships, 1973, as modified by the Protocol of 1978 related thereto (hereinafter referred to as "the Convention") under the authority of the Government of:

. . . .

(full designation of the country)

by . . . .

(full designation of the competent person or organization authorized under the provisions of the Convention)

Particulars of ship*

Name of ship . . . .

Distinctive number or letters . . . .

IMO Number† . . . .

Port of registry . . . .

Gross tonnage . . . .

THIS IS TO CERTIFY:

1 That the ship has been surveyed in accordance with regulation 5 of Annex VI of the Convention; and

2 That the survey shows that the equipment, systems, fittings, arrangements and materials fully comply with the applicable requirements of Annex VI of the Convention.

This Certificate is valid until (dd/mm/yyyy)‡ . . . .
subject to surveys in accordance with regulation 5 of Annex VI of the Convention.

Completion date of the survey on which this Certificate is based (dd/mm/yyyy) . . . .

Issued at. . . . .

(place of issue of Certificate)

Date (dd/mm/yyyy) . . . .

(date of issue) *(signature of duly authorized official issuing the Certificate)*

(seal or stamp of the authority, as appropriate)

* Alternatively, the particulars of the ship may be placed horizontally in boxes.

† In accordance with the IMO ship identification number scheme (resolution A.1078(28)).

‡ Insert the date of expiry as specified by the Administration in accordance with regulation 9.1 of Annex VI of the Convention. The day and the month of this date correspond to the anniversary date as defined in regulation 2.3 of Annex VI of the Convention, unless amended in accordance with regulation 9.8 of Annex VI of the Convention.

ENDORSEMENT FOR ANNUAL AND INTERMEDIATE SURVEYS

THIS IS TO CERTIFY that, at a survey required by regulation 5 of Annex VI of the Convention, the ship was found to comply with the relevant provisions of that Annex:

Annual survey

Signed. . . . .
(signature of duly authorized official)

Place . . . .

Date (dd/mm/yyyy) . . . .

(seal or stamp of the authority, as appropriate)

Annual/Intermediate* survey

Signed. . . . .
(signature of duly authorized official)

Place . . . .

Date (dd/mm/yyyy) . . . .

(seal or stamp of the authority, as appropriate)

Annual/Intermediate* survey

Signed. . . . .
(signature of duly authorized official)

Place . . . .

Date (dd/mm/yyyy) . . . .

(seal or stamp of the authority, as appropriate)

Annual survey

Signed. . . . .
(signature of duly authorized official)

Place . . . .

Date (dd/mm/yyyy) . . . .

(seal or stamp of the authority, as appropriate)

ANNUAL/INTERMEDIATE SURVEY IN ACCORDANCE WITH REGULATION 9.8.3

THIS IS TO CERTIFY that, at an annual/intermediate* survey in accordance with regulation 9.8.3 of Annex VI of the Convention, the ship was found to comply with the relevant provisions of that Annex:

Signed. . . . .
(signature of duly authorized official)

Place . . . .

Date (dd/mm/yyyy) . . . .

(seal or stamp of the authority, as appropriate)

ENDORSEMENT TO EXTEND THE CERTIFICATE IF VALID FOR LESS THAN 5 YEARS WHERE REGULATION 9.3 APPLIES

The ship complies with the relevant provisions of the Annex, and this Certificate shall, in accordance with regulation 9.3 of Annex VI of the Convention, be accepted as valid until (dd/mm/yyyy) . . . .

Signed. . . . .
(signature of duly authorized official)

Place . . . .

Date (dd/mm/yyyy) . . . .

(seal or stamp of the authority, as appropriate)

* Delete as appropriate.

ENDORSEMENT WHERE THE RENEWAL SURVEY HAS BEEN COMPLETED AND REGULATION 9.4 APPLIES

The ship complies with the relevant provisions of the Annex, and this Certificate shall, in accordance with regulation 9.4 of Annex VI of the Convention, be accepted as valid until (dd/mm/yyyy) . . . .

Signed . . . .
(signature of duly authorized official)

Place . . . .

Date (dd/mm/yyyy) . . . .

(seal or stamp of the authority, as appropriate)

ENDORSEMENT TO EXTEND THE VALIDITY OF THE CERTIFICATE UNTIL REACHING THE PORT OF SURVEY OR FOR A PERIOD OF GRACE WHERE REGULATION 9.5 OR 9.6 APPLIES

This Certificate shall, in accordance with regulation 9.5 or 9.6* of Annex VI of the Convention, be accepted as valid until (dd/mm/yyyy) . . . .

Signed . . . .
(signature of duly authorized official)

Place . . . .

Date (dd/mm/yyyy) . . . .

(seal or stamp of the authority, as appropriate)

ENDORSEMENT FOR ADVANCEMENT OF ANNIVERSARY DATE WHERE REGULATION 9.8 APPLIES

In accordance with regulation 9.8 of Annex VI of the Convention, the new anniversary date is (dd/mm/yyyy) . . . .

Signed . . . .
(signature of duly authorized official)

Place . . . .

Date (dd/mm/yyyy) . . . .

(seal or stamp of the authority, as appropriate)

In accordance with regulation 9.8 of Annex VI of the Convention, the new anniversary date is (dd/mm/yyyy) . . . .

Signed . . . .
(signature of duly authorized official)

Place . . . .

Date (dd/mm/yyyy) . . . .

(seal or stamp of the authority, as appropriate)

* Delete as appropriate.

SUPPLEMENT TO
INTERNATIONAL AIR POLLUTION PREVENTION CERTIFICATE
(IAPP CERTIFICATE)

RECORD OF CONSTRUCTION AND EQUIPMENT

Notes:

1 This Record shall be permanently attached to the IAPP Certificate. The IAPP Certificate shall be available on board the ship at all times.

2 The Record shall be at least in English, French or Spanish. If an official language of the issuing country is also used, this shall prevail in case of a dispute or discrepancy.

3 Entries in boxes shall be made by inserting either: a cross (x) for the answers "yes" and "applicable"; or a dash (–) for the answers "no" and "not applicable", as appropriate.

4 Unless otherwise stated, regulations mentioned in this Record refer to regulations of Annex VI of the Convention and resolutions or circulars refer to those adopted by the International Maritime Organization.

1 Particulars of ship

1.1 Name of ship . . . .

1.2 IMO Number. . . . .

1.3 Date on which keel was laid or ship was at a similar stage of construction (dd/mm/yyyy) . . . .

1.4 Length (*L*)* metres . . . .

2 Control of emissions from ships

2.1 *Ozone-depleting substances* (regulation 12)

2.1.1 The following fire-extinguishing systems, other systems and equipment containing ozone-depleting substances, other than hydrochlorofluorocarbons (HCFCs), installed before 19 May 2005 may continue in service:

System or equipment	Location on board	Substance

2.1.2 The following systems containing HCFCs installed before 1 January 2020 may continue in service:

System or equipment	Location on board	Substance

* Completed only in respect of ships constructed on or after 1 January 2016 that are specially designed, and used solely for recreational purposes and to which, in accordance with regulation 13.5.2.1 or regulation 13.5.2.3, the NOx emission limit as given by regulation 13.5.1.1 will not apply.

2.2 *Nitrogen oxides (NO_x)* (regulation 13)

2.2.1 The following marine diesel engines installed on this ship are in accordance with the requirements of regulation 13, as indicated:

Applicable regulation of MARPOL Annex VI (NTC = NO_x Technical Code 2008) (AM = Approved Method)			Engine #1	Engine #2	Engine #3	Engine #4	Engine #5	Engine #6
1	Manufacturer and model							
2	Serial number							
3	Use (applicable application cycle(s) – NTC 3.2)							
4	Rated power (kW) (NTC 1.3.11)							
5	Rated speed (rpm) (NTC 1.3.12)							
6	Identical engine installed ≥ 1/1/2000 exempted by 13.1.1.2		☐	☐	☐	☐	☐	☐
7	Identical engine installation date (dd/mm/yyyy) as per 13.1.1.2							
8a	Major conversion (dd/mm/yyyy)	13.2.1.1 & 13.2.2						
8b		13.2.1.2 & 13.2.3						
8c		13.2.1.3 & 13.2.3						
9a	Tier I	13.3	☐	☐	☐	☐	☐	☐
9b		13.2.2	☐	☐	☐	☐	☐	☐
9c		13.2.3.1	☐	☐	☐	☐	☐	☐
9d		13.2.3.2	☐	☐	☐	☐	☐	☐
9e		13.7.1.2	☐	☐	☐	☐	☐	☐
10a	Tier II	13.4	☐	☐	☐	☐	☐	☐
10b		13.2.2	☐	☐	☐	☐	☐	☐
10c		13.2.2 (Tier III not possible)	☐	☐	☐	☐	☐	☐
10d		13.2.3.2	☐	☐	☐	☐	☐	☐
10e		13.5.2 (Exemptions)	☐	☐	☐	☐	☐	☐
10f		13.7.1.2	☐	☐	☐	☐	☐	☐
11a	Tier III (ECA-NO_x only)	13.5.1.1	☐	☐	☐	☐	☐	☐
11b		13.2.2	☐	☐	☐	☐	☐	☐
11c		13.2.3.2	☐	☐	☐	☐	☐	☐
11d		13.7.1.2	☐	☐	☐	☐	☐	☐
12	AM[*]	installed	☐	☐	☐	☐	☐	☐
13		not commercially available at this survey	☐	☐	☐	☐	☐	☐
14		not applicable	☐	☐	☐	☐	☐	☐

[*] Refer to 2014 Guidelines on the approved method process (resolution MEPC.243(66)).

2.3 *Sulphur oxides (SO_x) and particulate matter* (regulation 14)

2.3.1 When the ship operates outside of an emission control area specified in regulation 14.3, the ship uses:

.1 fuel oil with a sulphur content as documented by bunker delivery notes that does not exceed the limit value of:

- 4.50% m/m (not applicable on or after 1 January 2012); or. . . . . ☐
- 3.50% m/m (not applicable on or after 1 January 2020); or . . . . ☐
- 0.50% m/m, and/or. . . . . ☐

.2 an equivalent arrangement approved in accordance with regulation 4.1 as listed in 2.6 that is at least as effective in terms of SO_x emission reductions as compared to using a fuel oil with a sulphur content limit value of:

- 4.50% m/m (not applicable on or after 1 January 2012); or. . . . . ☐
- 3.50% m/m (not applicable on or after 1 January 2020); or . . . . ☐
- 0.50% m/m . . . . ☐

2.3.2 When the ship operates inside an emission control area specified in regulation 14.3, the ship uses:

.1 fuel oil with a sulphur content as documented by bunker delivery notes that does not exceed the limit value of:

- 1.00% m/m (not applicable on or after 1 January 2015); or. . . . . ☐
- 0.10% m/m, and/or . . . . ☐

.2 an equivalent arrangement approved in accordance with regulation 4.1 as listed in 2.6 that is at least as effective in terms of SO_x emission reductions as compared to using a fuel oil with a sulphur content limit value of:

- 1.00% m/m (not applicable on or after 1 January 2015); or. . . . . ☐
- 0.10% m/m . . . . ☐

2.4 *Volatile organic compounds (VOCs)* (regulation 15)

2.4.1 The tanker has a vapour collection system installed and approved in accordance with MSC/Circ.585.. . . . . ☐

2.4.2.1 For a tanker carrying crude oil, there is an approved VOC management plan. . . . . ☐

2.4.2.2 VOC management plan approval reference . . . .

2.5 *Shipboard incineration* (regulation 16)

The ship has an incinerator:

.1 installed on or after 1 January 2000 that complies with:

.1 resolution MEPC.76(40), as amended* . . . . ☐

.2 resolution MEPC.244(66) . . . . ☐

.2 installed before 1 January 2000 that complies with:

.1 resolution MEPC.59(33), as amended† . . . . ☐

.2 resolution MEPC.76(40), as amended* . . . . ☐

* As amended by resolution MEPC.93(45).

† As amended by resolution MEPC.92(45).

2.6 *Equivalents* (regulation 4)

The ship has been allowed to use the following fitting, material, appliance or apparatus to be fitted in a ship or other procedures, alternative fuel oils, or compliance methods used as an alternative to that required by this Annex:

System or equipment	Equivalent used	Approval reference

THIS IS TO CERTIFY that this Record is correct in all respects.

Issued at. . . . .
(place of issue of the Record)

Date (dd/mm/yyyy) . . . .
(date of issue) *(signature of duly authorized official issuing the Record)*

(seal or stamp of the authority, as appropriate)

Appendix II

Test cycles and weighting factors (regulation 13)

The following test cycles and weighting factors shall be applied for verification of compliance of marine diesel engines with the applicable NO_x limit in accordance with regulation 13 of this Annex using the test procedure and calculation method as specified in the revised NO_x Technical Code 2008.

.1 For constant-speed marine engines for ship main propulsion, including diesel-electric drive, test cycle E2 shall be applied;

.2 For controllable-pitch propeller sets test cycle E2 shall be applied;

.3 For propeller-law-operated main and propeller-law-operated auxiliary engines the test cycle E3 shall be applied;

.4 For constant-speed auxiliary engines test cycle D2 shall be applied; and

.5 For variable-speed, variable-load auxiliary engines, not included above, test cycle C1 shall be applied.

Test cycle for *constant-speed main propulsion* application
(including diesel-electric drive and all controllable-pitch propeller installations)

Test cycle type E2	Speed	100%	100%	100%	100%
	Power	100%	75%	50%	25%
	Weighting factor	0.2	0.5	0.15	0.15

Test cycle for *propeller-law-operated main* and *propeller-law-operated auxiliary engine* application

Test cycle type E3	Speed	100%	91%	80%	63%
	Power	100%	75%	50%	25%
	Weighting factor	0.2	0.5	0.15	0.15

Test cycle for *constant-speed auxiliary engine* application

Test cycle type D2	Speed	100%	100%	100%	100%	100%
	Power	100%	75%	50%	25%	10%
	Weighting factor	0.05	0.25	0.3	0.3	0.1

Test cycle for *variable-speed and variable-load auxiliary engine* application

Test cycle type C1	Speed	Rated				Intermediate			Idle
	Torque	100%	75%	50%	10%	100%	75%	50%	0%
	Weighting factor	0.15	0.15	0.15	0.1	0.1	0.1	0.1	0.15

In the case of an engine to be certified in accordance with paragraph 5.1.1 of regulation 13, the specific emission at each individual mode point shall not exceed the applicable NO_x emission limit value by more than 50% except as follows:

.1 The 10% mode point in the D2 test cycle.

.2 The 10% mode point in the C1 test cycle.

.3 The idle mode point in the C1 test cycle.

Appendix III

Criteria and procedures for designation of emission control areas (regulations 13.6 and 14.3)

1 Objectives

1.1 The purpose of this appendix is to provide the criteria and procedures to Parties for the formulation and submission of proposals for the designation of emission control areas and to set forth the factors to be considered in the assessment of such proposals by the Organization.

1.2 Emissions of NO_x, SO_x and particulate matter from ocean-going ships contribute to ambient concentrations of air pollution in cities and coastal areas around the world. Adverse public health and environmental effects associated with air pollution include premature mortality, cardiopulmonary disease, lung cancer, chronic respiratory ailments, acidification and eutrophication.

1.3 An emission control area should be considered for adoption by the Organization if supported by a demonstrated need to prevent, reduce and control emissions of NO_x or SO_x and particulate matter or all three types of emissions (hereinafter emissions) from ships.

2 Process for the designation of emission control areas

2.1 A proposal to the Organization for designation of an emission control area for NO_x or SO_x and particulate matter or all three types of emissions may be submitted only by Parties. Where two or more Parties have a common interest in a particular area, they should formulate a coordinated proposal.

2.2 A proposal to designate a given area as an emission control area should be submitted to the Organization in accordance with the rules and procedures established by the Organization.

3 Criteria for designation of an emission control area

3.1 The proposal shall include:

.1 a clear delineation of the proposed area of application, along with a reference chart on which the area is marked;

.2 the type or types of emission(s) that is or are being proposed for control (i.e. NO_x or SO_x and particulate matter or all three types of emissions);

.3 a description of the human populations and environmental areas at risk from the impacts of ship emissions;

.4 an assessment that emissions from ships operating in the proposed area of application are contributing to ambient concentrations of air pollution or to adverse environmental impacts. Such assessment shall include a description of the impacts of the relevant emissions on human health and the environment, such as adverse impacts to terrestrial and aquatic ecosystems, areas of natural productivity, critical habitats, water quality, human health, and areas of cultural and scientific significance, if applicable. The sources of relevant data including methodologies used shall be identified;

.5 relevant information, pertaining to the meteorological conditions in the proposed area of application, to the human populations and environmental areas at risk, in particular prevailing wind patterns, or to topographical, geological, oceanographic, morphological or other conditions that contribute to ambient concentrations of air pollution or adverse environmental impacts;

.6 the nature of the ship traffic in the proposed emission control area, including the patterns and density of such traffic;

.7 a description of the control measures taken by the proposing Party or Parties addressing land-based sources of NO_x, SO_x and particulate matter emissions affecting the human populations and environmental areas at risk that are in place and operating concurrent with the consideration of measures to be adopted in relation to provisions of regulations 13 and 14 of Annex VI; and

.8 the relative costs of reducing emissions from ships when compared with land-based controls, and the economic impacts on shipping engaged in international trade.

3.2 The geographical limits of an emission control area will be based on the relevant criteria outlined above, including emissions and deposition from ships navigating in the proposed area, traffic patterns and density, and wind conditions.

4 Procedures for the assessment and adoption of emission control areas by the Organization

4.1 The Organization shall consider each proposal submitted to it by a Party or Parties.

4.2 In assessing the proposal, the Organization shall take into account the criteria that are to be included in each proposal for adoption as set forth in section 3 above.

4.3 An emission control area shall be designated by means of an amendment to this Annex, considered, adopted and brought into force in accordance with article 16 of the present Convention.

5 Operation of emission control areas

5.1 Parties that have ships navigating in the area are encouraged to bring to the Organization any concerns regarding the operation of the area.

Appendix IV

Type approval and operating limits for shipboard incinerators (regulation 16)

1 Shipboard incinerators described in regulation 16.6.1 shall possess an IMO Type Approval Certificate for each incinerator. In order to obtain such certificate, the incinerator shall be designed and built to an approved standard as described in regulation 16.6.1. Each model shall be subject to a specified type approval test operation at the factory or an approved test facility, and under the responsibility of the Administration, using the following standard fuel/waste specification for the type approval test for determining whether the incinerator operates within the limits specified in paragraph 2 of this appendix:

Sludge oil consisting of:	75% sludge oil from heavy fuel oil (HFO);
	5% waste lubricating oil; and
	20% emulsified water.
Solid waste consisting of:	50% food waste;
	50% rubbish containing:
	approx. 30% paper,
	" 40% cardboard,
	" 10% rags,
	" 20% plastic.
	The mixture will have up to 50% moisture and 7% incombustible solids.

2 Incinerators described in regulation 16.6.1 shall operate within the following limits:

O_2 in combustion chamber:	6–12%
CO in flue gas maximum average:	200 mg/MJ
Soot number maximum average:	Bacharach 3 or Ringelman 1 (20% opacity) (a higher soot number is acceptable only during very short periods such as starting up)
Unburned components in ash residues:	Maximum 10% by weight
Combustion chamber flue gas outlet temperature range:	850–1,200°C

Appendix V

Information to be included in the bunker delivery note (regulation 18.5)

Name and IMO Number of receiving ship

Port

Date of commencement of delivery

Name, address and telephone number of marine fuel oil supplier

Product name(s)

Quantity in metric tonnes

Density at 15°C, kg/m^3*

Sulphur content (% m/m)†

A declaration signed and certified by the fuel oil supplier's representative that the fuel oil supplied is in conformity with the applicable paragraph of regulation 14.1 or 14.4 and regulation 18.3 of this Annex.

* Fuel oil shall be tested in accordance with ISO 3675:1998 or ISO 12185:1996.

† Fuel oil shall be tested in accordance with ISO 8754:2003.

Appendix VI

Fuel verification procedure for MARPOL Annex VI fuel oil samples (regulation 18.8.2)

The following procedure shall be used to determine whether the fuel oil delivered to and used on board ships is compliant with the sulphur limits required by regulation 14 of Annex VI.

1 General requirements

1.1 The representative fuel oil sample, which is required by paragraph 8.1 of regulation 18 (the "MARPOL sample") shall be used to verify the sulphur content of the fuel oil supplied to a ship.

1.2 An Administration, through its competent authority, shall manage the verification procedure.

1.3 The laboratories responsible for the verification procedure set forth in this appendix shall be fully accredited* for the purpose of conducting the tests.

2 Verification procedure stage 1

2.1 The MARPOL sample shall be delivered by the competent authority to the laboratory.

2.2 The laboratory shall:

- **.1** record the details of the seal number and the sample label on the test record;
- **.2** confirm that the condition of the seal on the MARPOL sample is that it has not been broken; and
- **.3** reject any MARPOL sample where the seal has been broken.

2.3 If the seal of the MARPOL sample has not been broken, the laboratory shall proceed with the verification procedure and shall:

- **.1** ensure that the MARPOL sample is thoroughly homogenized;
- **.2** draw two subsamples from the MARPOL sample; and
- **.3** reseal the MARPOL sample and record the new reseal details on the test record.

2.4 The two subsamples shall be tested in succession, in accordance with the specified test method referred to in appendix V (second footnote). For the purposes of this verification procedure, the results of the test analysis shall be referred to as "A" and "B":

- **.1** If the results of "A" and "B" are within the repeatability (*r*) of the test method, the results shall be considered valid.
- **.2** If the results of "A" and "B" are not within the repeatability (*r*) of the test method, both results shall be rejected and two new subsamples should be taken by the laboratory and analysed. The sample bottle should be resealed in accordance with paragraph 2.3.3 above after the new subsamples have been taken.

2.5 If the test results of "A" and "B" are valid, an average of these two results should be calculated thus giving the result referred to as "X":

- **.1** If the result of "X" is equal to or falls below the applicable limit required by Annex VI, the fuel oil shall be deemed to meet the requirements.
- **.2** If the result of "X" is greater than the applicable limit required by Annex VI, verification procedure stage 2 should be conducted; however, if the result of "X" is greater than the specification limit by 0.59*R* (where *R* is the reproducibility of the test method), the fuel oil shall be considered non-compliant and no further testing is necessary.

* Accreditation is in accordance with ISO 17025 or an equivalent standard.

3 Verification procedure stage 2

3.1 If stage 2 of the verification procedure is necessary in accordance with paragraph 2.5.2 above, the competent authority shall send the MARPOL sample to a second accredited laboratory.

3.2 Upon receiving the MARPOL sample, the laboratory shall:

.1 record the details of the reseal number applied in accordance with 2.3.3 above and the sample label on the test record;

.2 draw two subsamples from the MARPOL sample; and

.3 reseal the MARPOL sample and record the new reseal details on the test record.

3.3 The two subsamples shall be tested in succession, in accordance with the test method specified in appendix V (second footnote). For the purposes of this verification procedure, the results of the test analysis shall be referred to as "C" and "D":

.1 If the results of "C" and "D" are within the repeatability (*r*) of the test method, the results shall be considered valid.

.2 If the results of "C" and "D" are not within the repeatability (*r*) of the test method, both results shall be rejected and two new subsamples shall be taken by the laboratory and analysed. The sample bottle should be resealed in accordance with paragraph 3.2.3 above after the new subsamples have been taken.

3.4 If the test results of "C" and "D" are valid, and the results of "A", "B", "C", and "D" are within the reproducibility (*R*) of the test method then the laboratory shall average the results, which is referred to as "Y":

.1 If the result of "Y" is equal to or falls below the applicable limit required by Annex VI, the fuel oil shall be deemed to meet the requirements.

.2 If the result of "Y" is greater than the applicable limit required by Annex VI, then the fuel oil fails to meet the standards required by Annex VI.

3.5 If the results of "A", "B", "C" and "D" are not within the reproducibility (*R*) of the test method then the Administration may discard all of the test results and, at its discretion, repeat the entire testing process.

3.6 The results obtained from the verification procedure are final.

Appendix VII

Emission control areas (regulations 13.6 and 14.3)

1 The boundaries of emission control areas designated under regulations 13.6 and 14.3, other than the Baltic Sea and the North Sea areas, are set forth in this appendix.

2 The North American area comprises:

.1 the sea area located off the Pacific coasts of the United States and Canada, enclosed by geodesic lines connecting the following coordinates:

Point	Latitude	Longitude
1	32°32′.10 N	117°06′.11 W
2	32°32′.04 N	117°07′.29 W
3	32°31′.39 N	117°14′.20 W
4	32°33′.13 N	117°15′.50 W
5	32°34′.21 N	117°22′.01 W
6	32°35′.23 N	117°27′.53 W
7	32°37′.38 N	117°49′.34 W
8	31°07′.59 N	118°36′.21 W
9	30°33′.25 N	121°47′.29 W
10	31°46′.11 N	123°17′.22 W
11	32°21′.58 N	123°50′.44 W
12	32°56′.39 N	124°11′.47 W
13	33°40′.12 N	124°27′.15 W
14	34°31′.28 N	125°16′.52 W
15	35°14′.38 N	125°43′.23 W
16	35°44′.00 N	126°18′.53 W
17	36°16′.25 N	126°45′.30 W
18	37°01′.35 N	127°07′.18 W
19	37°45′.39 N	127°38′.02 W
20	38°25′.08 N	127°53′.00 W
21	39°25′.05 N	128°31′.23 W
22	40°18′.47 N	128°45′.46 W
23	41°13′.39 N	128°40′.22 W
24	42°12′.49 N	129°00′.38 W
25	42°47′.34 N	129°05′.42 W
26	43°26′.22 N	129°01′.26 W
27	44°24′.43 N	128°41′.23 W
28	45°30′.43 N	128°40′.02 W
29	46°11′.01 N	128°49′.01 W
30	46°33′.55 N	129°04′.29 W
31	47°39′.55 N	131°15′.41 W
32	48°32′.32 N	132°41′.00 W

Point	Latitude	Longitude
33	48°57′.47 N	133°14′.47 W
34	49°22′.39 N	134°15′.51 W
35	50°01′.52 N	135°19′.01 W
36	51°03′.18 N	136°45′.45 W
37	51°54′.04 N	137°41′.54 W
38	52°45′.12 N	138°20′.14 W
39	53°29′.20 N	138°40′.36 W
40	53°40′.39 N	138°48′.53 W
41	54°13′.45 N	139°32′.38 W
42	54°39′.25 N	139°56′.19 W
43	55°20′.18 N	140°55′.45 W
44	56°07′.12 N	141°36′.18 W
45	56°28′.32 N	142°17′.19 W
46	56°37′.19 N	142°48′.57 W
47	58°51′.04 N	153°15′.03 W

.2 the sea areas located off the Atlantic coasts of the United States, Canada and France (Saint-Pierre-et-Miquelon), and the Gulf of Mexico coast of the United States enclosed by geodesic lines connecting the following coordinates:

Point	Latitude	Longitude
1	60°00′.00 N	64°09′.36 W
2	60°00′.00 N	56°43′.00 W
3	58°54′.01 N	55°38′.05 W
4	57°50′.52 N	55°03′.47 W
5	57°35′.13 N	54°00′.59 W
6	57°14′.20 N	53°07′.58 W
7	56°48′.09 N	52°23′.29 W
8	56°18′.13 N	51°49′.42 W
9	54°23′.21 N	50°17′.44 W
10	53°44′.54 N	50°07′.17 W
11	53°04′.59 N	50°10′.05 W
12	52°20′.06 N	49°57′.09 W
13	51°34′.20 N	48°52′.45 W
14	50°40′.15 N	48°16′.04 W
15	50°02′.28 N	48°07′.03 W
16	49°24′.03 N	48°09′.35 W
17	48°39′.22 N	47°55′.17 W
18	47°24′.25 N	47°46′.56 W
19	46°35′.12 N	48°00′.54 W
20	45°19′.45 N	48°43′.28 W
21	44°43′.38 N	49°16′.50 W
22	44°16′.38 N	49°51′.23 W

Point	Latitude	Longitude
23	43°53′.15 N	50°34′.01 W
24	43°36′.06 N	51°20′.41 W
25	43°23′.59 N	52°17′.22 W
26	43°19′.50 N	53°20′.13 W
27	43°21′.14 N	54°09′.20 W
28	43°29′.41 N	55°07′.41 W
29	42°40′.12 N	55°31′.44 W
30	41°58′.19 N	56°09′.34 W
31	41°20′.21 N	57°05′.13 W
32	40°55′.34 N	58°02′.55 W
33	40°41′.38 N	59°05′.18 W
34	40°38′.33 N	60°12′.20 W
35	40°45′.46 N	61°14′.03 W
36	41°04′.52 N	62°17′.49 W
37	40°36′.55 N	63°10′.49 W
38	40°17′.32 N	64°08′.37 W
39	40°07′.46 N	64°59′.31 W
40	40°05′.44 N	65°53′.07 W
41	39°58′.05 N	65°59′.51 W
42	39°28′.24 N	66°21′.14 W
43	39°01′.54 N	66°48′.33 W
44	38°39′.16 N	67°20′.59 W
45	38°19′.20 N	68°02′.01 W
46	38°05′.29 N	68°46′.55 W
47	37°58′.14 N	69°34′.07 W
48	37°57′.47 N	70°24′.09 W
49	37°52′.46 N	70°37′.50 W
50	37°18′.37 N	71°08′.33 W
51	36°32′.25 N	71°33′.59 W
52	35°34′.58 N	71°26′.02 W
53	34°33′.10 N	71°37′.04 W
54	33°54′.49 N	71°52′.35 W
55	33°19′.23 N	72°17′.12 W
56	32°45′.31 N	72°54′.05 W
57	31°55′.13 N	74°12′.02 W
58	31°27′.14 N	75°15′.20 W
59	31°03′.16 N	75°51′.18 W
60	30°45′.42 N	76°31′.38 W
61	30°12′.48 N	77°18′.29 W
62	29°25′.17 N	76°56′.42 W
63	28°36′.59 N	76°48′.00 W
64	28°17′.13 N	76°40′.10 W

Point	Latitude	Longitude
65	28°17′.12 N	79°11′.23 W
66	27°52′.56 N	79°28′.35 W
67	27°26′.01 N	79°31′.38 W
68	27°16′.13 N	79°34′.18 W
69	27°11′.54 N	79°34′.56 W
70	27°05′.59 N	79°35′.19 W
71	27°00′.28 N	79°35′.17 W
72	26°55′.16 N	79°34′.39 W
73	26°53′.58 N	79°34′.27 W
74	26°45′.46 N	79°32′.41 W
75	26°44′.30 N	79°32′.23 W
76	26°43′.40 N	79°32′.20 W
77	26°41′.12 N	79°32′.01 W
78	26°38′.13 N	79°31′.32 W
79	26°36′.30 N	79°31′.06 W
80	26°35′.21 N	79°30′.50 W
81	26°34′.51 N	79°30′.46 W
82	26°34′.11 N	79°30′.38 W
83	26°31′.12 N	79°30′.15 W
84	26°29′.05 N	79°29′.53 W
85	26°25′.31 N	79°29′.58 W
86	26°23′.29 N	79°29′.55 W
87	26°23′.21 N	79°29′.54 W
88	26°18′.57 N	79°31′.55 W
89	26°15′.26 N	79°33′.17 W
90	26°15′.13 N	79°33′.23 W
91	26°08′.09 N	79°35′.53 W
92	26°07′.47 N	79°36′.09 W
93	26°06′.59 N	79°36′.35 W
94	26°02′.52 N	79°38′.22 W
95	25°59′.30 N	79°40′.03 W
96	25°59′.16 N	79°40′.08 W
97	25°57′.48 N	79°40′.38 W
98	25°56′.18 N	79°41′.06 W
99	25°54′.04 N	79°41′.38 W
100	25°53′.24 N	79°41′.46 W
101	25°51′.54 N	79°41′.59 W
102	25°49′.33 N	79°42′.16 W
103	25°48′.24 N	79°42′.23 W
104	25°48′.20 N	79°42′.24 W
105	25°46′.26 N	79°42′.44 W
106	25°46′.16 N	79°42′.45 W

Point	Latitude	Longitude
107	25°43′.40 N	79°42′.59 W
108	25°42′.31 N	79°42′.48 W
109	25°40′.37 N	79°42′.27 W
110	25°37′.24 N	79°42′.27 W
111	25°37′.08 N	79°42′.27 W
112	25°31′.03 N	79°42′.12 W
113	25°27′.59 N	79°42′.11 W
114	25°24′.04 N	79°42′.12 W
115	25°22′.21 N	79°42′.20 W
116	25°21′.29 N	79°42′.08 W
117	25°16′.52 N	79°41′.24 W
118	25°15′.57 N	79°41′.31 W
119	25°10′.39 N	79°41′.31 W
120	25°09′.51 N	79°41′.36 W
121	25°09′.03 N	79°41′.45 W
122	25°03′.55 N	79°42′.29 W
123	25°03′.00 N	79°42′.56 W
124	25°00′.30 N	79°44′.05 W
125	24°59′.03 N	79°44′.48 W
126	24°55′.28 N	79°45′.57 W
127	24°44′.18 N	79°49′.24 W
128	24°43′.04 N	79°49′.38 W
129	24°42′.36 N	79°50′.50 W
130	24°41′.47 N	79°52′.57 W
131	24°38′.32 N	79°59′.58 W
132	24°36′.27 N	80°03′.51 W
133	24°33′.18 N	80°12′.43 W
134	24°33′.05 N	80°13′.21 W
135	24°32′.13 N	80°15′.16 W
136	24°31′.27 N	80°16′.55 W
137	24°30′.57 N	80°17′.47 W
138	24°30′.14 N	80°19′.21 W
139	24°30′.06 N	80°19′.44 W
140	24°29′.38 N	80°21′.05 W
141	24°28′.18 N	80°24′.35 W
142	24°28′.06 N	80°25′.10 W
143	24°27′.23 N	80°27′.20 W
144	24°26′.30 N	80°29′.30 W
145	24°25′.07 N	80°32′.22 W
146	24°23′.30 N	80°36′.09 W
147	24°22′.33 N	80°38′.56 W
148	24°22′.07 N	80°39′.51 W

Point	Latitude	Longitude
149	24°19′.31 N	80°45′.21 W
150	24°19′.16 N	80°45′.47 W
151	24°18′.38 N	80°46′.49 W
152	24°18′.35 N	80°46′.54 W
153	24°09′.51 N	80°59′.47 W
154	24°09′.48 N	80°59′.51 W
155	24°08′.58 N	81°01′.07 W
156	24°08′.30 N	81°01′.51 W
157	24°08′.26 N	81°01′.57 W
158	24°07′.28 N	81°03′.06 W
159	24°02′.20 N	81°09′.05 W
160	24°00′.00 N	81°11′.16 W
161	23°55′.32 N	81°12′.55 W
162	23°53′.52 N	81°19′.43 W
163	23°50′.52 N	81°29′.59 W
164	23°50′.02 N	81°39′.59 W
165	23°49′.05 N	81°49′.59 W
166	23°49′.05 N	82°00′.11 W
167	23°49′.42 N	82°09′.59 W
168	23°51′.14 N	82°24′.59 W
169	23°51′.14 N	82°39′.59 W
170	23°49′.42 N	82°48′.53 W
171	23°49′.32 N	82°51′.11 W
172	23°49′.24 N	82°59′.59 W
173	23°49′.52 N	83°14′.59 W
174	23°51′.22 N	83°25′.49 W
175	23°52′.27 N	83°33′.01 W
176	23°54′.04 N	83°41′.35 W
177	23°55′.47 N	83°48′.11 W
178	23°58′.38 N	83°59′.59 W
179	24°09′.37 N	84°29′.27 W
180	24°13′.20 N	84°38′.39 W
181	24°16′.41 N	84°46′.07 W
182	24°23′.30 N	84°59′.59 W
183	24°26′.37 N	85°06′.19 W
184	24°38′.57 N	85°31′.54 W
185	24°44′.17 N	85°43′.11 W
186	24°53′.57 N	85°59′.59 W
187	25°10′.44 N	86°30′.07 W
188	25°43′.15 N	86°21′.14 W
189	26°13′.13 N	86°06′.45 W
190	26°27′.22 N	86°13′.15 W

Point	Latitude	Longitude
191	26°33′.46 N	86°37′.07 W
192	26°01′.24 N	87°29′.35 W
193	25°42′.25 N	88°33′.00 W
194	25°46′.54 N	90°29′.41 W
195	25°44′.39 N	90°47′.05 W
196	25°51′.43 N	91°52′.50 W
197	26°17′.44 N	93°03′.59 W
198	25°59′.55 N	93°33′.52 W
199	26°00′.32 N	95°39′.27 W
200	26°00′.33 N	96°48′.30 W
201	25°58′.32 N	96°55′.28 W
202	25°58′.15 N	96°58′.41 W
203	25°57′.58 N	97°01′.54 W
204	25°57′.41 N	97°05′.08 W
205	25°57′.24 N	97°08′.21 W
206	25°57′.24 N	97°08′.47 W

.3 the sea area located off the coasts of the Hawaiian Islands of Hawai'i, Maui, Oahu, Moloka'i, Ni'ihau, Kaua'i, Lāna'i and Kaho'olawe, enclosed by geodesic lines connecting the following coordinates:

Point	Latitude	Longitude
1	22°32′.54 N	153°00′.33 W
2	23°06′.05 N	153°28′.36 W
3	23°32′.11 N	154°02′.12 W
4	23°51′.47 N	154°36′.48 W
5	24°21′.49 N	155°51′.13 W
6	24°41′.47 N	156°27′.27 W
7	24°57′.33 N	157°22′.17 W
8	25°13′.41 N	157°54′.13 W
9	25°25′.31 N	158°30′.36 W
10	25°31′.19 N	159°09′.47 W
11	25°30′.31 N	159°54′.21 W
12	25°21′.53 N	160°39′.53 W
13	25°00′.06 N	161°38′.33 W
14	24°40′.49 N	162°13′.13 W
15	24°15′.53 N	162°43′.08 W
16	23°40′.50 N	163°13′.00 W
17	23°03′.20 N	163°32′.58 W
18	22°20′.09 N	163°44′.41 W
19	21°36′.45 N	163°46′.03 W
20	20°55′.26 N	163°37′.44 W
21	20°13′.34 N	163°19′.13 W

Point	Latitude	Longitude
22	19°39′.03 N	162°53′.48 W
23	19°09′.43 N	162°20′.35 W
24	18°39′.16 N	161°19′.14 W
25	18°30′.31 N	160°38′.30 W
26	18°29′.31 N	159°56′.17 W
27	18°10′.41 N	159°14′.08 W
28	17°31′.17 N	158°56′.55 W
29	16°54′.06 N	158°30′.29 W
30	16°25′.49 N	157°59′.25 W
31	15°59′.57 N	157°17′.35 W
32	15°40′.37 N	156°21′.06 W
33	15°37′.36 N	155°22′.16 W
34	15°43′.46 N	154°46′.37 W
35	15°55′.32 N	154°13′.05 W
36	16°46′.27 N	152°49′.11 W
37	17°33′.42 N	152°00′.32 W
38	18°30′.16 N	151°30′.24 W
39	19°02′.47 N	151°22′.17 W
40	19°34′.46 N	151°19′.47 W
41	20°07′.42 N	151°22′.58 W
42	20°38′.43 N	151°31′.36 W
43	21°29′.09 N	151°59′.50 W
44	22°06′.58 N	152°31′.25 W
45	22°32′.54 N	153°00′.33 W

3 The United States Caribbean Sea area includes:

.1 the sea area located off the Atlantic and Caribbean coasts of the Commonwealth of Puerto Rico and the United States Virgin Islands, enclosed by geodesic lines connecting the following coordinates:

Point	Latitude	Longitude
1	17°18′.37 N	67°32′.14 W
2	19°11′.14 N	67°26′.45 W
3	19°30′.28 N	65°16′.48 W
4	19°12′.25 N	65°06′.08 W
5	18°45′.13 N	65°00′.22 W
6	18°41′.14 N	64°59′.33 W
7	18°29′.22 N	64°53′.51 W
8	18°27′.35 N	64°53′.22 W
9	18°25′.21 N	64°52′.39 W
10	18°24′.30 N	64°52′.19 W
11	18°23′.51 N	64°51′.50 W
12	18°23′.42 N	64°51′.23 W

Point	Latitude	Longitude
13	18°23′.36 N	64°50′.17 W
14	18°23′.48 N	64°49′.41 W
15	18°24′.11 N	64°49′.00 W
16	18°24′.28 N	64°47′.57 W
17	18°24′.18 N	64°47′.01 W
18	18°23′.13 N	64°46′.37 W
19	18°22′.37 N	64°45′.20 W
20	18°22′.39 N	64°44′.42 W
21	18°22′.42 N	64°44′.36 W
22	18°22′.37 N	64°44′.24 W
23	18°22′.39 N	64°43′.42 W
24	18°22′.30 N	64°43′.36 W
25	18°22′.25 N	64°42′.58 W
26	18°22′.26 N	64°42′.28 W
27	18°22′.15 N	64°42′.03 W
28	18°22′.22 N	64°40′.60 W
29	18°21′.57 N	64°40′.15 W
30	18°21′.51 N	64°38′.23 W
31	18°21′.22 N	64°38′.16 W
32	18°20′.39 N	64°38′.33 W
33	18°19′.15 N	64°38′.14 W
34	18°19′.07 N	64°38′.16 W
35	18°17′.23 N	64°39′.38 W
36	18°16′.43 N	64°39′.41 W
37	18°11′.33 N	64°38′.58 W
38	18°03′.02 N	64°38′.03 W
39	18°02′.56 N	64°29′.35 W
40	18°02′.51 N	64°27′.02 W
41	18°02′.30 N	64°21′.08 W
42	18°02′.31 N	64°20′.08 W
43	18°02′.03 N	64°15′.57 W
44	18°00′.12 N	64°02′.29 W
45	17°59′.58 N	64°01′.04 W
46	17°58′.47 N	63°57′.01 W
47	17°57′.51 N	63°53′.54 W
48	17°56′.38 N	63°53′.21 W
49	17°39′.40 N	63°54′.53 W
50	17°37′.08 N	63°55′.10 W
51	17°30′.21 N	63°55′.56 W
52	17°11′.36 N	63°57′.57 W
53	17°05′.00 N	63°58′.41 W
54	16°59′.49 N	63°59′.18 W
55	17°18′.37 N	67°32′.14 W

Appendix VIII
Form of International Energy Efficiency (IEE) Certificate

INTERNATIONAL ENERGY EFFICIENCY CERTIFICATE

Issued under the provisions of the Protocol of 1997, as amended, to amend the International Convention for the Prevention of Pollution by Ships, 1973, as modified by the Protocol of 1978 related thereto (hereinafter referred to as "the Convention") under the authority of the Government of:

. . . .

(full designation of the Party)

by . . . .

(full designation of the competent person or organization authorized under the provisions of the Convention)

Particulars of ship[*]

Name of ship . . . .

Distinctive number or letters . . . .

IMO Number[†] . . . .

Port of registry . . . .

Gross tonnage . . . .

THIS IS TO CERTIFY:

1 That the ship has been surveyed in accordance with regulation 5.4 of Annex VI of the Convention; and

2 That the survey shows that the ship complies with the applicable requirements in regulation 20, regulation 21 and regulation 22.

Completion date of survey on which this Certificate is based . . . .(dd/mm/yyyy)

Issued at. . . . .

(place of issue of Certificate)

Date (dd/mm/yyyy) . . . .

(date of issue) *(signature of duly authorized official issuing the Certificate)*

(seal or stamp of the authority, as appropriate)

[*] Alternatively, the particulars of the ship may be placed horizontally in boxes.

[†] In accordance with the IMO ship identification number scheme (resolution A.1078(28)).

SUPPLEMENT TO INTERNATIONAL ENERGY EFFICIENCY CERTIFICATE (IEE CERTIFICATE)

RECORD OF CONSTRUCTION RELATING TO ENERGY EFFICIENCY

Notes:

1 This Record shall be permanently attached to the IEE Certificate. The IEE Certificate shall be available on board the ship at all times.

2 The Record shall be at least in English, French or Spanish. If an official language of the issuing Party is also used, this shall prevail in case of a dispute or discrepancy.

3 Entries in boxes shall be made by inserting either: a cross (x) for the answers "yes" and "applicable"; or a dash (–) for the answers "no" and "not applicable", as appropriate.

4 Unless otherwise stated, regulations mentioned in this Record refer to regulations in Annex VI of the Convention, and resolutions or circulars refer to those adopted by the International Maritime Organization.

1 Particulars of ship

1.1 Name of ship . . . .

1.2 IMO number . . . .

1.3 Date of building contract . . . .

1.4 Gross tonnage . . . .

1.5 Deadweight. . . . .

1.6 Type of ship* . . . .

2 Propulsion system

2.1 Diesel propulsion. . . . . ☐

2.2 Diesel-electric propulsion . . . . ☐

2.3 Turbine propulsion. . . . . ☐

2.4 Hybrid propulsion . . . . ☐

2.5 Propulsion system other than any of the above . . . . ☐

3 Attained Energy Efficiency Design Index (EEDI)

3.1 The Attained EEDI in accordance with regulation 20.1 is calculated based on the information contained in the EEDI Technical File which also shows the process of calculating the Attained EEDI . . . . ☐

The Attained EEDI is: . . . . grams-CO_2/tonne-mile

3.2 The Attained EEDI is not calculated as:

3.2.1 the ship is exempt under regulation 20.1 as it is not a new ship as defined in regulation 2.23 . . . . ☐

3.2.2 the type of propulsion system is exempt in accordance with regulation 19.3 . . . . ☐

3.2.3 the requirement of regulation 20 is waived by the ship's Administration in accordance with regulation 19.4 . . . . ☐

3.2.4 the type of ship is exempt in accordance with regulation 20.1 . . . . ☐

* Insert ship type in accordance with definitions specified in regulation 2. Ships falling into more than one of the ship types defined in regulation 2 should be considered as being the ship type with the most stringent (the lowest) required EEDI. If ship does not fall into the ship types defined in regulation 2, insert "Ship other than any of the ship type defined in regulation 2".

4 Required EEDI

4.1 Required EEDI is: . . . . grams-CO_2/tonne-mile

4.2 The required EEDI is not applicable as:

4.2.1 the ship is exempt under regulation 21.1 as it is not a new ship as defined in regulation 2.23. . . . . ☐

4.2.2 the type of propulsion system is exempt in accordance with regulation 19.3 . . . . ☐

4.2.3 the requirement of regulation 21 is waived by the ship's Administration in accordance with regulation 19.4 . . . . ☐

4.2.4 the type of ship is exempt in accordance with regulation 21.1 . . . . ☐

4.2.5 the ship's capacity is below the minimum capacity threshold in Table 1 of regulation 21.2 . . . . . . ☐

5 Ship Energy Efficiency Management Plan

5.1 The ship is provided with a Ship Energy Efficiency Management Plan (SEEMP) in compliance with regulation 22 . . . . ☐

6 EEDI Technical File

6.1 The IEE Certificate is accompanied by the EEDI Technical File in compliance with regulation 20.1. . . ☐

6.2 The EEDI Technical File identification/verification number . . . .

6.3 The EEDI Technical File verification date. . . . .

THIS IS TO CERTIFY that this Record is correct in all respects.

Issued at. . . . .
(place of issue of the Record)

Date (dd/mm/yyyy) . . . . *(date of issue)*

. . . . *(signature of duly authorized official issuing the Record)*

(seal or stamp of the issuing authority, as appropriate)

Appendix IX

Information to be submitted to the IMO Ship Fuel Oil Consumption Database

Identity of the ship

IMO number . . . .

Period of calendar year for which the data is submitted

Start date (dd/mm/yyyy). . . . .

End date (dd/mm/yyyy). . . . .

Technical characteristics of the ship

Ship type, as defined in regulation 2 of this Annex or other (to be stated) . . . .

Gross tonnage (GT)* . . . .

Net tonnage (NT)† . . . .

Deadweight tonnage (DWT)‡ . . . .

Power output (rated power)§ of main and auxiliary reciprocating internal combustion engines over 130 kW (to be stated in kW). . . . .

EEDI¶ (if applicable). . . . .

Ice class** . . . .

Fuel oil consumption, by fuel oil type in metric tonnes and methods used for collecting fuel oil consumption data . . . .

Distance travelled . . . .

Hours underway. . . . .

---

* Gross tonnage should be calculated in accordance with the International Convention on Tonnage Measurement of Ships, 1969.

† Net tonnage should be calculated in accordance with the International Convention on Tonnage Measurement of Ships, 1969. If not applicable, note "N/A".

‡ DWT means the difference in tonnes between the displacement of a ship in water of relative density of 1,025 kg/m^3 at the summer load draught and the lightweight of the ship. The summer load draught should be taken as the maximum summer draught as certified in the stability booklet approved by the Administration or an organization recognized by it. If not applicable, note "N/A".

§ Rated power means the maximum continuous rated power as specified on the nameplate of the engine.

¶ As defined in the 2014 Guidelines on the method of calculation of the Attained Energy Efficiency Design Index (EEDI) for new ships (resolution MEPC.245(66), as amended) or other (to be stated).

** Ice class should be consistent with the definition set out in the International Code for ships operating in polar waters (Polar Code) (resolutions MEPC.264(68) and MSC.385(94)). If not applicable, note "N/A".

Appendix X
Form of Statement of Compliance – Fuel Oil Consumption Reporting

STATEMENT OF COMPLIANCE – FUEL OIL CONSUMPTION REPORTING

Issued under the provisions of the Protocol of 1997, as amended, to amend the International Convention for the Prevention of Pollution by Ships, 1973, as modified by the Protocol of 1978 related thereto (hereinafter referred to as "the Convention") under the authority of the Government of:

. . . .

(full designation of the Party)

by . . . .

(full designation of the competent person or organization authorized under the provisions of the Convention)

Particulars of ship*

Name of ship . . . .

Distinctive number or letters . . . .

IMO Number† . . . .

Port of registry . . . .

Gross tonnage . . . .

THIS IS TO DECLARE:

1. That the ship has submitted to this Administration the data required by regulation 22A of Annex VI of the Convention, covering ship operations from (dd/mm/yyyy) through (dd/mm/yyyy); and
2. The data was collected and reported in accordance with the methodology and processes set out in the ship's SEEMP that was in effect over the period from (dd/mm/yyyy) through (dd/mm/yyyy).

This Statement of Compliance is valid until (dd/mm/yyyy). . . . .

Issued at: . . . .

(place of issue of Statement)

Date (dd/mm/yyyy) . . . .

(date of issue) *(signature of duly authorized official issuing the Statement)*

(seal or stamp of the authority, as appropriate)

* Alternatively, the particulars of the ship may be placed horizontally in boxes.

† In accordance with the IMO Ship Identification Number Scheme (resolution A.1078(28)).

Notes